Toughened Composites

This book covers micro and macro aspects of toughened composites covering polymer matrix, metal matrix, ceramic matrix and nanomatrix. It gives the reader understanding of composite fabrication, construction, and lightweight yet high crack resistance performance, macroscopic testing supported by microscopic bonding and debonding features, models of stress transfer, and commercial features of developing cheaper yet high-quality materials.

Features:

- Focuses on micro and macro aspects of toughening methods and principles of composite materials.
- Includes all types of composites including polymer matrix, metal matrix, ceramic matrix and nanomatrix.
- Covers corrosion resistance and oxidation resistance as well as solubility resistance.
- Discusses the use of recycled materials.
- Provides a good balance of long fibre, short fibre, nanoparticle and particulate modifiers.

This book aims at researchers and professionals in materials science, composite materials, fracture mechanics, materials characterization and testing, properties and mechanics, nanomaterials, aerospace and automotive engineering and structural engineering.

Toughened Composites

Micro and Macro Systems

Edited by

Sri Bandyopadhyay

Raghavendra Gujjala

CRC Press is an imprint of the
Taylor & Francis Group, an informa business

First edition published 2023
by CRC Press
6000 Broken Sound Parkway NW, Suite 300, Boca Raton, FL 33487-2742

and by CRC Press
4 Park Square, Milton Park, Abingdon, Oxon, OX14 4RN

CRC Press is an imprint of Taylor & Francis Group, LLC

ISBN: 9780367353070 (hbk)
ISBN: 9781032385495 (pbk)
ISBN: 9780429330575 (ebk)

DOI: 10.1201/9780429330575

Typeset in Times
by codeMantra

Contents

Editors

Dr Sri Bandyopadhyay, a former academic at UNSW Sydney, Australia – School of Materials Science and Engineering, and an earlier senior research scientist at Australian Defence DSTO MRL, Melbourne, has a research background in (i) polymer matrix, (ii) metal matrix and (iii) ceramic matrix composites. At DSTO MRL, Australia, his *in situ* SEM study of micro and macro fractures earned him the 'DSTO MRL Scientist of the year' award. Whilst at UNSW SMSE, he taught fracture mechanics, polymer engineering and composite technology to students in his school as well as in other schools at UNSW in the Faculty of Engineering. By the way, he completed his PhD from Monash University, Australia, on environmental stress cracking of polyethylene. His organized proceedings of *ACUN-5 International Composite Conference – Developments in Composites: Advanced, Infrastructural, Natural and Nano-composites*, 11–14 July 2006, UNSW Sydney, Australia, are available on the Internet using the website http://unsworks.unsw.edu.au/fapi/datastream/unsworks:417/SOURCE1?view=true.

Dr Raghavendra Gujjala, an assistant professor in the mechanical department at the National Institute of Technology Warangal, Telangana, India, completed his PhD from NIT Rourkela. He has a research background in composite materials, FGMs, nanomaterials and bioglasses. He has published more than 130 international and national journals and conferences. He published more than 12 chapters in different books. He taught composites, tribology and material characterization of materials to students. Website: https://www.nitw.ac.in/faculty/id/16533/.

Contributors

Aziz Ahmed
School of Civil, Mining and Environmental Engineering
University of Wollongong
Wollongong, Australia

Saeed M. Al-Zahrani
SABIC Polymer Research Center (SPRC), Chemical Engineering Department
King Saud University
Riyadh, Saudi Arabia

Arfat Anis
SABIC Polymer Research Center (SPRC), Chemical Engineering Department
King Saud University
Riyadh, Saudi Arabia

Shazed Aziz
School of Chemical Engineering
The University of Queensland
Brisbane, Australia

G. Chitti Babu
Department of Mechanical Engineering
National Institute of Technology Warangal
Warangal, India

Janen N. Bandyopadhyay
Department of Civil Engineering
Indian Institute of Technology
Kharagpur, India

Sri Bandyopadhyay
School of Materials Science and Engineering
UNSW Sydney, Australia
Australian Defence DSTO MRL
Melbourne, Australia

Zahir Bashir
SABIC Polymer Research Center (SPRC), Chemical Engineering Department
King Saud University
Riyadh, Saudi Arabia

A. Bhattacharya
University of Cincinnati
Cincinnati, Ohio

Chinmoy Bhattacharya
Department of Chemistry
Indian Institute of Engineering Science and Technology (IIEST), Shibpur
Howrah, India

J. Campbell
Pike Township Fire Department
Indianapolis, Indiana

Ch. Sri Chaitanya
Department of Mechanical Engineering
VR Siddhartha Engineering College
Kanuru, India

K. K. Chattopadhyay
School of Materials Science and Nanotechnology
Thin Film and Nano Science Laboratory, Department of Physics
Jadavpur University
Kolkata, India

Sambhu Nath Chattopadhyay
Chemical and Biochemical Processing Division
ICAR – National Institute of Natural Fibre Engineering and Technology
Kolkata, India

D. Chauhan
University of Cincinnati
Cincinnati, Ohio

Theodora K.-H. Cheng
School of Materials Science and Engineering
UNSW Sydney
NSW, Australia

M. Chitranshi
University of Cincinnati
Cincinnati, Ohio

Pannalal Choudhury
Department of Mechanical Engineering
National Institute of Technology Silchar
Silchar, India

Mahuya Das
Greater Kolkata College of Engineering and Management
Kolkata, India

Parnab Das
Department of Civil, Construction and Environmental Engineering
The University of Alabama
Tuscaloosa, Alabama

Subhankar Das
Department of Mechanical Engineering
University of Petroleum and Energy Studies
Dehradun, India

Tapas Das
LMATS Brisbane Laboratory
Archerfield, Australia

Shaikh N. Faisal
ARC Centre of Excellence for Electromaterials Science and Intelligent Polymer Research Institute
University of Wollongong
Wollongong, Australia
School of Electrical and Data Engineering
University of Technology Sydney
Sydney, Australia

S. Fialkova
North Carolina Agricultural and Technology
Greensboro, North Carolina

Dheeraj K. Gara
Department of Mechanical Engineering
National Institute of Technology
Warangal, India

Shrabani Ghosh
School of Materials Science and Nanotechnology
Jadavpur University
Kolkata, India

S. Grinshpun
University of Cincinnati
Cincinnati, Ohio

Raghavendra Gujjala
Department of Mechanical Engineering
National Institute of Technology
Warangal, India

Paramita Hajra
Department of Chemistry
Indian Institute of Engineering Science and Technology (IIEST), Shibpur
Howrah, India

Sudipta Halder
Department of Mechanical Engineering
National Institute of Technology Silchar
Silchar, India

Naomi Ho
School of Materials Science and Engineering
UNSW Sydney
Sydney, Australia

Md. Mokarrom Hossain
School of Chemistry
The University of New South Wales
Sydney, Australia

Mohammad S. Islam
School of Mechanical and Manufacturing Engineering
The University of New South Wales
Sydney, Australia

Satish Jain
Department of Mechanical Engineering
National Institute of Technology
Warangal, India

B. J. Jetter
Glendale Ohio Fire Department
Glendale, Ohio

I.I. Kabir
School of Materials Science and Engineering
UNSW Sydney
Sydney, Australia

F. Selcen Kilinc-Balci
National Institute for Occupational Safety and Health
Washington, District of Columbia

Pramod Koshy
School of Materials Science and Engineering
UNSW Sydney
Sydney, Australia

A. Kubley
University of Cincinnati
Cincinnati, Ohio

Subhajit Kundu
Department of Chemical Technology
University of Calcutta
Kolkata, India

Babar Pasha Mahammod
Department of Mechanical Engineering
National Institute of Technology Warangal
Warangal, India

Supratim Maity
Thin film and Nano Science Laboratory, Department of Physics
Jadavpur University
Kolkata, India

Tina Majidi
School of Materials Science and Engineering
UNSW Sydney
Sydney, Australia

A. Basu Mallick
Department of Metallurgy and Materials Engineering
Indian Institute of Engineering Science and Technology, Shibpur
Howrah, India

Harahari Mandal
Department of Chemistry
Indian Institute of Engineering Science and Technology (IIEST), Shibpur
Howrah, India

Debarati Mitra
Department of Chemical Technology
University of Calcutta
Kolkata, India

G.S. Mukherjee
DRDO
Delhi, India

Amar N. Nayak
Department of Civil Engineering
Veer Surendra Sai University of Technology
Burla, India

V. Ng
University of Cincinnati
Cincinnati, Ohio

R. Noga
University of Cincinnati
Cincinnati, Ohio

Shakuntala Ojha
Department of Mechanical Engineering
Kakatiya Institute of Technology and Science
Warangal, India

Omprakash
Department of Mechanical Engineering
Kakatiya Institute of Technology and Science
Warangal, India

Bhavya Parameswaran
Rubber Technology Centre
Indian Institute of Technology Kharagpur
Kharagpur, India

Bernadette Pudadera
School of Materials Science and Engineering
UNSW Sydney
Sydney, Australia

A. Pujari
University of Cincinnati
Cincinnati, Ohio

Md. Arifur Rahim
School of Chemical Engineering
The University of New South Wales
Sydney, Australia

R. S. Rajeev
Polymers and Special Chemicals Division, Propellant, Polymers, Chemicals and Materials Entity
Vikram Sarabhai Space Centre
Thiruvananthapuram, India

R. Narasimha Rao
Department of Mechanical Engineering
National Institute of Technology Warangal
Warangal, India

Bidita Salahuddin
ARC Centre of Excellence for Electromaterials Science and Intelligent Polymer Research Institute
University of Wollongong
Wollongong, Australia

Biswanath Samanta
Department of Chemistry
Indian Institute of Engineering Science and Technology (IIEST), Shibpur
Howrah, India

Kartick K. Samanta
Chemical and Biochemical Processing Division
ICAR – National Institute of Natural Fibre Engineering and Technology
Kolkata, India

G. Santhosh
Propellant Engineering Division, Propellant, Polymers, Chemicals and Materials Entity
Vikram Sarabhai Space Centre
Thiruvananthapuram, India

Debasis Sariket
Department of Chemistry
Indian Institute of Engineering Science and Technology (IIEST), Shibpur
Howrah, India

M. J. Schulz
University of Cincinnati
Cincinnati, Ohio

B. Sengupta
Department of Metallurgical Engineering, School of Mines and Metallurgy
Kazi Nazrul University
Asansol, India

V. N. Shanov
University of Cincinnati
Cincinnati, Ohio

Sanjib Shyamal
Department of Chemistry
Indian Institute of Engineering Science and Technology (IIEST), Shibpur
Howrah, India

Nikhil K. Singha
Rubber Technology Centre
Indian Institute of Technology Kharagpur
Kharagpur, India

Charles C. Sorrell
School of Materials Science and Engineering
UNSW Sydney
Sydney, Australia

M. Sreejith
Propellant Engineering Division, Propellant, Polymers, Chemicals and Materials Entity
Vikram Sarabhai Space Centre
Thiruvananthapuram, India

Selvin P. Thomas
Department of Chemical Engineering
Yanbu Industrial College
Royal Commission for Jubail and Yanbu
Yanbu Al-Sinaiyah, Saudi Arabia

W. Jon Williams
National Institute for Occupational Safety and Health
Washington, District of Columbia

Vienna C. Wong
School of Materials Science and Engineering
UNSW Sydney
Sydney, Australia

S. Yarmolenko
North Carolina Agricultural and Technology
Greensboro, North Carolina

G.H. Yeoh
School of Mechanical and Manufacturing Engineering
UNSW Sydney
Sydney, Australia

Xiaoran Zheng
School of Materials Science and Engineering
UNSW Sydney
Sydney, Australia

Vicki Zhong
School of Materials Science and Engineering
UNSW Sydney
Sydney, Australia

Preface

We, co-editors, Dr Sri Bandyopadhyay of Sydney, Australia, and Dr R Gujjala of NITW, India, thank CRC Press for encouraging us to develop, prepare the concept and create this book *Toughened Composites: Micro and Macro Systems*. Our special acknowledgement goes to Dr Gagandeep Singh and his CRC Press Team with whom we have co-worked on this book. Also, we are grateful to the international reviewers of this book who recommended this book to CRC Press.

Preface

1 Introduction
In situ SEM Identification of High-Resolution Crack Tip Micro-Deformation in Some Polymer Matrix Materials

Sri Bandyopadhyay
UNSW Sydney
Aus Defence DSTO MRL

CONTENTS

1.1 INTRODUCTION

Walker et al. (1981) used optical microscopy and some other indirect techniques to measure initiation and opening of crack tips under low or medium magnification level. Hong and Gurland (1981) conducted *in situ* electron microscope stage tests to observe crack propagation on the surface of notched specimens of metal matrix composite recording sequential pictures of crack initiation and propagation. Sequential pictures of the fracture process were found useful for studying the microstructural aspects of crack initiation and crack growth. Some limitations were that (i) optical microscopes at moderately high magnification display a poor depth of field and (ii) examining after-failure fracture surfaces has an effect that significant number of features of scanning electron microscope (SEM) fracture can get lost. The author's work on specific *in situ* SEM crack propagation study of selected polymers is in ref (Bandyopadhyay 1984). Kikukawa et al. (1982) reported crack tip features under fatigue crack extension in a 3% Si-steel. Smith et al. (1976) reported observation of ductile fracture in the polymer poly(vinyl chloride) from existing diamond cavities by the use of a video recorder at very low magnifications somewhat similar to the work done by other researchers (Cornes and Haward 1974; Haward and Owen 1977). The author of this chapter presents earlier study done by him (Bandyopadhyay 1984), recording the video of in situ SEM crack tip behaviour that can be observed at various magnifications during slow and fast crack propagation.

DOI: 10.1201/9780429330575-1

1.2 EXPERIMENTAL DETAILS

1.2.1 SEM Studies of Crack Tip Micromechanics Features in Polymer Matrix

The work was conducted at Australian Defence Science and Technology Organisation Materials Research Laboratory, Melbourne, by scheduling study of crack propagation in bulk matrices of a range of polymer specimens conducted inside the chamber of a SEM where notched polymer specimens were attached inside specially designed tensile stage, shown schematically in Figure 1.1 (Bandyopadhyay 1984) displaying that a dumbbell-shaped specimen containing a sharp razor-made edge crack was loaded under tension in a Cambridge Scientific Instruments manufactured 'displacement controlled' tensile specimen stage, within a Cambridge Stereoscan 250 MK 2 Electron Microscope. Under the experiment, the continuous observation and recording of the crack tip were carried out using a Sony U-matic Video Recorder possessing at 25 frames/second which received the crack tip signals directly from the Stereoscan SEM. In case of slow crack propagation, the under-load crack tips' still pictures were also recorded by transferring/reverting to the normal frame raster. Such momentarily stopped crosshead's movement for such recording had no effective effect on the stages of follow-up extension of the crack. In addition, the specimen's simultaneous load–elongation plot during crack propagation was obtainable.

The used specimens had 25–30 mm gauge length, 4–6 mm width and up to 1.5 mm thickness. Specimens with and without coating were used. In majority of experiments, the SEM voltage used was maintained at a low value of 2.5 kV ensuring a minimum damage to the polymer surface and interior due to the irradiation caused by the electron beam. Observation of the crack tip was conducted using secondary electrons and tilting the specimen with a tilt of 45°. The movement speed of the crosshead was kept in the range between 0.5 and 2 mm/minute.

The crack tip situations were studied in (i) semicrystalline ductile polymer 'linear low-density polyethylene' (LLDPE), (ii) engineering rigid polymer 'polymethyl methacrylate' (PMMA) and (iii) 'carbon black-filled vulcanised natural rubber'. This selection of the structurally and modulus-wide different selected polymers demonstrated the potential of the in situ SEM methodical technique to achieve the selective objectives. Worth noting that as from the video record – if a feature was

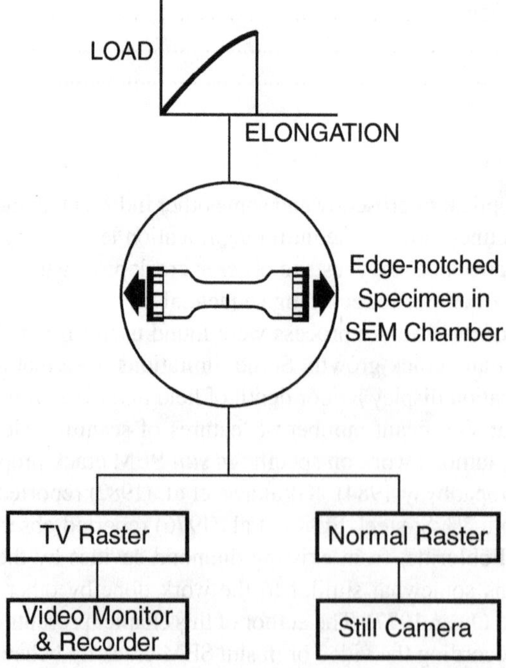

FIGURE 1.1 Experimental set-up diagram for in situ SEM fracture test (Bandyopadhyay 1984).

reproduced, that was obtained by 'freezing the frame on the video monitor thereby subsequent photographing it on the sheet film'. Thus, the technique can be explored in obtaining new information on structure–property relationship aspects as well as fracture resistance parameters of these ranges of materials where, notably, structure variation is achievable through varying (i) composition, (ii) molecular weight, (iii) cross-link density or (iv) processing parameters.

1.3 DUCTILE POLYETHYLENE

Figure 1.2 (Bandyopadhyay 1984) shows picture of an LLDPE specimen with a sharp perpendicular razor crack before loading in tension. It can then be observed that upon tensile loading of the crack containing LLDPE specimen, yielding happened at the crack tip and the loading makes the crack tip start blunting (Figure 1.3).

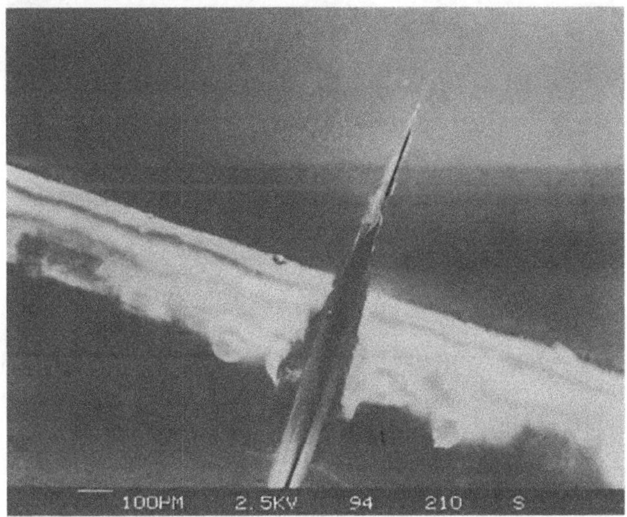

FIGURE 1.2 An LLDPE specimen with a sharp razor cut in SEM; scale: 100 μm (Bandyopadhyay 1984).

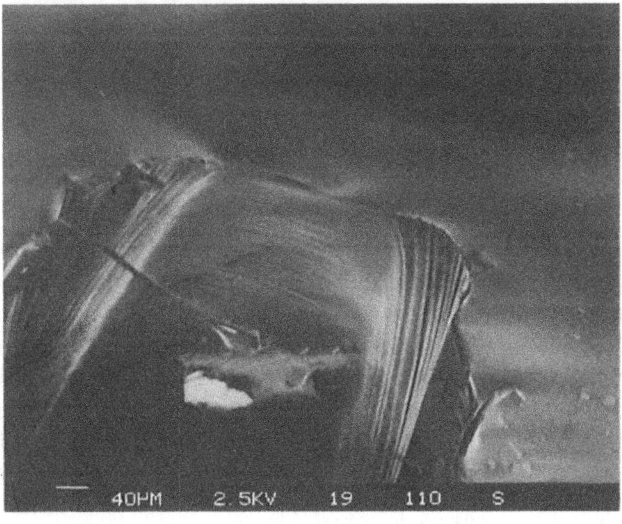

FIGURE 1.3 Crack tip blunting in high-molecular-weight LLDPE specimen under tensile loading (Bandyopadhyay 1984).

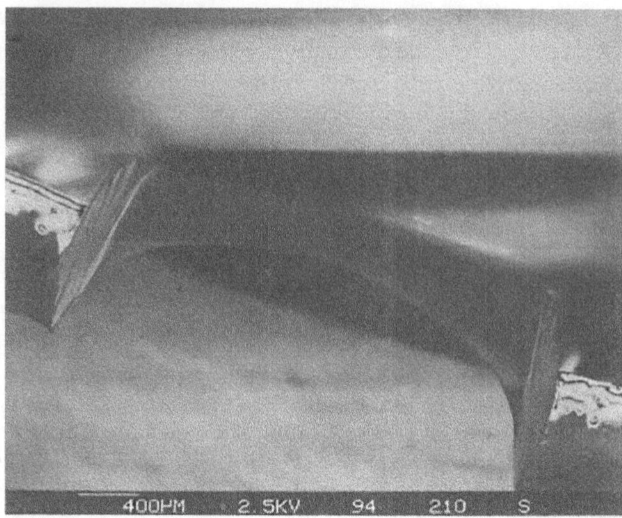

FIGURE 1.4 *In situ* SEM crack tip view in high-molecular-weight LLDPE specimen after long stretching (Bandyopadhyay 1984).

FIGURE 1.5 *In situ* crack tip blunting in low-molecular-weight LLDPE specimen upon tension; scale: 40 µm (Bandyopadhyay 1984).

Figure 1.3 (Bandyopadhyay 1984) shows the appearance of the crack tip at this stage, and Figure 1.4 (Bandyopadhyay 1984) represents a subsequent crack tip situation after substantial stretching in high-molecular-weight materials (as well as in quenched specimens).

On the other hand, Figures 1.5–1.7 (Bandyopadhyay 1984) (taken from *in situ* SEM video) in a slow cooled specimen of a low-molecular-weight LLDPE starting with a similar sharp razor cut as shown in Figure 1.2 (Bandyopadhyay 1984) highlight in sequence (i) the crack tip blunting (Figure 1.5), (ii) transformation of the material at the crack tip into fibrils (Figure 1.6) and then followed by breakdown and propagation of the crack by this means (Figure 1.7) until final failure happened.

Figure 1.8 displays two separated halves of the LLDPE specimen immediately after failure, making clear that fibrillation was confined to a very narrow region.

FIGURE 1.6 Fibrillation ahead of crack tip of low-molecular-weight LLDPE as in Figure 1.5; scale: 40 μm (Bandyopadhyay 1984).

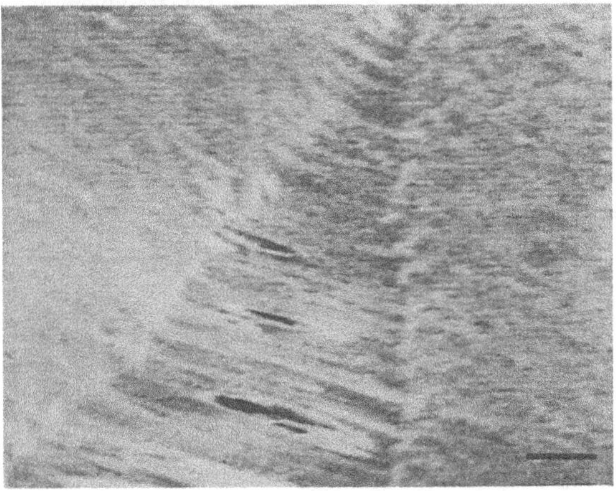

FIGURE 1.7 Crack propagation by fibril formation/breakdown in low-molecular-weight LLDPE; scale 40 μm (Bandyopadhyay 1984).

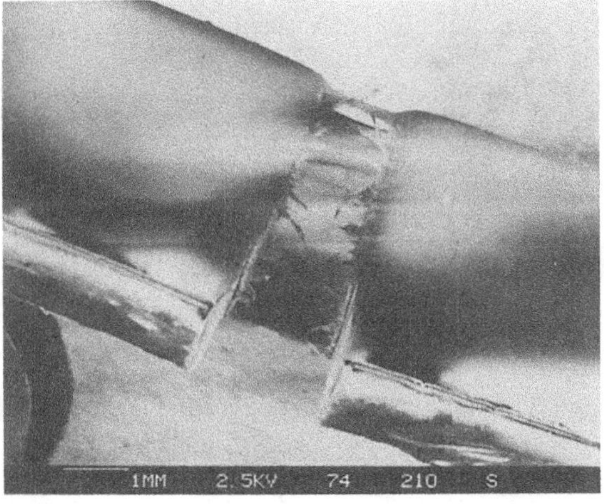

FIGURE 1.8 Two separated halves of the low-molecular-weight LLDPE specimen after failure; scale 1 mm (Bandyopadhyay 1984).

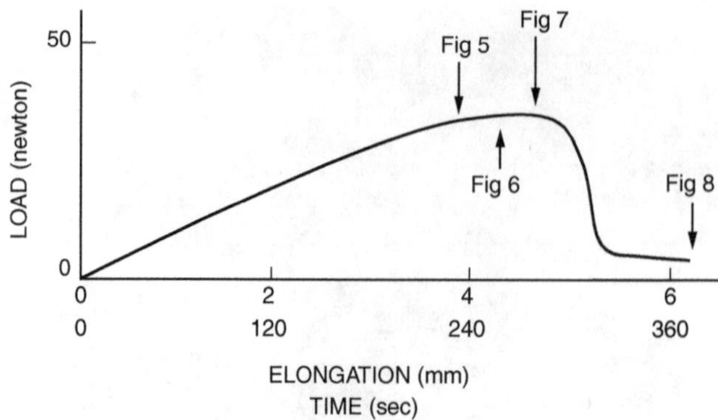

FIGURE 1.9 Load–elongation graph for crack tips shown in Figures 1.5–1.8 (Bandyopadhyay 1984).

Figure 1.9 (Bandyopadhyay 1984) shows four stages of the tensile load–elongation plots of LLDPE sharp crack specimen as shown in Figures 1.5–1.8 (Bandyopadhyay 1984). The two different types of crack tip behaviour shown in Figures 1.3–1.8 (Bandyopadhyay 1984) may be due to the difference in the crystalline texture, number and configuration of the tie molecules and work hardening ability of the corresponding materials (Haward and Owen 1977).

1.4 BRITTLE POLYMETHYL METHACRYLATE (PMMA)

Figures 1.10–1.12 (Bandyopadhyay 1984) (taken from video) highlight the crack tip situations during crack propagation in a brittle PMMA specimen where Figures 1.10 and 1.11 (Bandyopadhyay 1984) present the crack tips in early stage of crack propagation (after initiation took place from a very shallow razor cut) and Figure 1.12 (Bandyopadhyay 1984) depicts the critical crack tip (just 0.02 seconds before the specimen failed catastrophically). Figures 1.10 and 1.11 highlight that crack tip opening during crack advancement is just a few microns. It is quite apparent that the Dugdale-type plastic zone (Dugdale 1960) at the crack tip seen in Figure 1.10 did not exist when the critical crack propagation initiated as observed in Figure 1.12. Interestingly, Figure 1.10 seems to provide good experimental support to the crack tip model developed by Williams (1972) who suggested that crack tip opening displacement and plastic zone size in PMMA be 2 and 10 μm, respectively. Figure 1.13 (Bandyopadhyay 1984) shows the two separated halves which indicate that a clean brittle fracture had happened in the PMMA specimen.

Figure 1.13 shows two halves of the PMMA specimen immediately after fracture.

FIGURE 1.10 Early stage of crack propagation in brittle PMMA specimen; scale 20 μm (Bandyopadhyay 1984).

FIGURE 1.11 Advancement of crack in PMMA from Figure 1.10; scale 20 μm (Bandyopadhyay 1984).

FIGURE 1.12 Critical crack tip situation in PMMA specimen; scale 20 μm (Bandyopadhyay 1984).

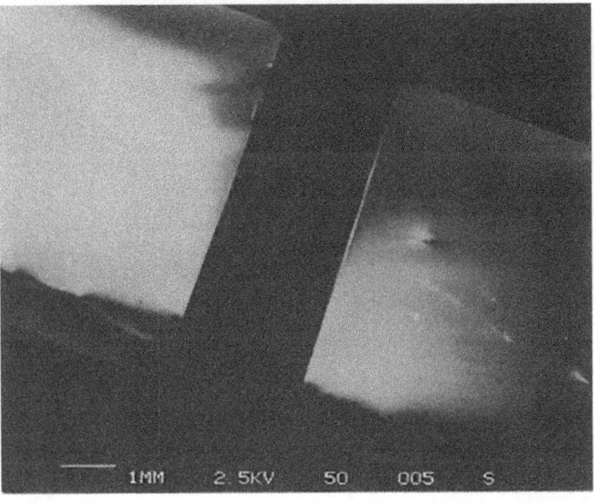

FIGURE 1.13 Two halves of immediately fractured brittle PMMA after Figure 1.12 (Bandyopadhyay 1984).

1.5 NATURAL RUBBER: A SUPER-ELASTIC MATERIAL

A crack tip situation during slow crack propagation in a natural rubber specimen has been shown in the next two figures to provide an interesting example of crack tip behaviour in a filled elastomer. The low magnification picture shown in Figure 1.14 (Bandyopadhyay 1984) indicates crack propagation by tearing. In Figure 1.15 (Bandyopadhyay 1984), the crack tip has been observed at a higher magnification, where the deviation in the crack path may have occurred when the crack tip met a hard filler particle – such as carbon black.

Figures 1.1–1.15 provide an indication of the potential of the in situ SEM method in understanding the crack tip micromechanics during crack initiation and propagation in different polymeric materials possessing ductile, brittle and super-elastic features. Further, elongation plots of the specimens which are simultaneously obtained during crack propagation do identify the different stages of crack propagation (including the point of incipient crack opening) on these plots as illustrated in Figure 1.9, thus showing it a useful and valid tool/methodology understanding and assessing the

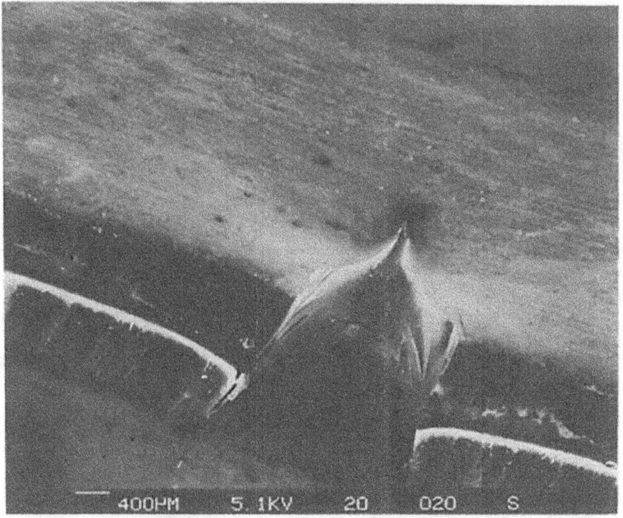

FIGURE 1.14 Crack growth in carbon black-filled vulcanised natural rubber; scale 400 μm (Bandyopadhyay 1984).

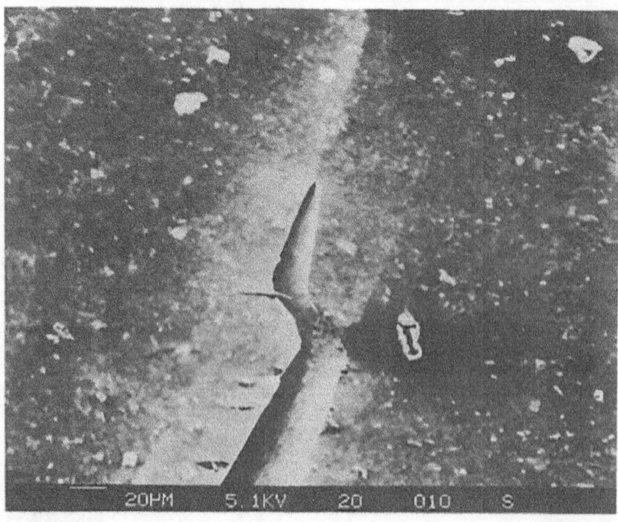

FIGURE 1.15 High magnification picture of crack tip in Figure 1.14; scale 20 μm (Bandyopadhyay 1984).

fracture characteristics of these types of polymer materials which form useful matrices in composites (Markström 1977; Begley and Landes 1972).

1.6 CRACK PROPAGATION VIDEO PICTURES IN NEAT AND CARBOXYL-TERMINATED ACRYLONITRILE-BUTADIENE (CTBN) RUBBER-MODIFIED EPOXY

In a subsequent publication (Bandyopadhyay 1990), Bandyopadhyay released his further research on crack propagation studies, where a grid was placed on the fracture test specimen by vapour depositing gold through a 17 μm copper mesh on the test specimen as shown in Figures 1.16 and 1.17 where Figure 1.16 shows deformation in 828 – 5 phr (parts per 100 resin) piperidine brittle polymer single-edge-notched (SEN) specimen and Figure 1.17 shows crack tip deformation in a *CTBN rubber-modified 828 – 5 phr piperidine* composite specimen.

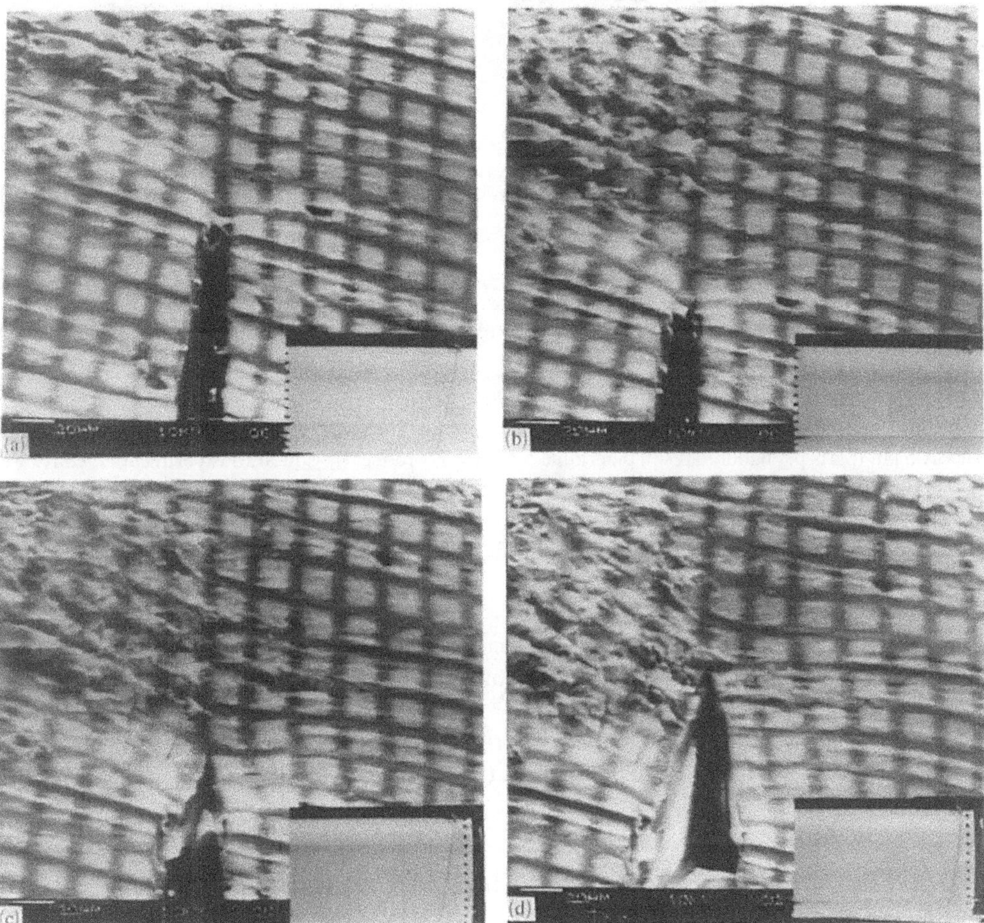

FIGURE 1.16 17 μm copper mesh through gold-coated view of online video crack growth in brittle 828 – 5 phr piperidine single-edge-notched specimen in SEM. SEM online video Crack tip photographs reproduced from a video record of in-situ crack propagation in a brittle (unmodified) 828-5 phr (parts per hundred) piperidine single edge notch specimen in the scanning electron microscope. The discrete grid around the crack tip was obtained by vapour depositing gold through a 17 micro-meter copper mesh : picture (a) shows crack tip opening picture under tensile load; picture (b) shows initiation of a sharp natural crack in the specimen upon loading; picture (c) shows advancement of the sharp natural crack, with micro-cracks appearing near the crack tip; picture (d) shows catastrophic stage of failure of the specimen. The increasing load at each of these stages should be noted in the load displacement diagrams at the bottom right corner of each SEM picture (Bandyopadhyay 1990).

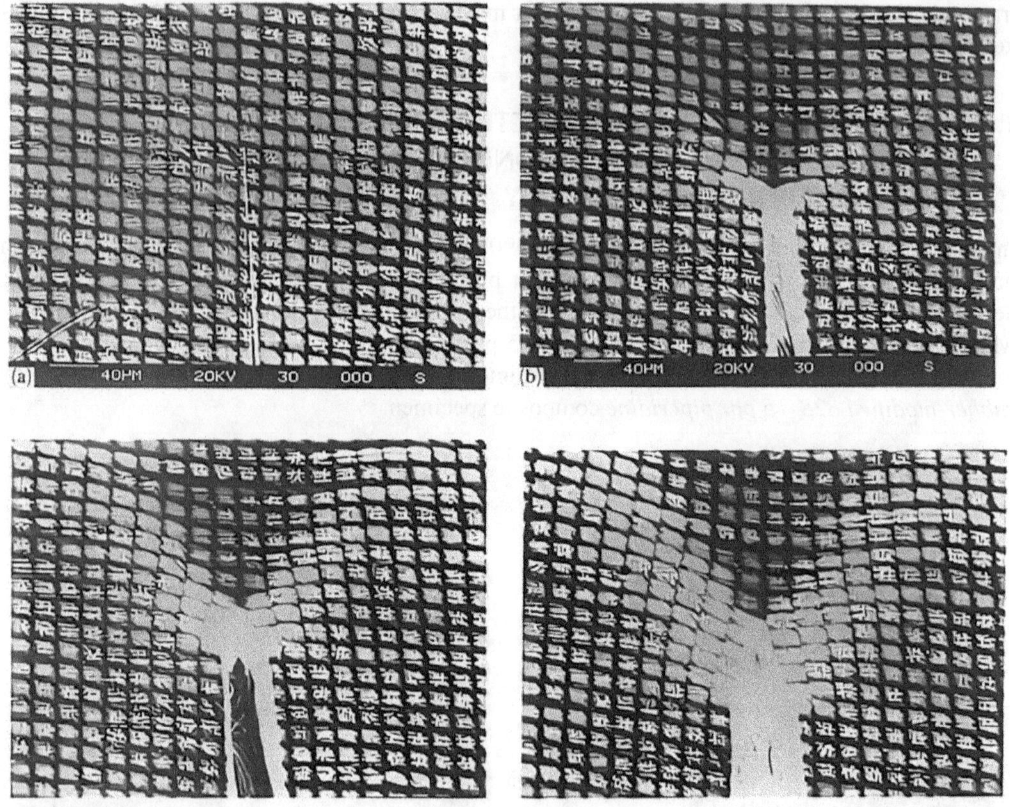

FIGURE 1.17 SEM online crack tip opening/ductile crack growth in rubber-toughened epoxy (Bandyopadhyay 1990). SEM on-line video pictures of Crack tip opening and slow crack growth sequence in a specimen of DGEBA-5 phr (parts per hundred resin) piperidine-5 phr CBTN rubber (in-situ SEM experiment). Pictures were taken in the still mode by stopping the cross-head movement : picture (a) top-left showing sharp starter crack; picture (b) top-right showing maximum crack tip opening (crack tip blunting); picture (c) bottom-left initiation of a crack in the specimen, and picture (d) bottom-right showing slow growth of the crack.

Figure 1.17 shows *(in situ* SEM experiment) online video pictures of crack tip opening and slow crack growth sequence in a ductile specimen of *DGEBA-5 phr piperidine-reinforced by 5 phr CTBN rubber.*

Figure 1.18 shows sequence of crack tip deformation in a commercial rubber-toughened epoxy containing 5 parts (per hundred) of asbestos filler (SEN specimen, *in situ* SEM experiment; pictured in still mode) (Bandyopadhyay 1990).

1.7 EFFECTS OF STRAIN RATE, AND FRACTURE UNDER IMPACT CONDITIONS IN RUBBER-MODIFIED EPOXY RESIN SPECIMEN

Fracture energy G_{ic} decreases with increased strain rate: (i) with stick-slip behaviour at lower strain rates and (ii) with continuous crack propagation at higher strain rate. These aspects are presented in Figure 1.19 (Bandyopadhyay 1990) in SEN rubber-modified epoxy resin specimen.

An elaborated SEM-based fracture study research on microscopic and macroscopic aspects of fracture of unmodified and modified epoxy resins involving research co-supervised by the author can be seen in references (Bandyopadhyay 1990; Cholake et al. 2016). A fibre–matrix 'de-bonding' model for multi-fibre-reinforced composites is displayed in Figure 1.20 (Bandyopadhay and Murthy 1975).

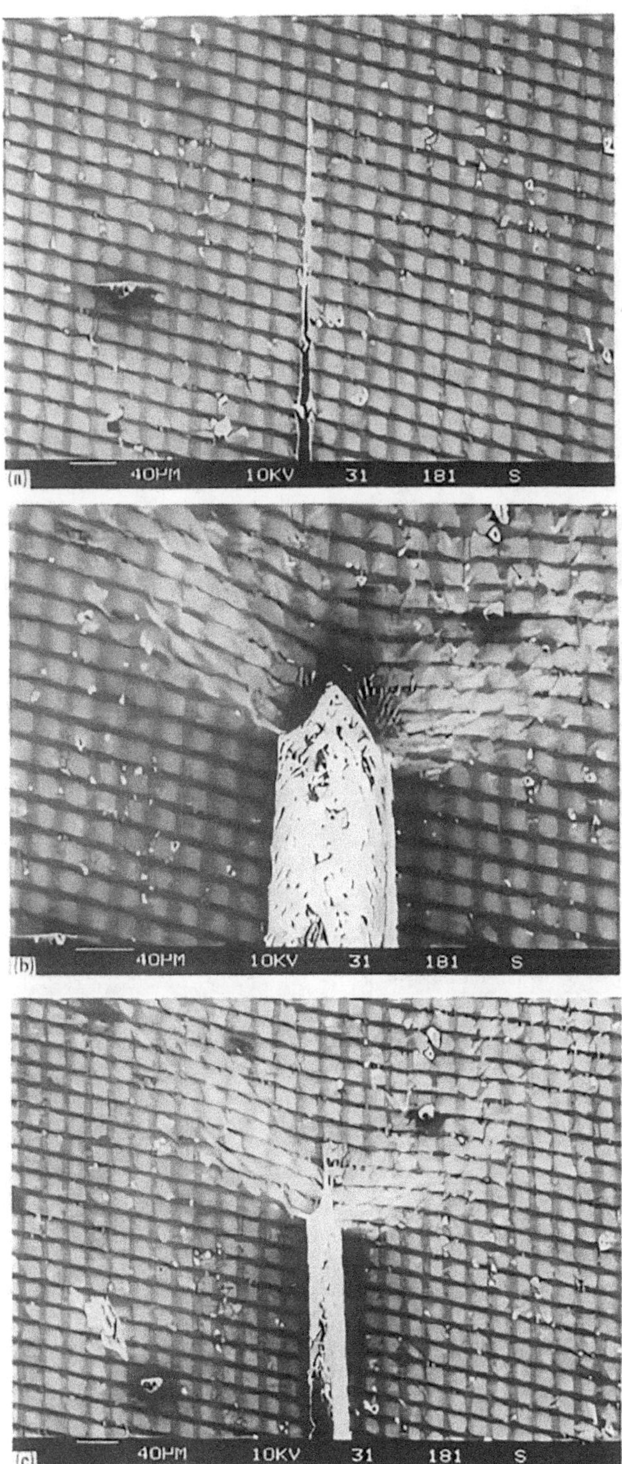

FIGURE 1.18 Crack tips in commercial rubber-toughened epoxy containing ceramic fibres. Still mode pictures: (a) shows sharp crack tip prior to loading, no sign of voids; (b) shows slow crack advance where formation of voids ahead of crack tip can be seen, and asbestos fibres are also visible across the crack edges; (c) shows relaxed crack tip upon unloading, meaning crack tip closure, residual deformation and unclosed voids (Bandyopadhyay 1990).

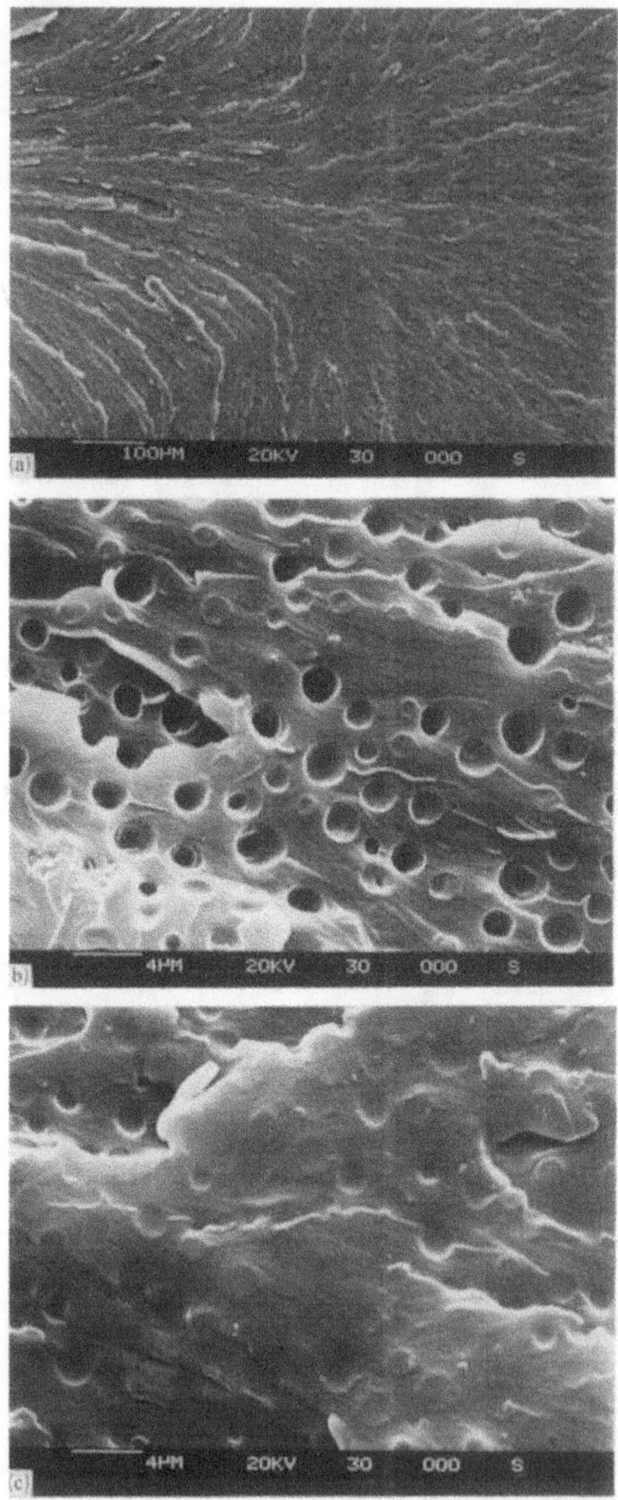

FIGURE 1.19 Fracture surface of SEN rubber-toughened epoxy sample at various speeds. Fracture surfaces of an SEN rubber-modified epoxy resin: (a) top micrograph : transition from slow speed crack advance (left) to fast growth; (b) middle micrograph : high magnification view of crack propagation through the rubber particles in the slow growth region; (c) bottom micrograph: crack propagation mostly through the epoxy matrix, around the rubber particles in the fast growth region (Bandyopadhyay 1990).

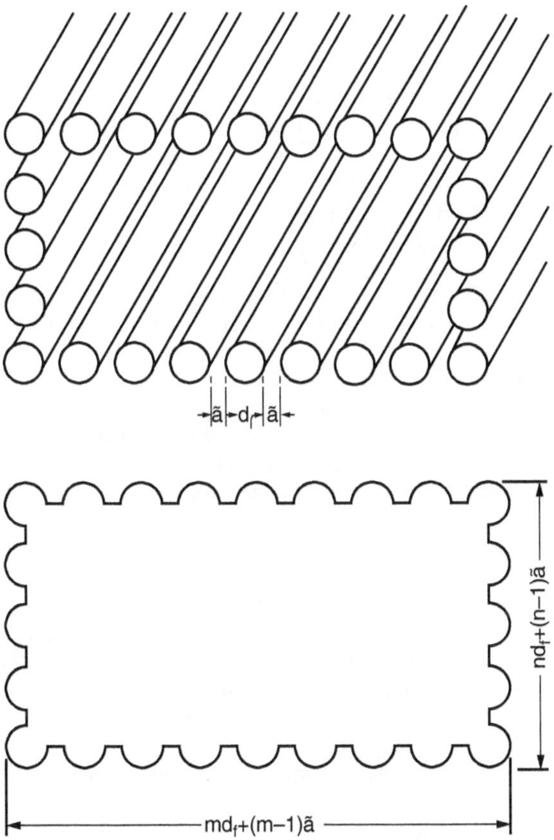

FIGURE 1.20 One-row models of shear surfaces for multiple fibre system (Bandyopadhay and Murthy 1975).

1.8 SUMMARY

The features presented here have given an indication of the potential of this in situ SEM method in understanding the crack tip micromechanics during crack propagation in different materials. Further, as the load–elongation plots of the specimens are simultaneously obtained during crack propagation, it is possible to identify the different stages of crack propagation (including the point of incipient crack opening) on these plots as illustrated in Figure 1.9 (Bandyopadhyay 1984), thereby a useful tool in the evaluation of their fracture properties (Williams 1972; Markström 1977; Begley and Landes 1972; Srinivasan and Bandyopadhyay 2016; Hauptstein et al. 2011; Nath and Bandyopadhyay 2011; Naval Materials Science and Engineering Course Notes n.d.; Bandyopadhyay and University of New South Wales 2006).

REFERENCES

Bandyopadhay S, and PN Murthy. 1975. Experimental studies on interfacial shear strength in glass fibre reinforced plastics systems. *Materials Science and Engineering* 19(1): 139–45.

Bandyopadhyay S. 1984. Crack propagation studies of bulk polymeric materials in the scanning electron microscope. *Journal of Materials Science Letters* 3(1): 39–43.

Bandyopadhyay S. 1990. Review of the microscopic and macroscopic aspects of fracture of unmodified and modified epoxy resins. *Materials Science and Engineering: A* 125(2): 157–84.

Bandyopadhyay S. and University of New South Wales. 2006. *Developments in Composites : Advanced, Infrastructural, Natural & Nano-Composites : ACUN-5 International Composites Conference, 11–14 July, 2006, University of New South Wales, Sydney, Australia*. University of New South Wales, 705.

Begley JA and JD Landes. 1972. The J Integral as a Fracture Criterion. Fracture Toughness: Part II. ASTM International, West Conshohocken, PA, 1–23.

Cholake ST, G Moran, B Joe, Y Bai, RK Singh Raman, XL Zhao, S Rizkalla, and S Bandyopadhyay. 2016. Improved mode I fracture resistance of CFRP composites by reinforcing epoxy matrix with recycled short milled carbon fibre. *Construction and Building Materials* 111(May): 399–407.

Cornes PL and RN Haward. 1974. Ductile fracture of rigid poly(vinyl chloride). *Polymer* 15(3): 149–56.

Dugdale DS. 1960. Yielding of steel sheets containing slits. *Journal of the Mechanics and Physics of Solids* 8(2): 100–104. doi:10.1016/0022-5096(60)90013-2.

Hauptstein A, S Bandyopadhyay, and D Lambino. 2011. *E-Beam Deformation of Ceramic Particles*. LAP Lambert Academic Publishing Germany.

Haward RN and DRJ Owen. 1977. The detergent stress-cracking of polyethylene. *Proceedings of the Royal Society of London. A. Mathematical and Physical Sciences* 352(1671): 505–21.

Hong J and J Gurland. 1981. Direct observation of surface crack initiation and crack growth in the scanning electron microscope. *Metallography* 14(3): 225–36.

Kikukawa M, M Jono, M Iwahashi, M Ichikawa, and N Uesugi. 1982. Fatigue testing apparatus assembled in a field-emission scanning electron microscope and direct observation of fatigue crack propagation. *Journal of the Society of Materials Science, Japan* 31(346): 669–74.

Markström K. 1977. Experimental determination of jc data using different types of specimen. *Engineering Fracture Mechanics* 9(3): 637–46.

Nath D and S Bandyopadhyay. 2011. *High Strength Polymer Fly Ash Composites: Fundamental Understanding*. Lambert Academic Publishing Germany.

Naval Materials Science and Engineering Course Notes. n.d. U.S. Naval Academy. https://www.usna.edu/NAOE/_files/documents/Courses/EN380/Course_Notes/Ch11_Fracture.pdf.

Smith K, MG Hall, and JN Hay. 1976. Observation of ductile fracture in poly(vinyl chloride) with a scanning electron microscope. *Journal of Polymer Science: Polymer Letters Edition* 14(12): 751–55.

Srinivasan A and S Bandyopadhyay. 2016. *Advances in Polymer Materials and Technology*. First ed. CRC Press, Boca Raton.

Walker N, RN Haward, and JN Hay. 1981. The growth rates of diamond cavities in polycarbonate. *Journal of Materials Science* 16(3): 817–24.

Williams JG. 1972. Visco-elastic and thermal effects on crack growth in PMMA. *International Journal of Fracture Mechanics* 8(4): 393–401.

2 Polystyrene-Based Composites and Their Toughening Mechanisms

Selvin P. Thomas
Yanbu Industrial College

CONTENTS

2.1 INTRODUCTION

Polystyrene (PS) is a brittle material even though it is having good mechanical properties such as tensile and flexural properties, clarity and processability. The material's impact properties and toughness are not that superior. Therefore, many attempts were made to improve its toughness properties.

PS was first produced commercially in the 1930s, and the ready availability of styrene feedstock has helped it to grow. Styrene will polymerize spontaneously on heating in an oxygen-free atmosphere, but catalysts are added to ensure complete polymerization at lower temperatures. Processes have been designed to aid heat transfer from the exothermic reaction, which can lead to low-molecular-weight polymers being formed if not controlled. Two main types of polymers are produced: crystal which is a clear, amorphous resin with good stiffness and electrical properties; and impact which contains varying levels of polybutadiene to improve toughness and impact resistance. There are three types of processes generally used: suspension, solution and bulk polymerization. The advantages of the solution route, which can be continuous or batch operation, are low residual monomer content and high purity polymers. The suspension route produces polymers of different molecular weights and can make specialist crystal and high impact grades of PS.

2.2 TOUGHENING IN POLYSTYRENE

PS generally has good stiffness, transparency, processability and dielectric qualities that make it useful. However, its low impact resistance at low temperatures makes catastrophic fracture failure (Zhang et al. 2013). The most widely used version of toughened PS is called high impact polystyrene or HIPS. Being cheap and easy to thermoform, it is utilized for many everyday uses. HIPS is made by polymerizing styrene in a polybutadiene rubber solution. After the polymerization reaction begins, the PS and rubber phases separate. When phase separation begins, the two phases compete

DOI: 10.1201/9780429330575-2

for volume until phase inversion occurs and the rubber can distribute throughout the matrix. The alternative emulsion polymerization with styrene–butadiene–styrene (SBS) or styrene–butadiene copolymers allows fine-tuned manipulation of particle size distribution. This method makes use of the core–shell architecture (Rovere et al. 2008).

The generation of vast quantities of waste rubber from car tires has sparked interest in finding uses for this discarded rubber. The rubber can be turned into a fine powder, which can then be used as a toughening agent for PS. However, poor miscibility between the waste rubber and PS weakens the material. This problem requires the use of a compatibilizer in order to reduce interfacial tension and ultimately make rubber toughening of PS effective. A PS/styrene–butadiene copolymer acts to increase the adhesion between the dispersed and continuous phases.

PS is toughened with various means such as the addition of rubbery materials, particulates and other polymers. Table 2.1 provides the details of toughening of PS by different agents and the results achieved.

The important methods are described in the following section.

2.2.1 Toughening by Rubbery Materials

Rubber toughening of PS is a prominent research area dating back to the 1970s. The main target was to improve the fracture resistance along with the retention of inherent properties of the parent polymer such as appearance, processing ease and stiffness. Even though some of the properties showed reduction, a balance of properties was achieved. Bucknall systematically reviewed the use of rubber as toughening agent and compiled the results in 1982 (Bucknall 1982). There were several reports on the toughening of PS by blending with rubbers. Examples of such rubbers are styrene–butadiene (SBR) (Martinez et al. 2000), natural rubber (NR) (Neoh and Hashim 2004), polybutadiene (BR) (Mathur and Nauman 1999), ethylene-propylene-diene terpolymer (EPDM) (Crevecoeur et al. 1995; Shaw and Singh 1990, 1990a,b), acrylonitrile–butadiene rubber (Mathew and Thomas 2003) and nitrile rubber (NBR) (Sreenivasan and Kurian 2007). It is generally accepted that the rubber particles in the PS/rubber blend work as stress concentrators to initiate a massive number of crazes in the PS matrix. The crazes will keep growing in the direction perpendicular to the principal applied stress until they encounter and are stabilized by the neighboring rubber particles. A number of factors, related to the rubber component, have been identified as affecting the toughness of these systems, including the volume fraction of the rubber phase, its chemical composition, particle morphology, level of adhesion to the matrix, type of rubber and, most importantly, the rubber particle size and its size distribution (Gopalakrishnan and Kutty 2015). HIPS toughened with rubber particles was studied by many authors in order to explain the toughening, and it was established that nucleation and growth of crazes in the matrix play major roles (Bucknall and Smith 1965). The blending of PS with emulsion polymerized core–shell rubber particles of 0.1–5 μm was reported as a means of toughening of the matrix. The advantage of this method is that the particles retain the morphology even during the blending process due to the cross-linking nature (Schneider et al. 1997; Vázquez et al. 1995).

Emre Tekay did a significant work on developing tough PS blends by melt compounding technique recently (Tekay 2020). PS/poly(styrene–b-isoprene–b-styrene) (SIS)/organophilic halloysite nanotube (Org-HNT) blend nanocomposites having 20%–40% SIS elastomer showed improvements in both toughness and impact strength compared to neat PS. All the nanocomposites exhibited continuous/fibrillar morphologies with smaller elastomer domains and higher tensile modulus and toughness. Among them, the nanocomposite having 7 phr Org-HNT and 30% SIS phase (7H-30SIS) exhibited the highest impact strength with enhanced tensile properties. The same nanocomposite exhibited about 21% and 100% increments in the modulus and toughness in comparison with its blend, respectively.

Styrene–butadiene rubber (SBR) was used as a toughening agent for PS/SBR blends to overcome the brittleness of the matrix material (Veilleux and Rodrigue 2016). Luna et al. studied the

TABLE 2.1
Examples of Toughening in Polystyrene-Based Blends and Composites

Toughening Strategy	Polystyrene Type	Toughening Agent	Compatibilizer Used	Results	Ref
Adding compatibilizers	Styron D685	EPR	SEBS	Toughness has not improved appreciably for the blends	Cigana et al. (1997)
	PS GP and Styron® 484	PB	PS-b-PB	Toughness improved on lower rubber concentrations	Mathur and Nauman (1999)
	PS (666D)	NBR	SBS	The graft reaction between PS and EPDM increased the toughness for the PS/EPDM/SBS blends	Fang et al. (2004)
	PS (GPPS 525)	PB	PB-g-PS	The notched impact strength of HIPS/PS/PB-g-PS was higher than expected as a linear average of HIPS and PS/PB-g-PS due to synergism of micrometer salami rubber particles and submicron-sized core–shell PB-g-PS particles.	Zhu et al. (2013)
Grafting a polymer on the surface of preformed rubber particles	General-purpose PS	PB	PS	Toughening mechanism of PS/PB-g-PS blends is shear yielding of the matrix promoted by cavitation	Gao et al. (2006)
	PS	P(B-co-MMA)	PS	A BA component in the core of about 70 wt % and St in the shell of 30 wt % was the optimum composition of core–shell particles that showed excellent toughening effect	Guo et al. (2003)
	PS (GPPS 525) and HIPS (POLYREXPH-88)	PB and PB-g-PS copolymers	-	When the rubber particle size was above 300 nm, the notched impact strength was 208 J/m, nearly two times that of toughened PS with particle size of 100 nm, and multiple crazing seems to be the toughening mechanism.	Deng et al. (2013)
Processing methodologies	HIPS	PB	-	The increase in toughness was attributed to the special parallel-arranged orientation structure formed by both the pebble-like salami rubber particles with larger surface area and the nanoscale PS filled in the PB phase.	Peng et al. (2014)
	PS (666D) and HIPS (Styron A)	SBR	-	Nanopowdered SBR is more suitable to toughen HIPS than PS at low rubber content.	Li et al. (2007)
	GPPS and HIPS	PB and PS-PB diblock copolymer	-	By the microdispersion of rubber particles that differ in morphology from standard HIPS particles, while keeping the conventional HIPS matrix, toughness showed substantial improvement	Alfarraj and Bruce Nauman (2004)
	HIPS (PH-888G)	-	-	An alternative mechanism for toughening of rubbery PS is proposed	Razavi et al. (2020)
	PS	SEBS		A scalable approach of 'casting–annealing–stretching' to concurrently reinforce and toughen PS is proposed	Zeng et al. (2021)
Adding inorganic materials or fibers	PS (Bycolene®158K)	Carbon nanofibers	SEBS	Simultaneous addition of carbon nanofiber and SEMS improved toughness significantly	Sun et al. (2019)
	PS (SC 206 GPPS)	NR (ISNR-5) and nylon-six fibers	-	Toughened composites based on polystyrene and natural rubber at a ratio of 85/15 were prepared by melt mixing with nylon-6 fibers	Gopalakrishnan and Kutty (2015)
	Styrene polymerized	POSS	-	Toughness showed improvement along with fire retardancy	Blanco and Bottino (2015)
	PS	Multilayer GO (mGO), graphite oxide (GrO) and graphite (Gr)	-	It was observed that the agglomerates toughen the PS polymer matrix. Well-controlled agglomerates can be extremely desirable, and the superlubricity state can also be observed in the polymer in the solid state.	Ferreira et al. (2020)

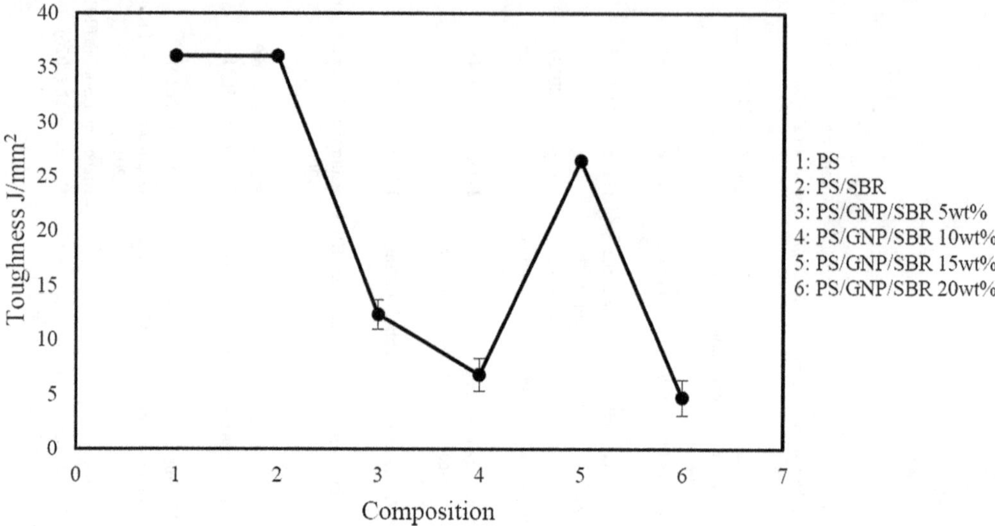

FIGURE 2.1 Impact properties of SBR-toughened PS composites (Ghani et al. 2020).

toughening of PS using different contents of white vulcanized styrene–butadiene rubber (SBRr) waste produced in the shoe industry (Luna et al. 2018). The composites were prepared by using twin screw extruder, followed by injection molding technique. The increase in SBRr concentration showed an increase in viscosity as per rheological studies, which meant that the stability of the PS got enhanced. The property enhancements indicated that these kinds of elastomeric waste materials can be utilized for hardening and toughening agents for PS materials.

Ab Ghani et al. studied the toughening effect of SBR in PS graphene nanoplatelet (PS/GNP) nanocomposites and showed that the increases in SBR content led to the decrease in toughness and tensile strength as shown in Figure 2.1 (Ghani et al. 2020). PS reinforced with a combination of styrene–butadiene–acrylonitrile and polyethylene as the minor phase (PS/SBR/PE) showed interesting results with respect to toughening (Luzinov et al. 1999). The PS content was a constant 70%, and the minor phase was varied in different ratios. With increase in PE content, a core–shell morphology is formed by the minor phase, which affects the properties of the ternary blends. The earliest report of using recycled rubber crumb for toughening PS is reported by Pittolo and Burford in 1985 (Pittolo and Burford 1985). The toughening of the brittle PS is increased on account of rubber-to-polymer matrix interaction and reduction in particle size of the crumb rubber. The rubber particles rupture during the measurements, leading to the improvement in elongation at break of the polymer. Some more examples for rubber toughening of PS are given in Table 2.1.

2.2.2 Toughening by Compatibilization

As the PS is not compatible with many of the polymers or fillers, suitable compatibilizers were employed to make the materials miscible. There were several reports of PS toughening with rubbery materials such as styrene–butadiene rubber (SBR), SBS, styrene–butylene/ethylene–styrene (SEBS) and ethylene propylene diene (EPDM) available in the literature (Araújo et al. 1997; Ibrahim and Kadum 2012; Wen-Dong et al. 2020; Libio et al. 2012; Luna et al. 2018). In order to improve the toughening properties along with appreciable tensile properties, compatibilizing with PS blends is investigated (Parameswaranpillai et al. 2015; Omonov et al. 2007; Halimatudahliana and Nasir 2002). Luna et al. utilized polypropylene (PP) waste from industrial containers as PS impact modifier using SEBS as a compatibilizing agent recently (Luna et al. 2020). The composition with 20 wt% of recycled PP and 10 wt% of SEBS showed remarkable impact properties and toughness for PS. The morphology development in the blend is clearly due to the compatibilizing effect of SEBS, and

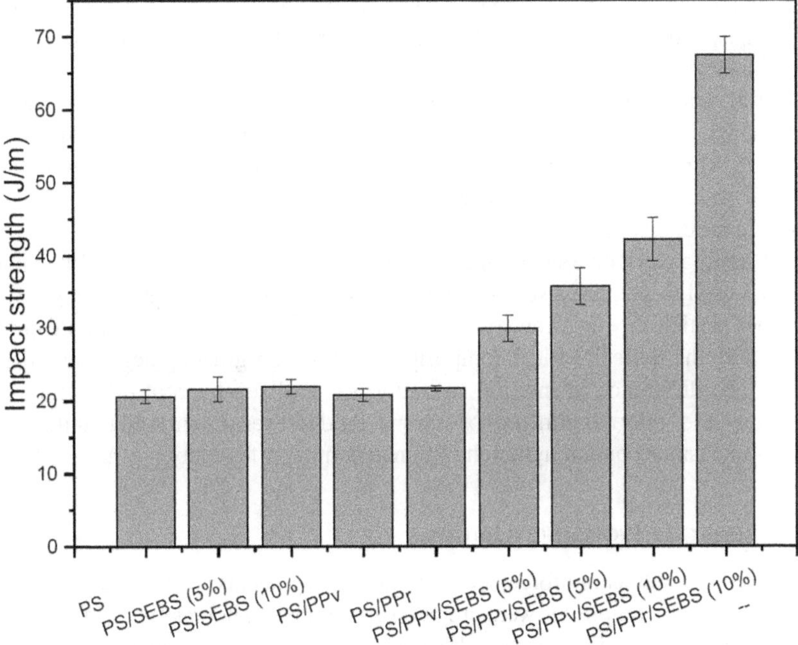

FIGURE 2.2 Impact strength of PS, binary and compatibilized blends (Luna et al. 2020).

it is well explained by morphological studies. The effect on impact strength is shown in Figure 2.2, and the SEBS copolymer is effective in making the PS/PP and PS/PPr mixtures compatibilized with the 10% weight content, because it helped to improve the PS toughening.

An interesting study reported by Sun et al. showed simultaneous improvement in thermal conductivity and toughness for polystyrene/carbon nanofiber composites (PS/CNF) incorporated with elastomer, styrene–ethylene/butylene–styrene (SEBS) (Sun et al. 2019). Both the CNF and SEBS were added in different ratios. The addition of SEBS improved the fracture toughness of the composite largely. For example, the composite with 20 wt% of CNF and 30 wt% of SEBS showed an impact strength of 45.2 kJ/m^2, while the composite with 20 wt% of CNF alone has a corresponding value of 8.9 kJ/m^2 only. The huge increase of almost 410% is due to the enhancement of dispersion of CNF tailored by SEBS, which promotes the assemblage of CNF in the polymer matrix.

Toughening of PS with ethylene propylene diene rubber (EPDM) is a classic example of rubber toughening techniques in polymers. Libio et al. reported that incorporation of EPDM to PS can enhance the mechanical properties and fracture toughness when used with styrenic compatibilizers too (Libio et al. 2012). Even 10 wt% of SBS compatibilization of PS/EPDM blends showed tremendous improvement in impact strength compared to uncompatibilized blend and HIPS.

PS/vulcanized rubber waste blends, compatibilized with SBS, were compared with HIPS in terms of photooxidative stability and toughness characteristics by Luna et al. (2019a). The effect of UV radiation on the mechanical properties on different aging times was surprising, and they were correlated with the interaction of the mineral fillers present in the rubber residue with the matrix in the blends. Impact strength values were higher than the uncompatibilized blends and HIPS for the whole aging days from 15 to 60 days. The higher toughness even for higher loading of compatibilizer is believed to be due to the diffusion of it into the interface and also good dispersion in the matrix.

Luna et al. attempted to incorporate recycled rubber compound from the shoe industry along with SBS compatibilizer in different weight percentages (Luna et al. 2019b). The effect of SBS on the mechanical properties and toughness characteristics of the composites was compared with HIPS. The toughness that was measured based on the impact strength values was similar to that of HIPS for the 5 and 7.5 wt% of compatibilizer. When the compatibilizer is increased to 10 wt%, the impact strength improved around 80% in relation to HIPS. These results are very important in terms of technological aspects with respect to HIPS as it contains a large amount of recycled rubber waste and the composites were prepared at room temperature.

Coutinho et al. made an interesting investigation on the toughening of PS with three types of polybutadiene (PB): low-cis polybutadiene (PBl), high-cis polybutadiene (PBh) and styrene–butadiene block (SBS) copolymer (PBco) (Coutinho et al. 2007). The impact strength increased 138%, 208% and 823% when low-cis polybutadiene, high-cis polybutadiene and styrene–butadiene block copolymers were used, respectively.

Another interesting work on toughening of PS is given by Zhou et al. (2015). Emulsion polymerization has been used to prepare a series of polybutadiene-graft-polystyrene (PB-g-PS) core–shell structural rubber particles by grafting styrene onto polybutadiene seeds. All the PB-g-PS particles were designed with the same chemical composition and similar grafting degree but different internal structures. The difference in internal structure was realized by controlling the allocation of 'external grafting' and 'internal grafting' of styrene. Izod test results showed that the PB-g-PS with large-scale sub-inclusions could toughen the PS matrix more effectively.

2.2.3 Toughening by Inorganic Materials

Inorganic fillers were used as reinforcement and toughening agents for PS recently. Calcium phosphate-reinforced PS nanocomposites were prepared and studied (Thomas et al. 2009). The significant interaction between the nanofiller and PS showed better mechanical properties and toughness and led to the proposal of the formation of the rigid amorphous fraction in the composites (Figure 2.3).

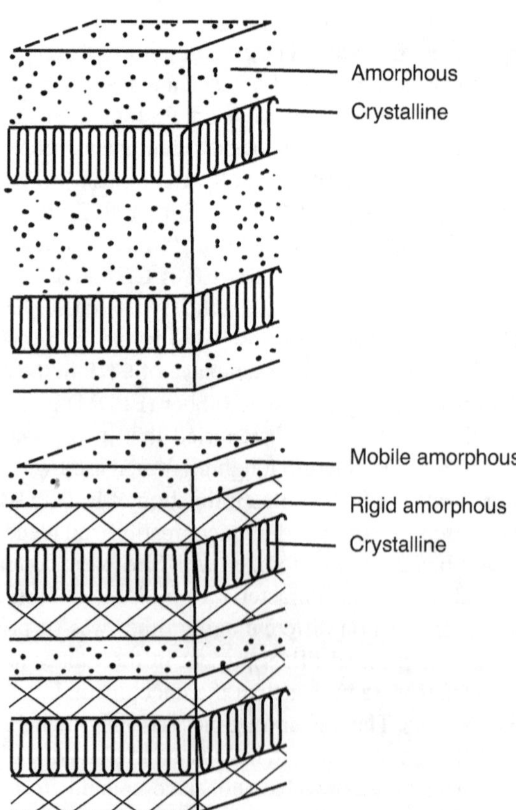

FIGURE 2.3 Simple model showing the formation of rigid amorphous fraction in amorphous polymers (Thomas et al. 2009).

Molybdenum disulfide (MoS_2) in very low quantities was used recently by Rodriguez et al. and reported interesting reports on toughening of PS (Rodriguez et al. 2021). The mechanical property measurements showed that the composite with 0.002 wt% of the two-dimensional filler improved the toughness to 100% than the neat PS. However, increase in weight percentage led to the agglomeration of the fillers, and toughness values decreased due to the initiation of fracture in the matrix. The toughness mechanism of the behavior in these types of PS composites is explained by Vaziri et al. (2011). The toughness mechanism, crack propagation and mechanical behavior of the composites were explained by Vaziri et al. by incorporating silica particles in PS matrix.

Figure 2.4 shows the transmitted optical microscopy (TOM) images of neat PS, 0.75 and 10 wt% PS nanocomposites without and with silica particles. It can be seen from (a) that the neat PS has craze propagation with little deviation with branches. By loading nanosilica to 0.75 wt%, a totally different propagation process is observed, and the deviation of craze propagation increases. With the good dispersion of silica particles at low loadings, more and more craze paths are formed in the sample. When the loading of the silica is increased to 10 wt%, the morphology clearly changed and it is similar to that of the virgin PS. The craze paths are less in number with practically no deviation, which can be accounted for with the formation of aggregates and increase in particle size. This was one of the pathbreaking papers describing the toughening mechanism in PS and rigid fillers.

Another important filler that can be coupled with rubbery materials for toughening of PS is GNP even in very low loading due to the high aspect ratio (Chakraborty et al. 2016). One of the main challenges is the fair distribution of GNP in the matrix by avoiding the agglomeration or stacking during processing. In order to avoid the agglomeration or restacking of the platelets, many methods are proposed, and adding a core–shell rubber nanoparticle with a thin layer of PS-compatible shell surrounding the rubbery core is used in this work. The core is made up of polyisobutadiene (PIB),

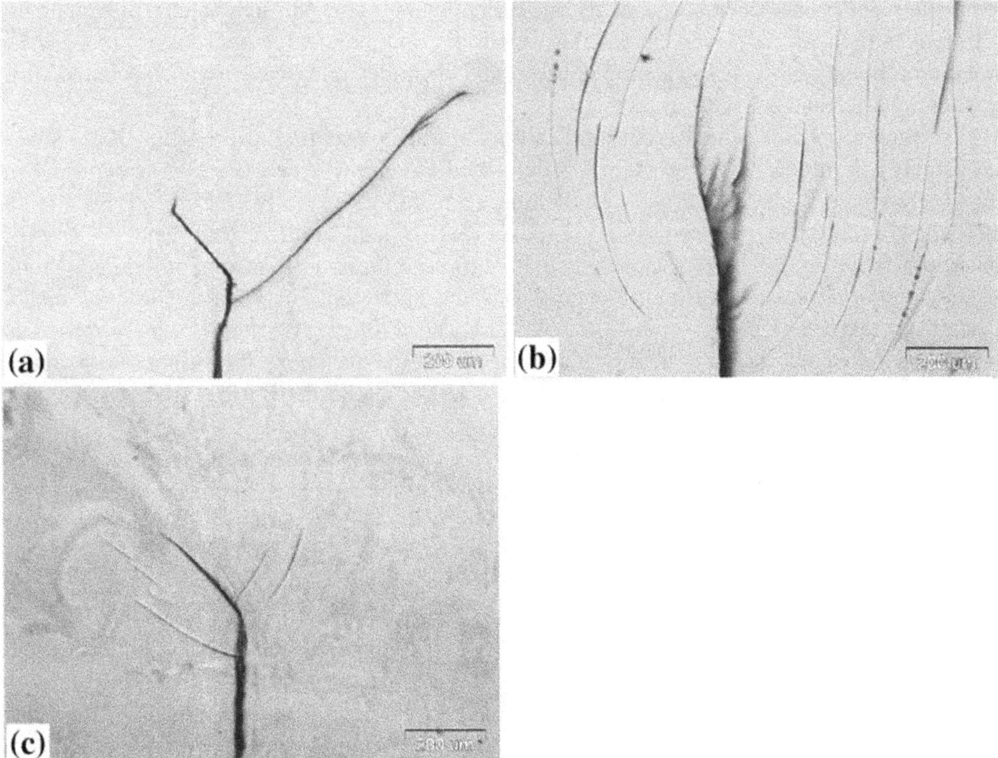

FIGURE 2.4 TOM micrograph of fracture surface and craze propagation of the nanocomposite samples (Salehi Vaziri et al. 2011).

and the shell is made up of methacrylate/styrene/acrylate copolymer. The addition of core–shell rubber to the PS/CNF composites improved the toughness and flexural strength, and it acted as a dispersion aid.

2.3 TOUGHENING MECHANISMS

Different theories describe how a dispersed phase toughens a polymeric substance; most employ methods of dissipating energy throughout the matrix. These theories include microcrack theory, shear-yielding theory, multiple-crazing theory, shear band and crazing interaction theory, and more recently those including the effects of critical ligament thickness, critical plastic area, voiding and cavitation, and damage competition. Fracture mechanics plays a vital role to understand the ability of the materials to toughen the polymer. The stress intensity factor, $K_{i,}$ depends on applied stress, specimen geometry and crack length, which will provide an idea about the linear elastic fracture mechanics of the polymeric systems. The measurement of *Kic* is very important to understand the fracture mechanics and thereby the toughening (Bucknall 1982). The higher Poisson's ratio of polymers compared to metals also affects the toughening mechanisms and contributes to plane strain effects and plastic zones in toughened polymers. The fracture mechanism of rubber-modified PS over a range of temperatures employing the surface notch method was reported by Parvin and Williams (1976).

The toughening depends on various aspects of structure–property relationships of polymer matrix and reinforcing fillers. The particle size of the filler has a profound effect on the toughening of rubber compounds. Another important factor to consider is the particle morphology as it can help the matrix to easily follow the crazing behavior. Adhesion at the interface between matrix and particle also plays a vital role in determining the toughening mechanism in polymer composites. If the particle debonds easily, the formation of cracks through holes will happen that will weaken the materials. Also, molecular weight distribution, degree of cross-linking, crystallinity, morphology of the crystals, and physical and chemical aging have to be considered while toughening different matrices.

The major toughening mechanisms are crazing and shear yielding, which allow PS to absorb more energy at impact since the matrix is deformed more extensively. Park et al. reported the use of acetylated cyclodextrin as a toughening agent for PS recently (Park et al. 2020). The host–guest interactions between the PS and cyclodextrin lead to interesting results of fracture toughness. Three to four wt% of cyclodextrin showed optimum properties of Young's modulus, elongation at break and fracture energy. Based on different analytical techniques such as mechanical tests, SAXS measurements and SEM test, a toughening model was proposed for the PS–cyclodextrin systems. The model envisaged that the cyclodextrins form a round domain initially, which helps the host–guest interactions between the matrix and filler as shown in Figure 2.5.

Toughening of PS component in composites is also done by various researchers. Very recently, the fused filament fabrication (FFF) technique is used to prepare PP/PS composites with in situ microfibrils and shish–kebab structure (Jiang et al. 2020). The shish–kebab structure is usually highly compacted and can provide excellent strength to the composites. The structure leads to a phenomenon called self-reinforcing material due to the arrangement of the molecular chains and ordered crystalline features. PP is a semicrystalline polymer, and PS is having higher tensile strength than PP, which will strengthen it, so that the blend material is suitable for 3D printing applications. Filament failure and interlayer failure are reported in the samples, which led to the increase in elongation at break and thereby toughness by the multiple energy dissipation routes. The toughness of PS is reported based on the impact strength values as 1.63 times greater than the virgin material. The schematic representation of the morphology evolution leading to the improvement in toughness is shown in Figure 2.6.

Core–shell structures with rubbery content are suitable for improving toughness properties of PS. The core shell is reported to be synthesized by an emulsion polymerization using $K_2S_2O_8$ as an

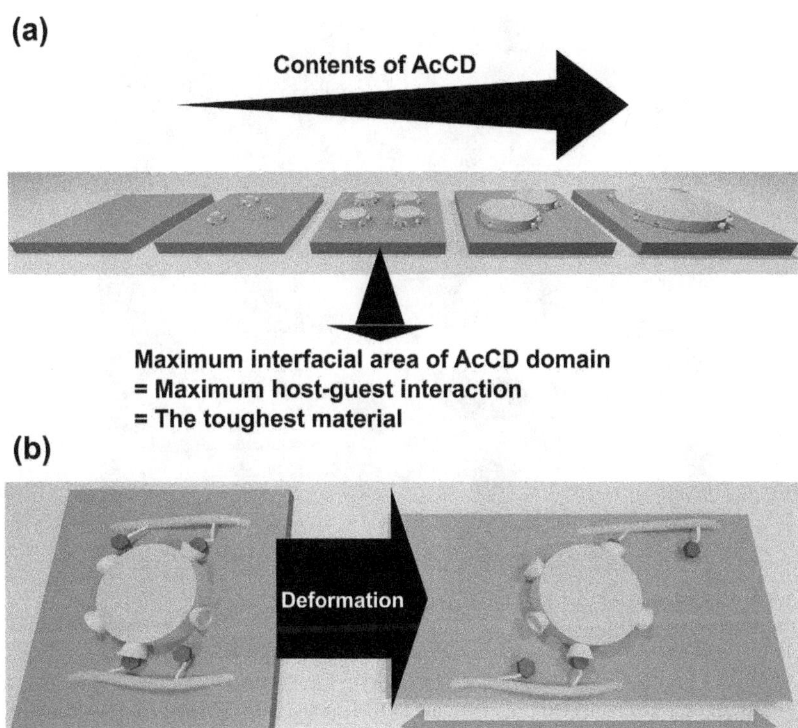

FIGURE 2.5 Suggested toughening mechanism by host-guest interaction between AcCD domains and PS. (a) Morphologies of domains with different contents of AcCD. (b) Recombination of host-guest interactions during deformation (Park et al. 2020).

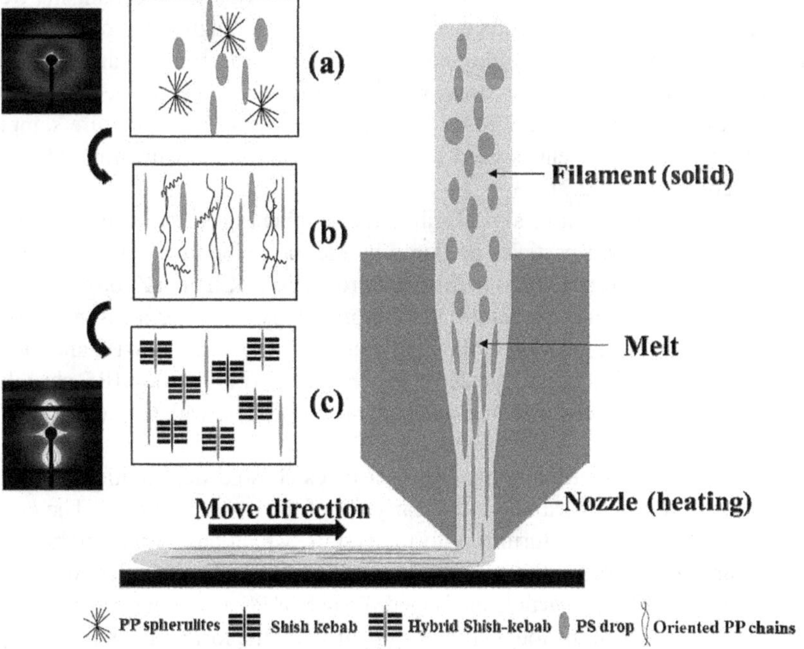

FIGURE 2.6 The schematic illustration of the morphology evolution during FFF, which shows composite at three stages, (a) as filament, (b) melting extrusion and (c) deposition on platform (Jiang et al. 2020).

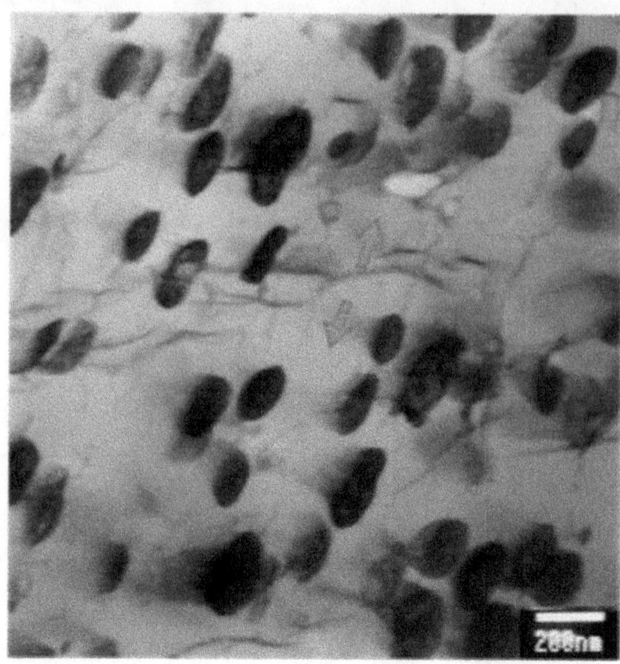

FIGURE 2.7 The transmission electron microscope (TEM) image of fractured stress-whitening zone of the PS/PB-g-PS60 blend in an impact test (Cai et al. 2013).

initiator (Cai et al. 2013). PS/PB-g-PS is prepared by blending the core–shell rubber particles with PS. It is noted that the optimum particle size of 0.3–0.5 mm significantly improved the toughness properties. In order to understand the toughness behavior of the formed materials, various properties such as the mechanical properties, morphologies and deformation mechanisms were studied. The toughening mechanism of the blends is explained based on crazing that occurred from rubber particles and its extension in a bridge-like manner to neighboring rubber particles parallel to the equatorial direction. The formation of stress-whitened zones because of the deformation energy indicates the fracture toughness in PS-based blends and composites. A fracture sample obtained from impact strength measurement is used to understand the stress-whitening zone beneath the surface of the fracture (Figure 2.7).

A very recent article by Zhou et al. described the toughening mechanism in polyamide 1012 with the addition of core–shell particles prepared by the melt blending of poly(methyl methacrylate)–polybutadiene–polystyrene (MBS) in different proportions (Chenxu Zhou et al. 2021). A study of the microstructure evolution upon deformation, morphology and interfacial interaction between MBS and polyamide led to explanation of the high toughness provided by MBS and the toughening mechanism in the system. The stress transfer between the matrix and the MBS particle led to the improvement in interfacial interaction and mechanical properties, which were established by the morphological studies.

During stretching of the materials, the MBS particles showed deformation, and cavitation is formed, which helped for the distribution of shear yield of the matrix material. The crystallization studies showed the transition of α-form to γ-form crystals, which has better toughness properties. Also, the MBS particles restricted the orientation of the PA1012 crystal lattice, which reduced the loading stress on the matrix. The article envisaged the use of PS as a toughening agent along with other polymers and established the toughening mechanism by the formation of hard core–shell particle. The softening of the shell is reduced due to the presence of PS part, and the strength remained constant for the polyamide matrix.

2.4 CONCLUSION

PS-based materials are used to manufacture rigid and brittle artifacts for various applications. The low toughness of the PS is a serious drawback, and various methods are employed to improve the toughness of this polymer. The recent developments in the field of toughening of PS are carefully revisited in this chapter. A short explanation of the mechanism of toughening by various means is also given. From the covered literature, it is imperative that the toughening of PS is still a topic to be explored and new mechanisms/theories could be developed.

REFERENCES

Alfarraj, A and E Bruce Nauman. 2004. Super HIPS: Improved high impact polystyrene with two sources of rubber particles. *Polymer* 45(25): 8435–42.

Araújo, EM, LH Carvalho, and MVL Fook. 1997. Mechanical properties of PS/rubber waste blends: Effects of concentration, particle size and moulding procedures. *Polímeros* 7(3): 45–52.

Blanco, I and FA Bottino. 2015. The influence of the nature of POSSs cage's periphery on the thermal stability of a series of new bridged POSS/PS nanocomposites. *Polymer Degradation and Stability* 121: 180–86.

Bucknall, CB and RR Smith. 1965. Stress-whitening in high-impact polystyrenes. *Polymer* 6(8): 437–46.

Bucknall, CB. 1982. The effective use of rubbers as toughening agents for plastics. *Journal of Elastomers & Plastics* 14(4): 204–21.

Cai, GD, HY Yang, LD Zhu, H Liu, GF Wu, MY Zhang, C Zhou, GH Gao, and HX Zhang. 2013. Toughening polystyrene by core–shell grafting copolymer polybutadiene-graft-polystyrene with potassium persulfate as initiator. *Journal of Industrial and Engineering Chemistry* 19(3): 823–28.

Chakraborty, I, A Shukla, and A Bose. 2016. Core–shell rubbery fillers for massive electrical conductivity enhancement and toughening of polystyrene–graphene nanoplatelet composites. *Journal of Materials Science* 51(23): 10555–60.

Cigana, P, BD Favis, C Albert, and T Vu-Khanh. 1997. Morphology–interface–property relationships in polystyrene/ethylene–propylene rubber blends. 1. Influence of triblock copolymer interfacial modifiers. *Macromolecules* 30(14): 4163–69.

Coutinho, F, Marcia PM Costa, MJOC Guimarães, and BG Soares. 2007. Comparative study of different types of polybutadiene on the toughening of polystyrene. *Polímeros* 17(4): 318–24.

Crevecoeur, JJ, LNIH Nelissen, MCM Van der Sanden, PJ Lemstra, HJ Mencer, and AH Hogt. 1995. Impact strength of reactively extruded polystyrene/ethylene-propylene-diene rubber blends. *Polymer* 36(4): 753–57.

Deng, Y, G Gao, Z Liu, C Cao, and H Zhang. 2013. Co-toughened polystyrene by submicrometer-sized core–shell rubber particles and micrometer-sized salami rubber particles. *Industrial & Engineering Chemistry Research* 52(14): 5079–84.

Fang, Z, Z Guo, and L Zha. 2004. Toughening of polystyrene with ethylene-propylene-diene terpolymer (EPDM) compatibilized by styrene-butadiene-styrene block copolymer (SBS). *Macromolecular Materials and Engineering* 289(8): 743–48.

Ferreira, EHC, LP de Lima, and GJM Fechine. 2020. The 'superlubricity state' of carbonaceous fillers on polymer composites. *Macromolecular Chemistry and Physics* 221(16): 2000192.

Gao, G, J Zhang, H Yang, C Zhou, and H Zhang. 2006. Deformation mechanism of polystyrene toughened with sub-micrometer monodisperse rubber particles. *Polymer International* 55(11): 1215–21.

Ghani, NFA, MSZ Mat Desa, M Yusop, M Bijarimi, A Ramli, and MF Ali. 2020. Mechanical properties of styrene butadiene rubber toughened graphene reinforced polystyrene. *{IOP} Conference Series: Materials Science and Engineering* 736(March): 52010.

Gopalakrishnan, J, and SKN Kutty. 2015. Mechanical, thermal, and rheological properties of dynamically vulcanized natural rubber-toughened polystyrene. *Journal of Elastomers & Plastics* 47(2): 153–69.

Guo, TY, GL Tang, GJ Hao, MD Song, and BH Zhang. 2003. Toughening modification of PS with N-BA/MMA/styrene core–shell structured copolymer from emulsifier-free emulsion polymerization. *Journal of Applied Polymer Science* 90(5): 1290–97.

Halimatudahliana, HI and M Nasir. 2002. The effect of various compatibilizers on mechanical properties of polystyrene/polypropylene blend. *Polymer Testing* 21(2): 163–70.

Ibrahim, BA and KM Kadum. 2012. Morphology studies and mechanical properties for PS/SBS blends. *International Journal of Engineering & Technology* 12(3): 19–27.

Jiang, Y, J Wu, J Leng, L Cardon, and J Zhang. 2020. Reinforced and toughened PP/PS composites prepared by fused filament fabrication (FFF) with in-situ microfibril and shish-kebab structure. *Polymer* 186: 121971.

Li, D, H Xia, J Peng, M Zhai, G Wei, J Li, and J Qiao. 2007. Radiation preparation of nano-powdered styrene-butadiene rubber (SBR) and its toughening effect for polystyrene and high-impact polystyrene. *Radiation Physics and Chemistry* 76(11): 1732–35.

Libio, IC, VG Grassi, MF Dal Pizzol, and SM Bohrz Nachtigall. 2012. Toughened polystyrene with improved photoresistance: Effects of the compatibilizers. *Journal of Applied Polymer Science* 126(1): 179–85.

Luna, CBB, DD Siqueira, E da Silva Barbosa Ferreira, WA da Silva, JA dos Santos Nogueira, and EM Araújo. 2020. From disposal to technological potential: Reuse of polypropylene waste from industrial containers as a polystyrene impact modifier. *Sustainability* 12(13).

Luna, CBB, DD Siqueira, EM Araújo, and MV Lia Fook. 2019a. Photodegradation of polystyrene/rubber waste blends compatibilized with SBS copolymer. *Journal of Elastomers & Plastics* 52(4): 356–79.

Luna, CBB, DD Siqueira, EM Araújo, DD de Souza Morais, and EB Bezerra. 2018. Toughening of polystyrene using styrene-butadiene rubber (SBRr) waste from the shoe industry. *REM-International Engineering Journal* 71(2): 253–60.

Luna, CBB, EM Araújo, DD Siqueira, DD de Souza Morais, EA dos Santos Filho, and MV Lia Fook. 2019b. Incorporation of a recycled rubber compound from the shoe industry in polystyrene: Effect of sbs compatibilizer content. *Journal of Elastomers & Plastics* 52(1): 3–28.

Luzinov, I, K Xi, C Pagnoulle, G Huynh-Ba, and R Jérôme. 1999. Composition effect on the core–shell morphology and mechanical properties of ternary polystyrene/styrene–butadiene rubber/polyethylene blends. *Polymer* 40(10): 2511–20.

Martinez, G, F Vazquez, A Alvarez-Castillo, R López-Castañares, and VM Castano. 2000. Mechanical and processing properties of polystyrene-(styrene butadiene) blends. *International Journal of Polymeric Materials* 46(1–2): 27–40.

Mathew, M, and S Thomas. 2003. Compatibilisation of heterogeneous acrylonitrile–butadiene rubber/polystyrene blends by the addition of styrene–acrylonitrile copolymer: Effect on morphology and mechanical properties. *Polymer* 44 (4). Elsevier: 1295–1307.

Mathur, D, and E B Nauman. 1999. Impact strength of bulk PS/PB blends: Compatibilization and fracture studies. *Journal of Applied Polymer Science* 72(9): 1151–64.

Neoh, SB and AS Hashim. 2004. Highly grafted polystyrene-modified natural rubber as toughener for polystyrene. *Journal of Applied Polymer Science* 93(4): 1660–65.

Omonov, TS, C Harrats, G Groeninckx, and P Moldenaers. 2007. Anisotropy and instability of the co-continuous phase morphology in uncompatibilized and reactively compatibilized polypropylene/polystyrene blends. *Polymer* 48(18): 5289–5302.

Parameswaranpillai, J, G Joseph, R Veliyath Chellappan, AK Zahakariah, and N Hameed. 2015. The effect of polypropylene-graft-maleic anhydride on the morphology and dynamic mechanical properties of polypropylene/polystyrene blends. *Journal of Polymer Research* 22(2): 2.

Park, J, S Murayama, M Osaki, H Yamaguchi, A Harada, G Matsuba, and Y Takashima. 2020. Reinforced polystyrene through host-guest interactions using cyclodextrin as an additive. *European Polymer Journal* 134: 109807.

Parvin, M and JG Williams. 1976. The effect of temperature on the fracture of rubber modified polystyrene. *Journal of Materials Science* 11(11): 2045–50.

Peng, X, J Chen, T Kuang, P Yu, and J Huang. 2014. Simultaneous reinforcing and toughening of high impact polystyrene with a novel processing method of loop oscillating push–pull molding. *Materials Letters* 123: 55–58.

Pittolo, M and RP Burford. 1985. Recycled rubber crumb as a toughener of polystyrene. *Rubber Chemistry and Technology* 58(1): 97–106.

Razavi, M, D Huang, S Liu, H Guo, and S-Q Wang. 2020. Examining an alternative molecular mechanism to toughen glassy polymers. *Macromolecules* 53(1): 323–33.

Rodriguez, CLC, MABS Nunes, PS Garcia, and GJM Fechine. 2021. Molybdenum disulfide as a filler for a polymeric matrix at an ultralow content: Polystyrene case. *Polymer Testing* 93: 106882.

Rovere, J, CA Correa, V Galhard Grassi, and MF Dal Pizzol. 2008. Role of the rubber particle and polybutadiene cis content on the toughness of high impact polystyrene. *Journal of Materials Science* 43(3): 952–59.

Salehi Vaziri, H, M Abadyan, M Nouri, IA Omaraei, Z Sadredini, and M Ebrahimnia. 2011. Investigation of the fracture mechanism and mechanical properties of polystyrene/silica nanocomposite in various silica contents. *Journal of Materials Science* 46(17): 5628–38.

Schneider, M, T Pith, and M Lambla. 1997. Toughening of polystyrene by natural rubber-based composite particles: Part I impact reinforcement by PMMA and PS grafted core-shell particles. *Journal of Materials Science* 32(23): 6331–42.

Shaw, S and R P Singh. 1990b. Studies on impact modification of polystyrene (PS) by ethylene–propylene–diene (EPDM) rubber and its graft copolymers. III. PS/EPDM-g-(styrene-co-maleic anhydride) blends and its relative performance. *Journal of Applied Polymer Science* 40(5–6): 701–7.

Shaw, S and RP Singh. 1990. Studies on impact modification of polystyrene (PS) by ethylene–propylene–diene (EPDM) rubber and its graft copolymers. II. PS/EPDM-g-(styrene-co-methylmethacrylate) blends. *Journal of Applied Polymer Science* 40 (5–6): 693–99.

Shaw, S and RP Singh. 1990a. Studies on impact modification of polystyrene (PS) by ethylene–propylene–diene (EPDM) rubber and its graft copolymers. I. PS/EPDM and PS/EPDM-g-styrene blends. *Journal of Applied Polymer Science* 40(5–6): 685–92.

Sreenivasan, PV and P Kurian. 2007. Mechanical properties and morphology of nitrile rubber toughened polystyrene. *International Journal of Polymeric Materials* 56(11): 1041–50.

Sun, D-x, Q-q Bai, X-z Jin, X-d Qi, J-h Yang, and Y Wang. 2019. Simultaneously enhanced thermal conductivity and fracture toughness in polystyrene/carbon nanofiber composites by adding elastomer. *Composites Science and Technology* 184: 107864.

Tekay, E. 2020. Preparation of tough, high modulus, and creep-resistant PS/SIS/halloysite blend nanocomposites. *Journal of Thermoplastic Composite Materials* 33(8): 1125–44.

Thomas, SP, S Thomas, and S Bandyopadhyay. 2009. Polystyrene–calcium phosphate nanocomposites: Preparation, morphology, and mechanical behavior. *Journal of Physical Chemistry C* 113(1): 97–104.

Vázquez, F, H Cartier, K Landfester, G-H Hu, T Pith, and M Lambla. 1995. Reactive blends of thermoplastics and latex particles. *Polymers for Advanced Technologies* 6(5): 309–15.

Veilleux, J and D Rodrigue. 2016. Properties of recycled PS/SBR blends: Effect of SBR pretreatment. *Progress in Rubber Plastics and Recycling Technology* 32(3): 111–28.

Wen-Dong, T, H Guang-Jian, H Wei-Tao, Z Xin-Liang, C Xian-Wu, and Y Xiao-Chun. 2020. The reactive compatibilization of PLA/PP blends and improvement of PLA crystallization properties induced by in situ UV irradiation. *CrystEngComm* 23(4): 864-875.

Zeng, Y, Q-C Yang, Y-T Xu, G-Q Ma, H-D Huang, J Lei, G-J Zhong, and Z-M Li. 2021. Durably ductile, transparent polystyrene based on extensional stress-induced rejuvenation stabilized by styrene–butadiene block copolymer nanofibrils. *ACS Macro Letters* 10(1): 71–77.

Zhang, J, H Chen, Y Zhou, C Ke, and H Lu. 2013. Compatibility of waste rubber powder/polystyrene blends by the addition of styrene grafted styrene butadiene rubber copolymer: Effect on morphology and properties. *Polymer Bulletin* 70(10): 2829–41.

Zhou, C, P Zhu, X Liu, X Dong, and D Wang. 2021. The toughening mechanism of core-shell particles by the interface interaction and crystalline transition in polyamide 1012. *Composites Part B: Engineering* 206: 108539.

Zhou, C, S Wu, B Yang, Y Gao, G Wu, and H Zhang. 2015. Toughening polystyrene by core-shell rubber particles: Analysis of the internal structure and properties. *Polymers and Polymer Composites* 23(5): 317–24.

Zhu, LD, H Yu Yang, GD Cai, C Zhou, G Feng Wu, MY Zhang, GH Gao, and HX Zhang. 2013. Submicrometer-sized rubber particles as 'craze-bridge' for toughening polystyrene/high-impact polystyrene. *Journal of Applied Polymer Science* 129(1): 224–29.

3 Metal Matrix Composites
An Overview

B. Sengupta
Kazi Nazrul University

A. Basu Mallick
Indian Institute of Engineering Science and Technology, Shibpur

CONTENTS

3.1 INTRODUCTION

Composite materials consist of a combination of two or more materials or phases, which possess the ability to provide the desired blend of properties that are otherwise unattainable in materials having only one type of constituent. Metal matrix composites (MMCs) contain at least two physically and chemically distinct phases. The continuous metal phase is the matrix, and the other phase that can be a metal or ceramic material that remains distributed in the matrix is the reinforcement. Incorporating reinforcement in metals provides the opportunity to achieve tailor-made properties, viz. high specific modulus, strength, thermal stability, better wear resistance and lower coefficient of thermal expansion (CTE), particularly critical for automotive, aerospace, sports goods, electrical, electronic applications and constructions. Popularly used metals as the matrix materials include Al, Ti, Mg and Cu (Cyriac 2011). Recently, with the incorporation of nanosized reinforcement in metallic matrices through advanced processing techniques, remarkable improvement in mechanical properties of MMCs has been achieved, which has opened up still further areas of their potential applications. Markets and Markets predicted the growth of the global MMCs market from USD 467 million in 2020 to USD 787 million by 2025, at a compound annual growth rate of 11.0% during the forecast period. To date, the lightweight of MMCs leading to the increased fuel efficiency of the components of automotive and aircraft parts remains a catalyst for increased usage of MMCs in this sector (Suresh, Mortensen, and Needleman 1993).

Considering the above views, in this chapter, highlights of the development of MMCs right from the historical background to the recent advancements are illustrated in a nutshell.

DOI: 10.1201/9780429330575-3

3.2 HISTORICAL BACKGROUND

The quest for materials that are stiffer materials that are stronger and lighter led to the development of composite materials. People were aware of composites, as evident from the usage of bricks made from mud reinforced with straw in ancient Egyptian civilizations to improve the structural integrity of the building. Egyptians also used linen or papyrus soaked in plaster for making death masks around 2181–2055 BC.

Throughout history, humankind had used composite-type materials for various purposes. In about 1200 AD, the Mongols invented the first bows made from a combination of wood, bamboo, bone, cattle tendons, horns and silk bonded with natural pine resin. These bows were small, powerful, extremely accurate and considered the most feared weapons on earth until the invention of firearms in the 14th century.

MMC, which uses metal as the matrix, was also known to humankind before the birth of Christ. An iron plate excavated from the Great Pyramid of Giza, Egypt, had numerous laminates of wrought iron welded together through hammering. After the Second World War and the launch of the Soviet satellite Sputnik in 1957, scientists looked for materials, which are lighter, more robust and resistant to high temperature. Researchers from different countries, particularly the USA, Germany and the erstwhile USSR, tried various combinations of metals with different reinforcements for developing composite materials with tailored properties. The recession of the 1970s and the West's oil embargo triggered the demand for energy-efficient systems, which led to Al matrix composite development with boron and carbon fibres as the reinforcements. However, research progress on the development of MMCs was slow due to the recession.

The interest in MMCs was rekindled in the 1980s. Several research works have been undertaken in the USA to fabricate various components of spacecraft with MMCs. The space shuttles developed by USSR in these periods contained different parts manufactured from the MMCs. In India, Defence Research and Development Laboratories initiated the Light Combat Aircraft programme development where extensive MMC parts usage is contemplated. In the same decade, Japanese companies started using MMCs in automotive parts.

In the early 1990s, research and development of lightweight Al matrix composites gained considerable worldwide importance. Ti matrix composites were also developed, and Mg matrix composites started receiving considerable attention. It is in this period that the use of MMCs in electronic packaging began. The start of this century saw a remarkable development in the processing techniques of MMCs. Friction stir welding, spark plasma sintering, squeeze casting and other casting methods are employed for processing MMCs. The main aim was to better distribute the reinforcements in the matrix and attain improved physical properties, e.g. wear behaviour, welding and dry sliding characteristics. At the same time, Cu matrix composites are also gaining importance besides Al and Mg matrix composites. Studies on MMCs reinforced with ceramic foams, sandwich panels and hybrid composites are also reported.

In the late 2000s, scientists' research interest shifted towards developing metal matrix nanocomposites (MMNCs). It was shown that compared to the micron-sized particle-reinforced MMCs, the nanoparticle-reinforced MMNCs possess superior strength, ductility, wear resistance, creep resistance and good elevated temperature properties. Like ordinary MMCs, Al, Mg, Ti, Cu and their alloys are used as matrix materials in MMNCs and nanosized oxides like Al_2O_3, Y_2O_3, nitrides (Si_3N_4, AlN), carbides (TiC, SiC), borides (TiB_2), carbon nanotubes (CNTs) and nanofibres are used as the reinforcements (Srivastava 2017).

3.3 MICROSTRUCTURE OF METAL MATRIX COMPOSITES

Microstructurally, there are three components present in MMC; metal, reinforcement and the interface between the matrix and reinforcement. Here, one should note that although the MMCs consist of two or more constituent phases separated by a distinct interface, they cannot be called an alloy. In alloys, one or more phases can appear or disappear as a function of temperature by phase transformation, which does not happen in MMCs during secondary processing treatments. The microstructural features of MMCs that are important from the viewpoint of their effect on the properties are (i) the metal or alloy, forming the host matrix, (ii) the reinforcement(s) type and (iii) the interface between the reinforcement(s) and the matrix.

The properties of MMCs primarily depend to a great extent on selecting the metal/alloys as the matrix and their grain size. The matrix–reinforcement interface also has a predominant role in controlling the strength property of the MMC. The reinforcements that are used can be ceramic, metallic or intermetallic material. However, their shape, size, volume fraction, composition and spatial distribution strongly influence the final properties conferred to the composites. Most of the structural components made up of MMCs have ceramic reinforcements due to their high strength and hardness, high melting point, high wear resistance, low CTE, high elastic modulus and good high-temperature properties.

The interface's importance lies in the fact that it is the interface through which load is transferred from the matrix to the reinforcement. The interface may be a simple row of atomic bonds, e.g. the interface between alumina and pure aluminium, or it can include matrix/reinforcement reaction products, e.g. aluminium carbide between aluminium and carbon fibres, or reinforcement coatings, e.g. interfacial coating between SiC monofilaments and titanium materials. Depending on the processing technique and condition, mechanical, electrostatic, chemical and reaction bonds may form at the interface.

3.4 CLASSIFICATIONS

Based on the reinforcement(s) type and geometry, MMCs can be broadly classified into (i) fibre-reinforced MMCs and (ii) particle-reinforced MMCs. The length of the fibres in the fibre-reinforced MMCs is much larger than their cross-sectional dimension. Fibre-reinforced composites are of two types, continuous fibre-reinforced composites and discontinuous fibre-reinforced composites.

Continuous fibre-reinforced composites are characterized by long fibres with a high aspect ratio, whereas discontinuous fibre-reinforced composites are short with a low aspect ratio. They may have a preferred or random orientation. Multilayered composites can be divided into two types: laminate and hybrid. Laminates are made up of several stacked layers of reinforcements, whereas the hybrids may contain a mixture of fibres or a mixture of fibrous and particulate reinforcements. Whiskers are also used as reinforcements. Whiskers possess a preferred shape and diameter, but lengths are much smaller as compared to fibres. Boron fibre-reinforced and SiC fibre-reinforced Al- and Ti-based MMCs are frequently used in the aerospace industry. It is a well-established fact that good elastic modulus can be achieved by the unidirectional incorporation of fibres or whiskers in the metal matrix.

Particle-reinforced MMCs are widely used in components where directional properties are not required. Particle reinforcement is typically used without any coating or pre-treatment; however, to improve the ease of incorporating the particles in the metal matrix, they may be pre-treated or coated. Particle reinforcements are often used to enhance the strength and other properties of inexpensive materials because they provide dispersion strengthening effects when the particles are finely sized. The dispersed second-phase particles may be metallic or non-metallic oxide materials. The strengthening effect arises when these fine particles offer resistance to the motion of the dislocations within the material. Thoria (ThO_2)-dispersed Ni alloys (TD Ni alloys) and sintered aluminium powder – where the aluminium matrix is dispersed with tiny flakes of alumina (Al_2O_3) – are some examples of this type of MMCs. WC or TiC particles embedded in cobalt or nickel matrix are used to make cutting tools. Aluminium alloy castings containing dispersed SiC particles are now widely used for automotive applications, including pistons and brake applications. Typical morphologies of some Al-based MMCs are indicated in Figure 3.1.

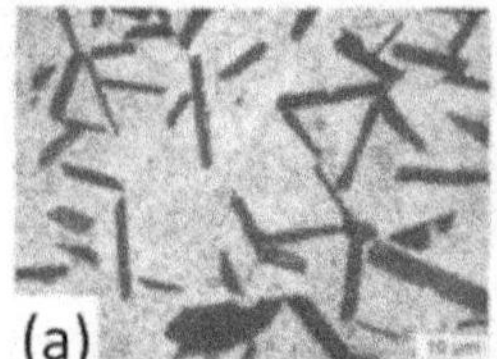

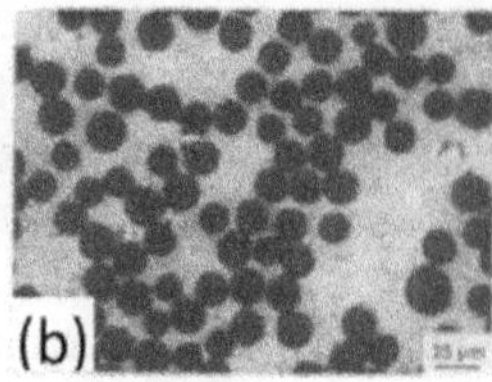

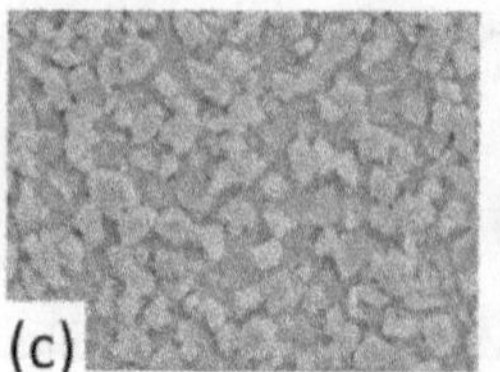

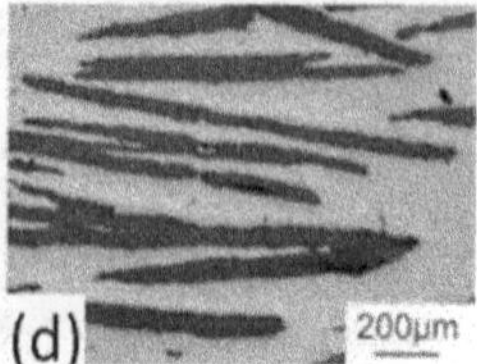

FIGURE 3.1 Microstructures of Al-based MMCs with (a) Al_2O_3 platelets, (b) Al_2O_3 continuous fibres, (c) SiC particles and (d) graphite flakes (Nturanabo, Masu, and Baptist Kirabira 2020).

Based on the method of preparation, MMCs can be also classified as (i) ex situ MMCs and (ii) in situ MMCs. In ex situ MMCs, the reinforcement materials are prepared separately before composite fabrication and incorporated into the host metal matrix without any reaction between the reinforcing materials and the host matrix. The distribution of the reinforcements often remains non-uniform throughout the matrix, as they are produced by mechanical mixing of the constituents. However, it impairs the mechanical properties of the composites and also causes property anisotropy. Conversely, the in situ MMCs contain reinforcements generated within the matrix by a reaction between one or more components. The in situ generated reinforcements are thermodynamically stable, are finer in size, and provide a cleaner interface, better distribution in the matrix and better mechanical and high-temperature properties. In most cases, the matrix materials selected for fabrication of in situ MMCs include Al, Ti, Cu, Ni and Fe. The reinforcements are borides, carbides, nitrides, oxides and their mixture.

3.5 PROCESSING TECHNIQUES

Major MMC processing techniques can be divided into two broad categories: liquid-state processing and solid-state processing.

3.5.1 Liquid-State Processing

This is the most widely used technique adopted by researchers and industry personnel. In this technique, the reinforcements are incorporated into the liquid metal. The liquid composite slurry so formed is either cast into various shapes by conventional casting techniques or cast into ingots through secondary processing. It is important to mention that since the matrix and the reinforcements are two separate entities, good bonding between the matrix and the reinforcement is essential. This can be achieved when the wettability of the reinforcement with the matrix is good. Thermodynamically, wetting occurs when it causes a decrease in the overall free energy of the system. The magnitude of the contact angle between the liquid and the solid surface denoted as "θ" is an important parameter. The conditions are $\theta = 180°$, no wetting, $\theta = 0°$, perfect wetting. When θ lies between 0° and 180°, the degree of wetting varies and is better when θ is less (Matthews and Rawlings 1994).

The role of oxygen is also important, particularly in MMCs reinforced with ceramic materials. Most metals in the liquid state react with oxygen to produce a thin layer of their respective oxides at the matrix and ceramic reinforcements' interface. It causes an increase in the contact angle θ. As a result, the wettability decreases, leading to poor bonding of the reinforcements with the matrix that is often remedied by using static atmospheres of neutral gas or by adding Ca or Mg. It is also essential to avoid adverse reaction products at the interface between the matrix and the second phase. Modification of the reinforcements is sometimes carried out by heat treatment or by applying a coating, so that improved wetting of the reinforcements can occur. Flux is also used for assisting the incorporation of the reinforcements into the matrix.

Highlights of some of the critical liquid-state processing techniques used for producing MMCs are as follows.

Stir Casting: One of the oldest variants in liquid-state processing is melt stirring. The metal matrix is melted, and the reinforcements are mixed with molten metal accompanied by stirring to ensure that the reinforcements get uniformly distributed in the matrix (Chen et al. 2015). However, non-uniform distribution and agglomeration of the reinforcements are severe limitations of this process. To overcome the problem, the melt is allowed to cool to a two-phase solid–liquid state to improve the stirring effect. This modified melt stirring process is known as compocasting or rheocasting. The conventional stir casting and the disintegrated melt deposition (DMD) technique are also used to produce MMCs. Although stir casting is the simplest and economical process, the reinforcements in the form of nanoparticles tend to form clusters and agglomerates due to high surface area and resulting in high van der Waals forces. Moreover, the wettability of nanoparticles is also poor due to either an oxide film on the melt surface or a gas layer that covers the nanoparticles. To surmount the limitation, alloying elements are added to the matrix alloy/metal. Ca, Ti, Mg and Zr are known to reduce surface tension. Ultrasonic wave-assisted stirring is also employed by some researchers. It has been shown that high-intensity ultrasonic waves generate an implosive impact that improve wettability and cause breaking of the clustered nanoparticles (Mirihanage et al. 2016).

In the DMD process, composite slurry produced by the stir casting process is passed through a pouring nozzle with an inert gas jet at a superheated temperature and deposited over a metallic substrate. This process is widely used for manufacturing Mg-based nanocomposites. The final product is an ingot which is then extruded to achieve the desired shape.

Squeeze Casting: The liquid matrix is infiltrated into a porous pre-form under applied pressure.

Spray Co-Deposition: The process involves atomization of the molten metal. The reinforcement particles are introduced into the spray of metal droplets, which is then followed by co-deposition of the metal and the reinforcement particles over a substrate.

Exothermic Dispersion: This process uses elemental constituent of a high-temperature ceramic phase, which is heated at a temperature above the melting point of the metal matrix phase to initiate an exothermic reaction between the elemental constituents and generate the ceramic phase. The master alloy so produced contains fine ceramic reinforcement particles in high volume. The master alloy is added and remelted with the base alloy to obtain the composite with desired volume per cent of reinforcement.

Flux-Assisted Synthesis: This is also a widely used liquid-state technique for producing MMCs, viz. Al-TiB_2, where the TiB_2 reinforcements are generated in situ in the molten Al matrix by the reaction between K_2TiF_6 and KBF_4 salt (Ramesh, Pramod, and Keshavamurthy 2011).

Other processes like selective laser melting, semisolid casting and friction stir processing are also used frequently for the production and processing of MMCs.

3.5.2 Solid-State Processing Techniques

Diffusion bonding, powder metallurgy, mechanical alloying, extrusion, rolling and hot compaction processes fall under this category. In diffusion bonding, a fibre material is sandwiched between two sheets or a foil of the matrix material. Sometimes, the fibres are coated by plasma spraying or ion plating for enhancing the bond strength. Ti, Ni, Cu and Al reinforced with boron fibres are examples of MMCs prepared by this method.

The powder metallurgy process is particularly suitable for producing whisker- or particulate-reinforced MMCs. The process involves mixing the matrix material and reinforcements in powder form and pressing the powder blend in cylindrical dies at room temperature or at an elevated temperature.

The hot compaction technique is a variant of the powder metallurgy process, where the ceramic reinforcement is formed through a chemical reaction between the elements or an element and compound while compacting the powder mix in a hot press. In many cases, mixing and pressing processes are followed by secondary working processes like forging and extrusion. These secondary forming processes help in breaking undesirable oxide layers.

Recently, equal channel angular pressing (ECAP) has emerged as an established method for fabricating ultrafine-grained metal alloys and composites from billets or powders by severe plastic deformation (Deb, Panigrahi, and Weiss 2018). In ECAP, the material is pressed through a die having equal channels that intersect at a specified angle. In this technique, the cross-sectional area of the material remains unchanged, and the grain size of the matrix is refined by inducing massive shear, which causes remarkable improvement in the mechanical and functional properties of the composites.

In the mechanical alloying process, the constituents are mixed and subjected to mechanical milling. MMNCs are produced by this technique. It has been reported that Al and Ni matrix reinforced with CNTs, Al-Si_3N_4 nanocomposites, Mg matrix, and Cu matrix nanocomposites reinforced with CNTs are prepared by this method (Malaki et al. 2019). Al-TiB_2 MMNC was also prepared by a two-step mechanical alloying process (Sadeghian, Enayati, and Beiss 2009). In the first step, Ti and B powders were mixed at a particular ratio and milled with a small amount of Al powder for 20 hours. This mixture was further milled for 20 hours with additional Al powder. Transmission electron microscopic studies revealed the formation of nanosized TiB_2 particles in the Al matrix (Nampoothiri et al. 2016).

3.6 PROPERTIES

One can achieve significant improvement in ultimate tensile strength (UTS), yield strength (YS), wear and hardness values by incorporating ceramic reinforcements in the metal matrix, but the ductility and toughness values of the MMCs decrease compared to the monolithic metals (Christy et al. 2010; Ramesh, Pramod, and Keshavamurthy 2011). For example, K_{IC} or the critical stress intensity factor for monolithic Al alloy is about 20–45 MPa $m^{1/2}$, which is reduced to a value of around 5–25 MPa $m^{1/2}$ for SiC-reinforced Al-MMC (Pramod, Bakshi, and Murty 2015). Recently, researchers indicated that the ductility and toughness of MMCs could also be improved significantly with a simultaneous increase in strength by reducing the reinforcement particle sizes to the nanometer range. It has been reported that in AlSiMgCu alloy containing TiB_2 nanoparticles, a high milestone hardness of 1.56 GPa, YS of 356 MPa, the tensile strength of 457 MPa and an industrially applicable ductility of around 5.5% could be achieved (Dong et al. 2019). It has also been shown that with the addition of small weight % of CNTs in Al alloy, the UTS and the YS increased considerably (Malaki et al. 2019)

In most cases, Orowan strengthening, CTE mismatch strengthening and Hall–Petch strengthening mechanisms are found to be the main reasons behind the improvement of the mechanical properties of the MMCs. Grain refinement and uniform distribution of the reinforcement(s) can also cause an improvement in the mechanical properties. The details of these strengthening mechanisms are discussed in later sections.

3.7 STRENGTHENING MECHANISMS AND ROLE OF SECOND-PHASE PARTICLES

The increase in hardness, YS and tensile strength in MMCs and MMNCs is attributed to the contribution of several strengthening mechanisms, namely Hall–Petch strengthening, Orowan strengthening, CTE mismatch strengthening and particle strengthening. Their contributions to the strength increase can be summed up according to the following equation as the total change in YS ($\Delta\sigma_{total}$) is:

$$\Delta\sigma_{total} = \Delta\sigma_{GR} + \Delta\sigma_{Oro} + \Delta\sigma_{CTE} + \Delta\sigma_{Mod} + \Delta\sigma_{Load} \tag{3.1}$$

where $\Delta\sigma_{GR}$ is the increase in YS due to grain refinement or Hall–Petch strengthening, $\Delta\sigma_{Oro}$ is the increase in YS due to Orowan strengthening effect, $\Delta\sigma_{CTE}$ represents the contribution from CTE mismatch strengthening, $\Delta\sigma_{Mod}$ is the increase in YS due to the creation of geometrically necessary dislocations (GNDs) to accommodate plastic deformation mismatch between matrix and reinforcement particles and $\Delta\sigma_{Load}$ is due to the load transfer effect from the soft matrix to the hard reinforcements under an applied external load.

Here, it would be worthwhile to mention that the strength of a crystalline solid is determined by the stress required to either generate dislocations or move dislocations across a span of the crystal lattice. Dislocation motion is controlled by either the dislocation itself or dislocation interaction with the dispersed solids or crystal defects, e.g. grain boundaries. Usually, MMCs are processed at high temperature and cooled to room temperature. Dislocations are generated during heating and cooling of the composites due to the mismatch in the CTE and contraction between the matrix and the reinforcement. Consequently, volumetric strains are generated in the composite and GNDs are produced, which cause strengthening (Gupta, Chaudhari, and Daniel 2018).

Grain refinement also plays an important role in increasing the YS of the composite according to the Hall–Petch equation, which is represented by the equation:

$$\sigma_y = \sigma_0 + k_y D^{-1/2} \tag{3.2}$$

where σ_y is the YS, σ_0 is the friction stress that allows dislocations to move on slip planes in a single crystal in the absence of any strengthening mechanisms, k_y is the stress concentration factor and D is the average grain size. It is to be noted that grain boundaries are highly disordered regions and offer considerable resistance to the movement of dislocations. The reinforcements, depending on their size, favour particle-stimulated nucleation and grain refinement. This causes an increase in the grain boundary area and an increase in the resistance to dislocation motion. In some cases, it has been reported that nanosized reinforcements cause grain refinement and strengthen the composite material (Zhao et al. 2019).

Orowan strengthening occurs when small particles present in the MMCs impede the dislocation motion, and the dislocations try to bypass the particles. The basic equation for the Orowan shear stress is:

$$\tau_0 = Gb/\lambda \tag{3.3}$$

where G is the shear modulus of the matrix material, b is the Burgers vector of the matrix material and λ is the interparticle spacing. However, it is suggested by several researchers that the Orowan strengthening mechanism cannot significantly operate when particle size is more than 1 µm since the interparticle separation becomes larger (Akbari et al. 2017).

Mismatch in CTE between the soft metallic matrix and hard reinforcement particles increases the YS of the MMCs by generating GNDs due to the mismatch of thermal expansion or due to the mismatch of Young's modulus. The increase in the YS can be evaluated by using the following equation (Pramod et al. 2015):

$$\sigma_{CTE} = \eta G b \rho^{1/2} \tag{3.4}$$

where η is a constant and ρ is the dislocation density.

Dislocations are also generated during the secondary forming processes like forging, rolling and extrusion. Stress increment due to the generation of GNDs is dependent on the average diameter of the nanoparticles, the Burgers vector of the dislocation and bulk strain of the composite. Finally, the contribution from the load transfer effect is written as

$$\sigma_{Load} = 0.5 s \sigma_m V_p$$

where s is the aspect ratio of the particles.

3.8 DESIGNATIONS OF MMCs

In view of the intense interest in the processing of Al MMCs, the American National Standards Institute specified the designations of Al-based MMCs, which provide accepted designation of the matrix, an abbreviation of the reinforcement designation and arrangement, and volume fraction in percentage with the symbol of type/form (shape) of the reinforcement, e.g. matrix/reinforcement/volume% form. For example, magnesium alloy matrix, AM10 reinforced by continuous carbon fibre-type T300, unidirectional oriented amounting to a volume fraction of 65% is designated as "AM10/C – T300/UD65f". Similarly, Al alloy, AA6061 reinforced by particulates of alumina of 22% by volume is designated as "AA6061/A1203/22p". Recently, due to the remarkable strength properties exhibited by MMCs containing nanosized reinforcements, researchers are recommending the inclusion of particle size in the designation.

3.9 IMPORTANT MMC SYSTEMS

A majority of the common MMC systems have Al, Mg, Ti, Cu and superalloys as matrices. Al-based MMCs are the most widely used MMCs having the largest segment in the global composite market. Al is a lightweight metal, which exhibits excellent thermal and electrical conductivity and superior corrosion resistance. Ceramic materials such as SiC, Al_2O_3, TiB_2 and B_4C are frequently used as reinforcements with the Al matrix.

Particulate-reinforced Al matrix composites (PAMC) are also used in fan exit guide vane (FEGV) in the gas turbine engine, as ventral fins and fuel access cover doors in military aircraft. Flight control hydraulic manifolds made of 40 vol% of SiC_p-reinforced aluminium composites have also been used successfully.

Al-SiC composites are generally produced by casting, co-deposition and powder metallurgy route and are used in manufacturing various automotive components, e.g. brake rotors. Control arms and wheel hubs made of SiC-reinforced Al nanocomposites have shown improved strength characteristics similar to cast iron and have a high potential for commercial use in lightweight applications.

The sports goods industry uses aluminium matrix reinforced with particles of SiC or BC. Al-Al_2O_3 MMC is another Al-based MMC that is also used in automotive parts (Bandyopadhyay, Ghosh, and Basumallick 2007). Short Al_2O_3 fibre-reinforced Al matrix composites are widely used in pistons. Al-TiB_2 is another important Al matrix composite system. The presence of TiB_2 reinforcements in the matrix increases the strength and wear resistance of Al. In recent years, Al-TiB_2 composites are increasingly being utilized in high-tech structural and functional applications, including aerospace, defence, automotive, and thermal management areas, as well as in sports and recreation (Kumar et al. 2016; Yadav and Bauri 2015).

Currently, new generation advanced integrated circuits are generating more heat than their previous counterparts. Therefore, the dissipation of heat is a major concern. Thermal fatigue may occur due to a small mismatch of the CTE between the Si substrate and the heat sink (normally molybdenum). It has been proposed that this problem can be solved by using MMCs with exactly matching coefficients (e.g. Al with boron or graphite fibres and Al with SiC particles). Besides a low CTE and high thermal conductivity, these Al-based MMCs also have a low density and a high elastic modulus. Magnesium matrix composites have similar advantages. However, they are not easy to fabricate and possess lower thermal conductivity compared to the Al-based MMCs. Magnesium matrix composites are mainly used in the space industry. Ti and Ti-based alloys have good strength at elevated temperatures and excellent corrosion resistance, which is considered to be an essential criterion for use in aircraft and missile structures. Continuously reinforced, SiC fibre/Ti matrix composite materials are used in a gas turbine engine. WC or TiC embedded in Co or Ni matrix is used to make cutting tools.

Superalloy composites reinforced with tungsten alloy fibres are being developed for the component in jet turbine engines that operate at temperatures above 550°C. Graphite/Cu composites have tailored properties and are useful in high-temperature applications in the air; they provide excellent mechanical characteristics, as well as high electrical and thermal conductivity. Ductile superconductors are fabricated with a matrix of Cu and superconducting filament of niobium–titanium. Cu reinforced with W particles is used as heat sinks in electronic packaging. Ti reinforced with SiC is now considered to be used as the skin material for space applications.

3.10 CONCLUDING REMARKS AND FUTURE DIRECTION

This chapter discussed some of the critical aspects of MMC systems, including their definition, processing, properties, reinforcements, interfaces, strengthening mechanisms, important MMC systems, including MMNCs and current developments. Researchers are continuously striving for developing an optimum combination of properties in MMCs by various modifications in conventional processing techniques like ultrasonic vibration treatment. Also, surface treatments of the reinforcements are carried out to improve wettability. Efforts have also been made to study the microstructure evolution during cold rolling of the composites (Dan et al. 2017). It was observed by researchers that secondary forming processes improve the properties of composites. However, more study regarding the optimization of these process parameters is required. MMNCs have huge potential as a futuristic material, and therefore, upscaling of various methods used in manufacturing MMNCs is necessary to make these materials commercially valuable for the industry.

REFERENCES

Akbari, MK, K Shirvanimoghaddam, Z Hai, S Zhuiykov, and H Khayyam. 2017. Al-TiB_2 micro/nanocomposites: particle capture investigations, strengthening mechanisms and mathematical modelling of mechanical properties. *Materials Science and Engineering: A* 682(January): 98–106.

Bandyopadhyay, NR, S Ghosh, and A Basumallick. 2007. *New Generation Metal Matrix Composites.* 22: 679–82, Taylor & Francis Group.

Chen, F, Z Chen, F Mao, T Wang, and Z Cao. 2015. TiB2 reinforced aluminum based in situ composites fabricated by stir casting. *Materials Science and Engineering: A* 625(February): 357–68.

Christy, TV, N Murugan, S Kumar, TV Christy, N Murugan, and S Kumar. 2010. A comparative study on the microstructures and mechanical properties of Al 6061 alloy and the MMC Al 6061/TiB_2/12p. *Journal of Minerals and Materials Characterization and Engineering* 9(1): 57–65.

Cyriac, AJ. 2011. *Metal Matrix Composites: History, Status, Factors and Future.* Oklahoma State University ProQuest Dissertations Publishing, 1500830.

Dan, CY, Z Chen, G Ji, SH Zhong, Y Wu, F Brisset, HW Wang, and V Ji. 2017. Microstructure study of cold rolling nanosized in-situ TiB_2 particle reinforced Al composites. *Materials & Design* 130(September): 357–65.

Deb, S, SK Panigrahi, and M Weiss. 2018. Development of bulk ultrafine grained Al-SiC nano composite sheets by a SPD based hybrid process: Experimental and theoretical studies. *Materials Science and Engineering: A* 738(December): 323–34.

Dong, X, H Youssef, Y Zhang, S Wang, and S Ji. 2019. High performance Al/TiB_2 composites fabricated by nanoparticle reinforcement and cutting-edge super vacuum assisted die casting process. *Composites Part B: Engineering* 177(November): 107453.

Gupta, R, GP Chaudhari, and BSS Daniel. 2018. Strengthening mechanisms in ultrasonically processed aluminium matrix composite with in-situ Al_3Ti by salt addition. *Composites Part B: Engineering* 140(May): 27–34.

Kumar, N, G Gautam, RK Gautam, A Mohan, and S Mohan. 2016. Synthesis and characterization of TiB_2 reinforced aluminium matrix composites: A review. *Journal of the Institution of Engineers (India): Series D* 97(2): 233–53.

Malaki, M, W Xu, AK Kasar, PL Menezes, H Dieringa, RS Varma, and M Gupta. 2019. Advanced metal matrix nanocomposites. *Metals* 9(3): 330.

Matthews, FL, and RD (Rees) Rawlings. 1994. *Composite Materials : Engineering and Science.* Chapman & Hall, 470.

Mirihanage, W, W Xu, J Tamayo-Ariztondo, D Eskin, M Garcia-Fernandez, P Srirangam, and P Lee. 2016. Synchrotron radiographic studies of ultrasonic melt processing of metal matrix nano composites. *Materials Letters* 164(February): 484–87.

Nampoothiri, J, RS Harini, SK Nayak, B Raj, and KR Ravi. 2016. Post in-situ reaction ultrasonic treatment for generation of Al–4.4Cu/TiB_2 nanocomposite: A route to enhance the strength of metal matrix nanocomposites. *Journal of Alloys and Compounds* 683(October): 370–78.

Nturanabo, F, L Masu, and JB Kirabira. 2020. Novel Applications of aluminium metal matrix composites. *Aluminium Alloys and Composites*. DOI:10.5772/intechopen.86225.

Pramod, SL, SR Bakshi, and BS Murty. 2015. Aluminum-based cast in situ composites: A review. *Journal of Materials Engineering and Performance* 24(6): 2185–2207.

Ramesh, CS, S Pramod, and R Keshavamurthy. 2011. A study on microstructure and mechanical properties of Al 6061–TiB_2 in-situ composites. *Materials Science and Engineering: A* 528(12): 4125–32.

Sadeghian, Z, MH Enayati, and P Beiss. 2009. In situ production of Al–TiB_2 nanocomposite by double-step mechanical alloying. *Journal of Materials Science* 44(10): 2566–72.

Srivastava, A. 2017. Metal matrix nanocomposites (MMCs): A review of their physical and mechanical properties. *International Journal of Nanotechnology in Medicine & Engineering* 2(8): 152–54.

Suresh, S, A Mortensen, and A Needleman. 1993. *Fundamentals of Metal-Matrix Composites*. Butterworth-Heinemann, 342.

Yadav, D, and R Bauri. 2015. Friction stir processing of Al-TiB_2 in situ composite: Effect on particle distribution, microstructure and properties. *Journal of Materials Engineering and Performance* 24(3): 1116–24.

Zhao, B, Q Yang, L Wu, X Li, M Wang, and H Wang. 2019. Effects of nanosized particles on microstructure and mechanical properties of an aged in-situ TiB_2/Al-Cu-Li composite. *Materials Science and Engineering: A* 742(January): 573–83.

4 Improvement of Characteristics of Composites Using Natural Fibre as Reinforcement by Different Pretreatments

Sambhu Nath Chattopadhyay and Kartick K. Samanta
ICAR - National Institute of Natural Fibre Engineering and Technology

CONTENTS

DOI: 10.1201/9780429330575-4

4.1 INTRODUCTION

Our country is blessed with plenty of renewable natural fibre resources obtainable from the plant kingdom. Many of these natural fibres hold out immense potential for utilization as raw materials for non-traditional applications (Chattopadhyay et al. 2021). Natural fibres can be classified into three main groups – plant, animal and mineral. The most suitable fibres for composite reinforcements are from plants, in particular bast, leaf and wood fibres. Bast fibres, such as flax, hemp and kenaf, are taken from the stem of the plant and are most commonly used as reinforcements because they have the longest length and the highest strength and stiffness, while being significantly lighter than conventional reinforcements, such as glass fibres, and they are relatively low cost and biodegradable (Samanta, Basak, and Chattopadhyay 2015; Pandey et al. 1993). These types of plant fibres are composed of cellulose, hemicellulose and lignin (Chattopadhyay et al. 2020). From the environmental perspective, these fibres are biodegradable, recyclable and 'carbon positive', since they absorb more carbon dioxide than they release.

Jute is being utilized for a long time as a flexible packaging material in the form of sacks and hessian bags for crops and commodities (Chattopadhyay et al. 2018; Chattopadhyay, Pan, and Day 2004). But nowadays synthetic packaging materials have replaced them to a great extent due to various advantages of synthetic packaging materials. So, alternative use of this golden fibre (jute) is a must, and full utilization of its potential to be utilized in non-traditional sectors is the only option (Chattopadhyay et al. 2010). One of the major areas for utilization of jute fibre is the production of composite materials (Ammayappan et al. 2016). A composite material consists of one or more discontinuous phases embedded in a continuous phase with distinct interfaces (Ganguly 2013). The discontinuous phase is usually stronger than the continuous phase and is known as the reinforcement, while the continuous phase is known as the matrix. The properties of composites are strongly influenced by the properties of the individual constituents. It has superior properties and is possibly unique in some specific respects to the properties of the individual components. Each constituent or phase has to be present in reasonable proportions, which according to one estimate is greater than 5%. Composites are gaining popularity day by day as substitute of metal or wood due to their light in weight and superior mechanical properties. Fibre-reinforced composite (FRC) is a composite with two or more distinct physical phases, one of which is a fibrous phase. If the fibrous phase is jute, it is called jute-reinforced composite (JRC). Biocomposites are composite materials comprising one or more phase(s) derived from biological origin, e.g., jute-reinforced polyester resin and jute-reinforced polypropylene. Green composites, on the other hand, are materials in which all the components or phases are biodegradable, e.g., jute-reinforced polylactic acid (PLA). The component might be either derived from the biological origin or may be synthesized.

Jute is a naturally occurring lignocellulosic fibre made up of nearly 60% cellulose, 24% hemicellulose, 13% lignin and 3% other minor constituents. The chains of cellulose and hemicellulose run almost parallel to the fibre axis. The hydrogen bonds and other linkages between the constituents provide high stiffness of jute fibre. It is observed that jute fibre required large quantities of resin to wet the fibres/fabric structure and a good portion of this resin is squeezed out by the application of pressure during curing. Moreover, hydroscopic nature of jute fibre poses another problem in case of JRC (Ammayappan et al. 2020). So, modification of jute fibre is an important operation on one hand and suitable identification of area, where it can replace glass fibre composite is another important factor on the other hand. Probably, jute fibre composite may well substitute glass fibre composite, where high strength and cost of glass fibre are not justified. A number of research works have been done in the recent past for modification of jute fibre to improve its several functional properties like hygroscopicity, rot resistance, photochemical degradation and wetting property. But the literature shows only marginal improvement in its strength and moisture absorption. In some cases, the modified fibre surfaces are very difficult to wet with organic matrix materials, resulting marginal improvement in strength.

Jute may be considered to be a composite with the anisotropic cellulose microfibrils acting as the load-bearing entity in an isotropic lignin matrix with hemicellulose acting as the coupling agent between the two. Morphological studies of jute fibre show the outer sheath of lignin that develops on the cellulose ultimates. It is the cellulose ultimates, where the bonding must take place, if an efficient composite is to be manufactured. So, proper chemical treatment of jute fibre may break the different linkages releasing lignin, exposing cellulose ultimates and hemicellulose coupling agent resulting in better and more stable bond. In the present paper, some important chemical treatment on jute fibre has been done to improve its compatibility with the resin matrix. Considering practical utility and ease of operation, following important points have been taken care of while making fibre modification: (i) simple and suitable process for small scale sector, (ii) low cost, (iii) without costly equipment involvement and (iv) use of safe chemicals.

Modification of jute fibre has been done either at fibre stage or at non-woven fabric format. Modification of jute at fibre stage has been done to evaluate some of its important properties, while the same modification has been done at non-woven stage for making composites. The TD-3 grade jute fibre (as per BIS standard) has been used for fibre modification, and jute non-woven mat of 300 gsm has also been used for modification at fabric stage. Unsaturated polyester resins were applied on jute non-woven mats along with catalyst and accelerator in the material to liquor ratio 1:2.5, using 120 kg/cm^2 pressure at ambient temperature. Modification of jute was done using the following techniques: (i) hydrogen peroxide bleaching, (ii) alkali treatment, (iii) enzyme treatment, (iv) reactive dyeing and (v) phenol-formaldehyde resin treatment of jute fibre.

4.2 PREPARATION OF COMPOSITE

The way natural fibres are introduced as reinforcing materials in polymer composites must be adapted to the available production techniques. It is based on the polymer type (thermoplastic or thermosetting), the status of the products after processing (semi- or finished product) and the post-processing fibre geometry. Initially, a mould has to be prepared either from iron or from any other suitable conductive metal or from glass fibre-reinforced plastic. Out of various moulding techniques, hand layup and extrusion moulding techniques in the form of open or closed mould are frequently used (Ganguly 2013).

4.2.1 Hand Laying Process

Jute fibre mat or fabric is laid manually on the mould. Resin along with catalyst and accelerator is applied on the reinforcement. This can be done with application of pressure or without any pressure. Depending on the requirement, temperature can also be applied as per the need during the curing operation. Mould releasers, colours and gel-coats, could also be applied, if required.

4.2.2 Compression Moulding

Compression moulding process includes sheet moulding, bulk moulding, and cold press and hot press techniques. The method is similar to the hand layup technique, with the exception that makes use of a matched dies, which are closed before cure taken place by the application of pressure. By this process, a much larger volume of fibre loading can be achieved comparing with hand layup process. Jute fibre before as well as after chemical modification was used for preparation of biocomposite sheet by hand laying process followed by compression moulding method. The preparation protocol follows the (i) preparation of jute reinforcement, (ii) preparation of matrix resin and (iii) preparation of biocomposite. Schematic diagram of compression moulding process is depicted in Figure 4.1.

4.2.3 Extrusion Moulding Process

Jute reed is cut into desired length and modified either chemically or otherwise. These are then compounded with thermoplastic polymer in the kneader, passed through extruder and then cut into granules, which are subsequently moulded into products.

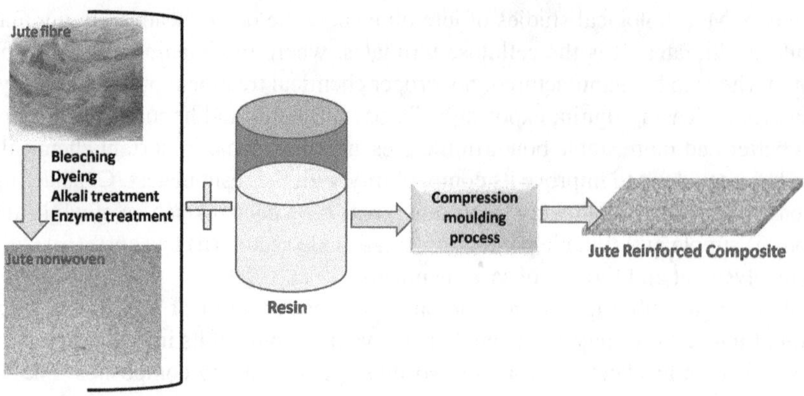

FIGURE 4.1 Schematic of compression moulding process.

4.3 MATERIAL AND METHODS

4.3.1 Materials

A commercial enzyme received from preparation Biocellulase ZK (Biocon India Ltd., Bangalore) was used. The following chemicals of analytical grade were used for study: hydrogen peroxide, trisodium phosphate, sodium hydroxide, sodium carbonate, sodium hydrosulphite, sodium silicate, non-ionic detergent (Ultravon JU), acetic acid and sodium acetate. Phenol-formaldehyde and polyester resin of commercial grades were also used. Three commercial reactive dyes, namely, Procion Yellow M4R (Cold brand, Substitution type), Procion Brown H4R (Hot brand, Substitution type) and Remazol Yellow FG (Vnylsulphone, addition type) were used for dyeing of jute fibre. Different reinforcement materials viz., jute fibre and non-woven fabric were used in the experiment.

4.3.2 Scouring, Bleaching and Enzyme Treatment

Grey jute fibre was scoured with sodium hydroxide (1%, o.w.f.) and Ultravon JU (2 mL/L) at boil for 60 minutes using 1:20 to liquor ratio. Bleaching of scoured fibre was done at 90°C for 90 minutes, keeping the material to liquor ratio at 1:20 with 1 vol. hydrogen peroxide, trisodium phosphate (5 g/L), sodium silicate (10 g/L), sodium hydroxide (1 g/L) and Ultravon JU (2 mL/L). The pH of the bath maintained at 10–11. The material was washed thoroughly after bleaching and then dried air. The scoured jute fibre was treated with 4% (o.w.f.) enzyme at 55°C for 120 minutes using 1:10 material to liquor ratio keeping the pH at 5. Then, samples were washed and dried.

4.3.3 Reactive Dyeing of Jute

4.3.3.1 Cold Brand Reactive Dye

Bleached jute fibre was dyed with three different reactive dyes with 4% shade with M:Lr ratio of 1:20. Dye bath was made with Glauber's salt (60 g/L). The samples are dipped in the bath and kept for 1 hour with stirring at ambient temp. Then, sodium carbonate (20 g/L) was added and kept for 45 minutes. The samples were then washed with cold water, soaped with Ultravon JU (2 g/L) for 15 minutes followed by washing in fresh water and drying in the air (Chattopadhyay, Pan, and Day 2009).

4.3.3.2 Hot Brand Reactive Dye

The samples are dipped in the dye bath with Glauber's salt (80 g/L) for 1 hour with stirring at 80°C temp. Then, alkali (NaOH) 2 g/L and sodium carbonate (20 g/L) were added and kept for 45 minutes in the same condition. The samples were washed as above.

4.3.3.3 Remazol FG Dye

The samples are dipped in the dye bath with Glauber's salt (80 g/L) for 1 hour with stirring at ambient temp. Then, sodium hydroxide (4 g/L) was added and kept for 45 minutes in the same condition. The samples were washed as above.

4.3.4 Alkali Treatment

Jute fibres were treated with different concentration of sodium hydroxide solution at 22°C for 60 minutes using material to liquor ratio at 1:10. The fibres were washed with water and then kept 5% acetic acid solution for 15 minutes and then washed and dried.

4.3.5 Determination of Physical Properties

Weight loss of the jute fibre was measured by gravimetric method. The samples were weighed carefully after drying in an oven at 100°C for 60 minutes before and after treatment and calculated using following expression, where W_1 and W_2 are the weights of the sample before and after treatment. Weight loss = $100(W_1 - W_2)/W1$. X-ray crystallinity values of finely powdered samples of control and enzyme-treated jute fibres were determined in Philips X-ray diffractometer (Model 1700). Whiteness index in the HUNTER scale of grey and bleached jute fibre samples was measured by the spectrascan-5100 computer colour matching system using relevant software. Moisture regain of different jute fabric samples was determined following standard method (ASTM-D-2654-76). Jute fibre samples were exposed in a hot air oven to a stream of air, heated to 105°C±5°C until no further loss of weight occurs. The moisture regain (MR) value was calculated using below expression, where 1 is weight of the fabric specimen and '*w*' weight of oven dry weight fabric.

$$\text{MR}(\%) = \frac{(1-w)\times 100}{w}$$

4.3.6 Determination of Mechanical Properties

Bundle strength of grey and processed jute fibre samples was determined as per IS: 7032 (Part-VII) – 1975 method. The fineness of grey and processed jute fibre samples was determined as per the IS: 7032 (Part VIII) – 1976 method. The tensile properties of jute non-woven fabrics have been evaluated in the Instron Tensile Tester following the ASTM standard (D683–86) keeping gauge length-50 mm, test width-6 mm and crosshead speed-5 mm/minute. Average of ten tests has been reported. A three-point loading system utilizing centre loading on a simply supported beam has been applied to evaluate flexural properties of composite samples in Instron Material Testing System following the ASTM Standard (D790-81) keeping sample dimension 80 × 10 mm, support span 64 mm, rate of crosshead speed 1.7 mm/minute. Average of ten tests has been reported.

4.4 RESULTS AND DIMENSIONS

4.4.1 Hydrogen Peroxide Bleaching of Jute

During the preparation of composites, various impurities present in the fibre inhibit good adhesion of the fibre with the resin matrix, which in turn affects composite properties. In order to expose the reactive groups of jute fibre, it has been treated with oxidising bleaching agent (Pandey et al. 1993). The average values of the physical properties of grey and modified fibres are as follows: whiteness index: 53 and 80, fibre fineness: 2.84 and 2.06 tex and bundle strength: 25 and 24 g/tex, respectively. It appears that whiteness index and brightness index of the fibre improve appreciably after fibre modification. In comparison to grey fibre, the fibre becomes finer after the treatment. But there is no appreciable change in the strength of the fibre (Chattopadhyay et al. 2010; Sengupta et al. 2008; Samajpati et al. 2005). Different non-woven fabrics have been prepared from this bleached and control fibres. These non-wovens were then treated with resin to make different composites. The properties of these composites and non-woven made from this modified and control fibres have been evaluated.

It is clear from Table 4.1 that resin uptake increases considerably, if composite is made from non-woven made from bleached jute fibre. The increased resin uptake in non-woven made from bleached jute fibre resulted in increased flexural and tensile properties. So, it is included that modification of jute fibre by bleaching method leads to better jute fibre-reinforced composite products with improved mechanical properties.

4.4.2 Alkali Treatment of Jute

In order to improve the quality of composite materials made from jute fibres, chemical modification of jute fibre is an important step. Jute fibre has been treated with different concentrations of sodium hydroxide solutions (0%, 2%, 5%, 9%, 12%, 14%, 16%, 18% and 20%) for 2 hours at room temperature to remove some ingredients for accessibility for chemicals. The fibres were washed thoroughly, neutralized and dried. The alkali-treated fibres has been evaluated, and the results are tabulated in Table 4.2 (Chattopadhyay, Sengupta, Samajpati, and Day 2006; Sengupta et al. 2008; Samajpati et al. 2005; Ammayappan et al. 2020).

It is evident from the table, that starting from 9% alkali treatment moisture regain value decreases. Alkali treatment leads to removal of hemicellulose, which is the main component of jute fibre responsible for high moisture affinity. So, jute fibre becomes less hygroscopic after 9% alkali treatment. Weight loss due to alkali treatment also starts increasing beyond 9% alkali treatment. It is also clear from the table that there is no loss of strength of fibre, if fibre is treated with 9% alkali solution.

Alkali treatment of jute fibre results in weight loss due to removal of xylan and lignin-carbohydrate complex. Quantitative analysis of extract after alkali treatment (9% NaOH) at room temperature was carried out in respect to 100 g of dewaxed jute. The values of different important parameters are: (i) total loss of cellulose by alkali treatment – 9.63, (ii) lignin-carbohydrate complex

TABLE 4.1
Effect of Different Reactive Dyeing of Jute Fibre on Flexural Properties

Samples	Tensile Stress (MPa)	Elongation (%)	Tensile Modulus (GPa)	Flexural Stress (MPa)	Resin Uptake (%)
Control	47.08	2.06	2.28	39.8	1.35
Bleached	63.85	3.16	2.01	55.7	3.20

TABLE 4.2
Effect of Alkali Treatment of Jute Fibre on Different Physical Properties

Sample Treated with Different Concentration of NaOH	Weight Loss (%)	Moisture Regain (%)	Bundle Strength (g/tex)
0% (Control)	–	12.18	26.75
2%	2.00	12.89	26.93
5%	6.52	12.98	25.58
9%	9.60	10.34	26.10
12%	11.76	10.27	20.11
14%	10.02	10.29	13.48
16%	12.36	10.36	15.03
18%	11.23	10.42	15.13
20%	12.56	10.72	14.26

– 2.78, (iii) xylan – 5.36 and (iv) soluble carbohydrate lignin breakdown products – 1.53. Removal of lignin-carbohydrate complex makes the structure loose, while xylan removal makes the fibre porous making room for the large resin molecules to enter the matrix for cross-linking. Actually, alkali-mediated pretreatment process helps to increase the binding of resin with jute fibre. So, pretreatment of jute fibre with alkali lead to improve jute fibre-reinforced plastic products.

The same alkali treatment has been given to jute non-woven mat, and composites have been prepared from them. They have been evaluated for tensile and flexible properties following ASTM standard. Detailed analysis of the data (Tables 4.3 and 4.4) reveals that the important points like alkali-treated fibres have better tensile properties than control. Tenacity values increase up to 9% after alkali treatment and then start falling, which can be explained from the fact that bundle strength of the fibres starts falling drastically after 9% alkali treatment. Breaking extension is also good at 9%, and then, tensile modulus drops substantially after 9%. Energy to break is also very high at 9% alkali treatment. Analysis of flexural property also suggests that the composites produced from 9% alkali-treated jute fibre mat shows good performance. So, it can be concluded that if jute is treated with 9% alkali solution and composite is produced, it results in good tensile and flexural property.

4.4.3 Enzyme Treatment of Jute

Enzyme used for improvement of softness and smoothness of jute fibre is lytic in nature, so there will be certain weight loss by enzymatic hydrolysis process (Chattopadhyay, Sanyal, Day, and Kundu 2000). The smooth fibres produced after enzyme treatments are having high porosity (improved by 30%) due to removal of cellulose, hemicellulose and lignin. Enzyme is found to attack preferentially the amorphous region of fibre, which is manifested by relative increase in % crystallinity, where it increased steadily from 57% to 66% from the grey to bleached and enzyme-treated samples. Properties of enzyme treated fibres have been tabulated in Table 4.5. Bundle strength of the fibre is marginally reduced after scouring and the strength further decreases. It is found that the fineness improves to a great extent after enzyme treatment (Chattopadhyay, Sengupta, Samajpati, and Day 2006; Sengupta et al. 2008; Samajpati et al. 2005).

TABLE 4.3
Effect of Alkali Treatment of Jute Non-woven on Tensile Properties of JRC

Samples	Tensile Stress (MPa)	Breaking Extension (%)	Tensile Modulus (GPa)	Energy to Break (J)
Control	47.1	2.06	2.28	0.248
2% Alkali	54.5	2.15	2.54	0.195
5% Alkali	58.9	2.16	2.72	0.234
9% Alkali	60.8	2.49	2.44	0.426
14% Alkali	46.9	3.12	1.50	0.445
18% Alkali	43.1	2.61	1.65	0.485

TABLE 4.4
Effect of Alkali Treatment of Jute Non-woven on Flexural Properties of JRC

Samples	Flexural Stress (MPa)	Flexural Strain	Flexural Modulus (MPa)	Energy to Break (mJ)
Control	39.8	0.031	1,254	74
2% Alkali	44.1	0.019	2,217	31
9% Alkali	64.1	0.039	1,634	214
18% Alkali	66.0	0.045	1,463	465

TABLE 4.5
Comparative Property of Jute Fibre before and after Enzyme Treatment

Sample	Bundle strength (g/Tex)	Fineness (Tex)
Control	25.0	2.40
Scoured jute fibre (1% NaOH)	23.0	2.20
Enzyme-treated fibre	21.8	1.83

TABLE 4.6
Effect of Enzyme Treatment of Jute Non-woven on Tensile Properties of JRC

Samples	Tensile Stress (MPa)	Breaking Extension (%)	Tensile Modulus (GPa)	Energy to Break (J)
Control	47.08	2.06	2.288	0.24889
1% Alkali + Enzyme	51.59	2.28	2.262	0.330532
2% Alkali + Enzyme	55.14	2.21	2.495	0.421317

TABLE 4.7
Effect of Enzyme Treatment of Jute Non-woven on Flexural Properties of JRP

Samples	Flexural Stress (MPa)	Flexural Strain	Flexural Modulus (MPa)	Energy to Break (mJ)
Control	39.8	0.031	1,254	74
1% Alkali + Enzyme	77.0	0.039	1,962	261
2% Alkali + Enzyme	67.1	0.031	2,147	197

Composites prepared from enzyme-treated jute non-woven mats have been evaluated for tensile and flexural properties following ASTM standard. It is clear from the analysis of Tables 4.6 and 4.7 that enzyme-treated non-woven composite samples have better tensile properties than control. Regarding flexural property, there is a substantial improvement. Improved crystallinity, softness, fineness and pliability of enzyme-treated fibre may be responsible for better flexural property of the composites. It is seen that enzyme treatment on 1% alkali-treated fibre produced better composites (Chattopadhyay et al. 2010; Sengupta et al. 2008; Samajpati et al. 2005).

4.4.4 Reactive Dyeing of Jute Non-woven

Pretreatment of jute fibre is a must for making composites, and it will be better if the pretreatment process is chosen in such a way that it will make value addition to the product apart from making fibres suitable for preparation of composites (Basü and Chattopadhyay 1996). With this aim in view, jute non-wovens were dyed with the three different types of reactive dyes (Chattopadhyay, Sengupta, Samajpati, and Day 2006; Samajpati et al. 2005). The details of the dyes are (i) Procion Yellow M4R- Cold brand, Substitution type, (ii) Procion Brown H4R- Hot brand, Substitution type and (iii) Remazol Yellow FG- Vinyl sulphone, addition type. The reason for choosing reactive dye is that it can react with fibre as well as resin to work as a coupling agent and it produces very bright product. Consideration of molecular structure of reactive dyes suggests that dichlorotriazinyl compound should be able to cross-link between adjacent cellulose molecules. This is proved by significant reduction of swelling in

TABLE 4.8
Effect of Different Reactive Dyeing of Jute Non-woven on Tensile Properties

Samples	Tensile Stress (MPa)	Breaking Extension (%)	Tensile Modulus (GPa)	Energy to Break (mJ)
Control	47.1	2.06	2.28	248
Vinyl sulphone reactive dye	55.8	1.94	2.87	111
Hot brand reactive dye	59.9	2.10	2.85	247
Cold brand reactive dye	75.1	2.16	3.48	239

TABLE 4.9
Effect of Different Reactive Dyeing of Jute Non-woven on Flexural Properties

Samples	Flexural Stress (MPa)	Flexural Strain	Flexural Modulus (MPa)	Energy to Break (mJ)
Control	39.8	0.031	1,254	074
Vinyl sulphone reactive dye	42.1	0.022	1,888	077
Cold brand reactive dye	49.5	0.025	1,971	132

cuprammonium solution. A recent development (Procion-Resin process) is based upon the capacity of reactive dyes to combine with resin precondensate during curing condition.

Composites have been prepared from these dyed mats, it shows very uniform application of resin, and the product is also uniform. The samples have been tested following ASTM standards with respect to its tensile and flexural properties. Analyses of the results show that all the dyed samples show better tensile and flexural properties than the control samples (Tables 4.8 and 4.9). Cold brand reactive dyed samples show the best result. Compared with improvement in flexural property, tensile property improvement is more.

4.4.4.1 Phenol Formaldehyde Resin Treatment on Jute

Jute fibre, in general, requires high percentage of resin to wet with polymeric matrices due to weak interfacial bonding between the fibre and the resin and due to hygroscopic nature of jute fibres. The α-cellulose content of jute fibre possesses three hydroxyl groups for each glucose residues, and the hemicellulose content of jute fibre possesses two hydroxyl groups for each pentose residues. These hydroxyl groups are the active centres to attract water molecules. Hydrogen bonds are formed between hydroxyl groups and water molecules. First, the direct attachment of water molecules through the active sites occurs, and then, the indirect attachment through directly attached water molecules takes place. Hence, similar to sizing of the synthetic fibres, attempts have been made to block partially the hydroxyl groups of cellulose and hemicellulose of jute fibre by the water repellent phenolic solutions. Jute non-wovens have been dipped in phenolic resin and squeezed to give 100% wet pick up. The samples are dried, and unsaturated polyester resin applied and cured at high temperature and pressure. Polyester resins have been applied in three different proportions to produce three different samples. Evaluations of their properties have been tabulated in Table 4.10. Similar result was also seen by other researcher (Ammayappan, Chakraborty, and Pan 2018).

Two samples have been found to have slightly higher tenacity values but have significantly higher tensile modulus. Similarly, with respect to flexural properties, two samples have higher stress value than control with substantial improvement of flexural modulus in one case. So, it can be concluded that composites prepared by following phenol formaldehyde (PF) & polyester (PE): 1 + 2.5 produced the best one. Some of the jute composites as well as their products are shown in Figure 4.2.

TABLE 4.10
Effect of PF Resin Pretreatment of Jute Non-woven on Tensile and Flexural Properties of JRC

Samples	Tensile Stress (MPa)	Flexural Stress (MPa)	Tensile Strain (%)	Flexural Strain (%)	Tensile Modulus (GPa)	Flexural Modulus (MPa)	Tensile Energy to Break (mJ)	Flexural Energy to Break (mJ)
Control	47.1	39.8	2.06	0.031	2.28	1,254	248	074
PF + PET (1 + 1.5)	30.2	42.2	1.12	0.019	2.69	2,164	045	057
PF + PET (1 + 2)	47.9	47.9	1.33	0.019	3.60	2,491	120	062
PF+PET (1+2.5)	51.2	66.9	1.39	0.018	3.68	3,674	127	069

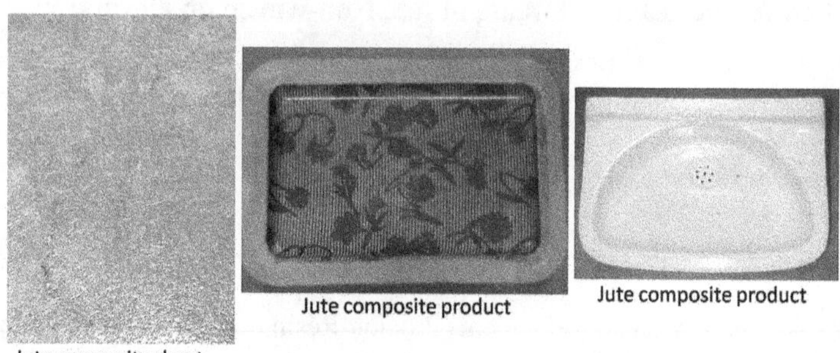

FIGURE 4.2 Pictures of jute composite and their products.

4.5 EFFECT OF VARIOUS IRRADIATIONS ON JUTE COMPOSITE

Various chemical methods, such as chlorination, alkali treatment, synthetic polymer coating and many more, are reported for textile modification. However, these processes are often found to be adverse, non-eco-friendly, modifying the bulk properties of materials and also generating noticeable effluents. On the other hand, fibre surface modification by physical means, i.e., using various irradiation technologies has also been explored to improve the textile properties. In this context UV rays, gamma rays, plasma and laser ray irradiations are being utilized in chemical processing of textiles since long (Teli, Pandit, and Samanta 2015; Gupta and Basak 2010; Samanta et al. 2016; Morgan, Tyrer, and Kane 2014). Among these, the plasma, ultraviolet (UV) and laser treatments are found to be clean, cheaper and multipurpose option, besides their advantages of environmentally friendly, dry and energy saving process over the traditional wet chemical process as discussed below (Sengupta, Ammayappan and Samanta, 2021).

4.5.1 Plasma Treatment

Atmospheric plasma, a potential environmentally friendly emerging technology, has been widely explored for surface modification of fibre at nanometre level without affecting the bulk properties (Teli et al. 2015a). Plasma-enhanced surface modification puts forward an alternative to wet processing with consequent benefits to the environment. The nature of the effects realized mostly depends on the type of feed gas and time of irradiation. The key importance of exploiting atmospheric plasma is to enhance the wettability of the natural fibre surface by removing non-cellulosic substances from the outermost surface that plays a key role in improving the composite properties. Textile surface modification using plasma technology is achieved with a various gases viz., oxygen, helium, air, argon, nitrogen or fluorine to improve/ introduce properties like water and oil absorption, adhesion, dye exhaustion, anti-static and anti-felting of wool (Teli, Pandit, and Samanta 2015;

Park et al. 2008; Ratnapandian et al. 2011; Teli et al. 2021). Likewise, plasma modification with a reactive molecule containing carboxyl, hydroxyl, carbonyl, vinyl, acrylate, silicone, phosphorous and fluorocarbon could lead to improvement in hydrophobic, oleophobic, flame retardant, UV protective, crease-resistant and antimicrobial properties of textiles (Samanta et al. 2008, 2010, 2012; 2014, 2015, 2021; Pandit, Samanta, and Teli 2020; Teli et al. 2015b; Vinogradov and Lunk 2005).

The jute woven fabric (320 g/m^2 with 270 tex warp and 242 tex weft yarns) was subjected to atmospheric pressure glow discharge plasma treatment in continuous manner with plasma power of 970 W and helium gas of 14 L/minute. Fabric samples of 200 mm^2 were cut and mixed with unsaturated polyester resin at a ratio of 3:1 (resin to jute) (Kafi et al. 2012). The plasma treatment of jute fabric improves the flexural properties of jute/polyester composites. Flexural properties were improved due to a plasma-induced increase in the strength of the fabric reinforcement material. A 45% increase in flexural strength (FS) and 141% increase in flexural modulus (FM) were seen in composites manufactured with 25 passes of plasma-treated fabric. The majority of the improvements in FS and FM can be directly related to the strength improvement in the fabric reinforcement material. Unbleached woven jute fabric was treated with atmospheric pressure glow discharge plasma using helium (He) (Flow rate: 14 L/minute), helium (He) and acetylene (Ac) (Flow rates: 14 and 0.7 L/minute, respectively), and helium (He) and nitrogen (N) (Flow rates: 14 and 0.7 L/minute, respectively) gases for different time. The power of the treatment plasma was 970 W, the frequency was 90 kHz and the process was carried out for 5, 25, 50 and 100 revolutions, keeping treatment time of 0.424 seconds per revolution. For all gases used, 10 seconds of plasma treatment was sufficient to significantly improve the wettability of the fabric and unsaturated polyester resin composites. There was a 55%, 62% and 40% improvement in the flexural strength, flexural modulus and inter-laminar shear stress, respectively, when the composites were produced from the plasma-treated fabrics. The storage modulus and glass transition temperature were also improved by up to 200% and 16°C, respectively (Kafi, Magniez, and Fox 2011; Sengupta, Ammayappan and Samanta, 2021).

In another study, jute fabrics were subjected to plasma irradiation with different powers of 60, 90 and 120 W for 1, 3 and 6 minutes. Tensile modulus of untreated jute/polyester composite was 5.2 GPa. When jute fibres were atmospheric air plasma treated, tensile modulus of jute/ polyester composites increases for each plasma power and exposure time. The ILSS values increased about 122%, 144% and 171% after atmospheric air plasma treatment for 1, 3 and 6 minutes at 120 W, respectively (Table 4.11). It is inferred that plasma treatment improves the interfacial adhesion between the jute fibre and polyester. The greatest tensile strength and flexural strength values were determined at 120 W for 1 minute and at 60 W for 3 minutes, respectively (Demir et al. 2011).

TABLE 4.11
Tensile and Flexural Properties of Jute Fabric Composites (Demir et al., 2011)

Plasma Power (W)	Treatment Time (minute)	Tensile Strength (MPa)	Flexural Strength (MPa)	Tensile Modulus (GPa)	Flexural Modulus (GPa)	Elongation at Break (%)
Untreated	0	56.8±2.4	76.5±3.2	5.2±0.7	5.12±0.54	1.19±0.06
60	1	71.5±4.3	83.1±6.4	5.6±0.9	6.34±0.97	0.97±0.04
	3	86.6±6.3	93.6±7.5	5.9±0.7	6.67±0.76	1.09±0.06
	6	79.6±5.8	89.7±7.8	5.7±0.4	6.56±0.45	0.85±0.02
90	1	78.4±5.6	81.5±8.6	5.6±0.9	6.78±1.23	0.97±0.04
	3	87.3±5.7	86.6±5.7	5.9±0.7	7.94±0.96	1.12±0.08
	6	82.6±3.5	83.7±7.3	5.6±0.5	6.72±0.67	0.79±0.05
120	1	91.2±4.6	88.1±5.7	6.0±0.5	6.27±0.45	0.95±0.06
	3	84.1±5.2	81.2±7.3	5.8±0.2	7.03±0.35	1.13±0.03
	6	72.3±3.4	72.1±4.5	6.1±0.4	6.87±0.31	0.76±0.01

4.5.2 UV and Laser Treatment

The surface modification, such as etching, ablation, deposition and evaporation by laser treatment, can be carried out in various ways depending on the purpose of such modification (Samanta, Basak, and Chattopadhyay 2017). The operating principle of laser is based on the radiative decomposition of excimer states created by a silent discharge in a high-pressure gas column. A typical lamp consists of two concentric quartz tubes, outer and inner metallic electrodes with the discharge gap of few millimetres and an external high-voltage generator (Basak et al. 2015). The discharge gap is filled with either a rare gas or a rare gas-halogen mixture. An alternative high voltage of 7–10 kV and frequency of 50 Hz to several MHz are adequate to run the arc discharge between the electrodes (Samanta, Basak, and Chattopadhyay 2014). Excimer lamps are used for surface modification of polymers and textile substrates for improving in anti-microbial, anti-soiling, anti-felting, anti-static, wettability, adhesion strength and cross-linking properties.

A laser is a device that could emit light through a process of optical amplification based on the stimulated emission of electromagnetic radiation. It works by creating an extensive heat, where the material within a very small region is subjected to a very intense heat that may also lead to melting of the material due to a phase transformation from solid to liquid. The optimal characteristics of the laser light have increased its application in many fields including medical, industrial and military (Nourbakhsh and Ashjaran 2012). The laser used in the textile industry is of great importance for discoloration of denim, cutting of fabrics and for wool-shrinkage controlling purpose (Kan, Yuen, and Cheng 2010). Laser finishing is a complete dry process and with the careful control of laser processing parameters, which it can provide a fast and accurate production with a good reproducibility and repeatability (Hung et al. 2011). Several commercial lasers are available for the industrial applications, including the laser of Nd:YAG, CO_2 and Excimer. Similar to water-free plasma processing technique, the CO_2 laser treatment could also be considered for denim processing without usage of water, as an alternative to the conventional wet processing viz., stone washing, sand-washing and bleaching for achieving the faded-look and worn-out effects.

Jute fibre (hessian cloth) and E-glass fibre (mat)-reinforced, unsaturated polyester (USP) resin, along with additives and initiator, composites are prepared by the hand layup technique at room temperature (25°C) (Abdullah-Al-Kafi et al. 2006). Among the various hybrid composites produced, the composite with jute to glass ratio of 1:3 demonstrates improved mechanical properties, such as tensile strength (TS) 125%, tensile modulus (TM) 49%, bending strength (BS) 162% and bending modulus (BM) 235% over untreated jute composite. To further improve the properties, the surface of jute and glass fibre was irradiated under UV radiation of different intensities. UV-pretreated jute and glass fibres (1:3) at optimum intensities show the highest mechanical properties, such as TS 70%, TM 33%, BS 40% and BM 43% compared with untreated jute- and glass-based hybrid composites with best Charpy impact strength of 40 kJ/m^2. In another experiment, jute yarns were initially dried at 105°C for 20 hours to remove absorbed moisture and stored in a desiccator (Idriss Ali et al. 1999). The dried yarns (15 cm long and weighing about 2 g) were then immersed for 30 minutes in HEMA+MeOH solutions mixed at different proportions (1%–20%) of HEMA with MeOH in test tubes that were placed in a rotatory rack that moved around a global UV lamp (100 W) for different periods. After the irradiation, the treated jute yarn was washed in acetone to remove unreacted monomer HEMA and gained about 10% polymer loading along with enhanced tensile strength (80%) and elongation (95%). The tenacity did not further improve by incorporation of a minute amount (1%) of novel additives into the HEMA+MeOH solutions, but elongation was found to increase up to 140%. The tenacity increases with irradiation time up to 15 minutes and then decreases as the irradiation increases beyond 15 minutes. The highest TS value is produced by the sample treated with 3% HEMA, followed by 2% HEMA.

4.5.3 Gamma Treatment

Gamma radiation is an electromagnetic ionizing radiation produced during the radioactive decay of the atomic nucleus. One of the key results of gamma irradiation is the excitation of the electron, resulting in ionization-induced bond breaking. It leads to formation of excited states, short-life radicals and finally formation of new bonds. In research, especially textile, gamma irradiation is used in optimization of textile dyeing, reducing the ecological impact and cultural heritage artefacts biological decontamination. Gamma rays are high-energy electromagnetic radiations having energies above 100 keV and wavelengths less than 10 pm. Surface modification of textiles using gamma ray is considered one of the promising techniques for various applications. Gamma radiation is known to deposit energy in solid cellulose by Compton scattering, and the rapid localization of energy within molecules produces trapped macrocellulosic radicals. The radicals thus generated are responsible for changing the physical, chemical and biological properties of cellulose fibres (Căpraru et al. 2018). For preparing the composites, PP sheets (0.20 – 0.35 mm) and bleached hessian jute fabrics were cut into small pieces (15 cm × 12 cm) (Zaman et al. 2010). Dried jute fabrics and also the PP sheets were treated with gamma radiation for different doses (250 – 1000 krad) at 350 krad/hour using Co-60 gamma source. Both the irradiated jute fabrics and PP sheets were used for the preparation of composites with 50% fibre content by compression moulding, and the mechanical properties were evaluated.

For irradiated jute/irradiated PP composite, TS and TM values increased up to 100 UV radiation dose then decreased (Table 4.12). The maximum TS and TM values were found to be 59.69 MPa and 1.32 GPa, respectively, which correspond to a 24% increase in TS and a 39% increase in TM as compared to the untreated composite. Investigation showed that BS, BM and IS of the composites were increased significantly compared with the non-treated one (Zaman et al. 2010). For irradiated jute/irradiated PP composite, TS and TM values increased up to 500 krad and then decreased. The maximum TS and TM values were found to be 57.9 MPa and 1.24 GPa, respectively, resulting in 21% enhancement of TS and 31% enhancement of TM compared with the untreated composite. The BS, BM and IS of the composites were increased proportionally compared with the untreated one, and those values were maximum of 62.4 MPa, 1.58 GPa and 24.2 kJ/m^2 at 500 krad gamma dose, respectively. It was seen that the mechanical properties of UV-treated jute fabrics/PP composites had increased TS, TM, BS, BM and IS compared with that of the gamma-treated jute fabrics/PP composites.

TABLE 4.12
Mechanical Properties of UV- and Gamma-Treated Jute Woven Fabrics/PP Composite (Zaman et al. 2010)

UV Pass	Tensile Strength (MPa)	Bending Strength (MPa)	Tensile Modulus (GPa)	Bending Modulus (GPa)	Impact Strength (kJ/m^2)
0	48.1	51.2	0.95	1.12	17.1
25	52.1	56.6	0.98	1.25	20.1
50	56.2	59.1	1.25	1.47	24.2
100	59.6	65.3	1.32	1.62	26.1
150	56.6	63.2	1.29	1.58	23.4
200	54.1	60.4	1.26	1.52	21.6
			Gamma Radiation		
0	48	51.2	0.95	1.12	17.1
250	51.1	56.2	0.99	1.35	20.2
500	57.9	62.4	1.24	1.58	24.2
750	54.2	59.7	1.14	1.51	23.1
1,000	51.8	57.1	1.1	1.42	21.9

4.6 SUMMARY

1. Modification of jute fibre by hydrogen peroxide helps to produce better jute fibre-reinforced plastic products. Bleached jute non-woven dyed with cold brand reactive dye not only produced beautiful coloured composite but also possessed very high tensile and flexural property than control.
2. Alkali treatment removes surface lignin as well as improves porosity. Composites made from jute non-woven modified by 9% alkali treatment produce best composite with improved tensile and flexural properties.
3. Enzyme treatment on 1% alkali-treated jute non-woven produces better composite with high tensile and flexural property.
4. Phenol formaldehyde-treated non-woven jute mat produced good composite product with unsaturated polyester resin, when used 2.5 times weight of the only non-woven mat.
5. It was observed that in all the irradiation jute composite samples, there was notable improvement in tensile and flexural properties.

REFERENCES

Abdullah-Al-Kafi, M. Abedin, M. Beg, K. Pickering, and M.A. Khan. 2006. Study on the mechanical properties of jute/glass fiber-reinforced unsaturated polyester hybrid composites: Effect of surface modification by ultraviolet radiation. *Journal of Reinforced Plastics and Composites* 25(6): 575–88.

Ammayappan, L., S. Chakraborty, I. Musthafa, and N.C. Pan. 2020. *Standardization of a Chemical Modification Protocol for Jute Fabric Reinforcement*. Taylor & Francis. https://doi.org/10.1080/15440478.2020.1758276.

Ammayappan, L., S. Chakraborty, and N.C. Pan. 2018. Effect of areal density and layering of jute nonwoven fabrics on the performance of biocomposite. *Indian Journal of Natural Fibres* 4(2): 25–32.

Ammayappan, L., S. Das, R. Guruprasad, D.P. Ray, and P.K. Ganguly. 2016. Effect of lac treatment on mechanical properties of jute fabric/polyester resin based biocomposite. *Indian Journal of Fibre & Textile Research* 41(3): 312–17.

Basak, S, K.K. Samanta, S.K. Chattopadhyay, and R. Narkar. 2015. Development of dual hydrophilic/ hydrophobic wool fabric by Q172 NM VUV irradiation. *Journal of Scientific and Industrial Research (JSIR)* 75(439): 439–43.

Basü, G., and S.N. Chattopadhyay. 1996. Ambient temperature bleaching of jute fibre - its effect on yarn properties and dyeing behaviour. *Indian Journal of Fibre and Textile Research* 21(3): 217–22.

Căpraru, O., C. Herman, B. Lungu, and I. Stănculescu. 2018. Application of gamma irradiation for the functionalization of textile materials the textile industry is a major branch of the world's economy, some of its biggest producers and exporters being China, India and Italy. Textiles have been found to orig. *The 7th International Conference on Advanced Materials and Systems*, 51–56.

Chattopadhyay, S.N., N.C. Pan, A. Day, S.B. Mondal, and A. Khan. 2010. Reactive dyeing of a pretreated jute fabric using minimum application technology. *The Journal of the Textile Institute* 97(6): 493–501. https://doi.org/10.1533/joti.2005.0202.

Chattopadhyay, S.N., N.C. Pan, A.N. Roy, and K.K. Samanta. 2018. Pretreatment of jute and banana fibre—Its effect on blended yarn and fabric. *Journal of Natural Fibers* 17(1): 75–83.

Chattopadhyay, S.N., N.C. Pan, A.N. Roy, K.K. Samanta, and A. Khan. 2020. Two-step bleaching of jute yarn and fabric using hydrogen peroxide and peracetic acid. *Journal of Natural Fibers* 19(3). https://doi.org/10.1080/15440478.2020.1821278.

Chattopadhyay, S.N., N.C. Pan, A.N. Roy, K.K. Samanta, and A. Khan. 2021. Hybrid bleaching of jute yarn using hydrogen peroxide and peracetic acid. *Indian Journal of Fibre & Textile Research (IJFTR)* 46(1): 78–82.

Chattopadhyay, S.N., N.C. Pan, and A. Day. 2004. A wet processing of jute at ambient temperature. *AATCC Review* 4(9): 27–27.

Chattopadhyay, S.N., N.C. Pan, and A. Day. 2009. Ambient-temperature bleaching and reactive dyeing of jute: The effects of pre-treatment, bleaching, and dyeing methods. *The Journal of the Textile Institute* 93(3): 306–15. https://doi.org/10.1080/00405000208630572.

Chattopadhyay, S.N., Sanyal, S.K., A. Day, and A.B. Kundu. 2000. Enzyme treatment on jute - mechanism of enhancement of enzyme activity due to pretreatment. *Man-Made Textiles in India* XLIII(5): 209–219.

Chattopadhyay, S.N., Sengupta, S., S. Samajpati, and A. Day. 2006. Modification of jute fibres for making composites. *Asian Textile Journal* 15(8): 42–51.

Demir, A., Y. Seki, E. Bozaci, M. Sarikanat, S. Erden, K. Sever, and E. Ozdogan. 2011. Effect of the atmospheric plasma treatment parameters on jute fabric: The effect on mechanical properties of jute fabric/polyester composites. *Journal of Applied Polymer Science* 121(2): 634–38.

Ganguly, P.K. 2013. Natural fibre reinforced polymeric composites: Some basic considerations. *Diversification of Jute and Allied Fibres* 175–202.

Gupta, D., and S. Basak. 2010. Surface functionalization of wool using 172 Nm UV Excimer lamp. *Journal of Applied Polymer Science* 117(6): 3448–53.

Hung, O.N., L.J. Song, C.K. Chan, C.W. Kan, and C.W.M. Yuen. 2011. Using artificial neural network to predict colour properties of laser-treated 100% cotton fabric. *Fibers and Polymers* 12(8): 1069–76.

Idriss Ali, K.M., M.A. Khan, M. Azam Ali, and K.S. Akhunzada. 1999. In situ jute yarn composite with HEMA via UV radiation. *Journal of Applied Polymer Science* 71: 841–846.

Kafi, A.A., C.J. Hurren, M.G. Huson, and B.L. Fox. 2012. Analysis of the effects of atmospheric helium plasma treatment on the surface structure of jute fibres and resulting composite properties. *Journal of Applied Polymer Science* 23(16): 2109–20. http://dx.doi.org/10.1163/016942409x12526743388006.

Kafi, A.A., K. Magniez, and B.L. Fox. 2011. A surface-property relationship of atmospheric plasma treated jute composites. *Composites Science and Technology* 71(15): 1692–98.

Kan, C.W., C.W.M Yuen, and C.W. Cheng. 2010. Technical study of the effect of CO_2 laser surface engraving on the colour properties of denim fabric. *Coloration Technology* 126(6): 365–71.

Morgan, L., J. Tyrer, and F. Kane. 2014. The effect of CO_2 laser irradiation on surface and dyeing properties of wool for textile design. *International Congress on Applications of Lasers & Electro-Optics* 2018(1): 1.

Nourbakhsh, S., and A. Ashjaran. 2012. Laser treatment of cotton fabric for durable antibacterial properties of silver nanoparticles. *Materials* 5(7): 1247–57.

Pandey, S.N., A. Day, S.N. Chattopadhyay, and N.C. Pan, 1993. Recent developments in bleaching of jute and allied fibres. *Colourage* XL(10): 29–34.

Pandey, S.N., A. Day, S.N. Chattopadhyay, N.C. Pan, and M.D. Mathew. 1993. Effect of aftertreatments on direct dyed jute fabrics. *Indian Journal of Fibre & Textile Research (IJFTR)* 18(2): 87–90.

Pandit, P., K.K. Samanta, and M.D. Teli. 2020. Optimization of atmospheric plasma treatment parameters for hydrophobic finishing of silk using box behnken design. *Journal of Natural Fibers* 19(2): 463–474

Park, D.J., M.H. Lee, Y.I. Woo, D.W. Han, J.B. Choi, J.K. Kim, S.O. Hyun, K.H. Chung, and J.C. Park. 2008. Sterilization of microorganisms in silk fabrics by microwave-induced argon plasma treatment at atmospheric pressure. *Surface and Coatings Technology* 202(22–23): 5773–78.

Ratnapandian, S., L. Wang, S.M. Fergusson, and M. Naebe. 2011. Effect of atmospheric plasma treatment on pad-dyeing of natural dyes on wool. *Journal of Fiber Bioengineering and Informatics* 4(3): 267–76.

Samajpati, S., S. Sengupta, S.N. Chattopadhyay, A. Day, and S.K. Bhattacharyya. 2005. Jute-based composite products. *Asian Textile Journal-Bombay* 14(7): 70–72.

Samanta, K.K., A.G. Joshi, M. Jassal, and A.K. Agrawal. 2012. Study of hydrophobic finishing of cellulosic substrate using He/1, 3-butadiene plasma at atmospheric pressure. *Surface and Coatings Technology* 213: 65–76.

Samanta, K.K., A.G. Joshi, M. Jassal, and A.K. Agrawal. 2021. Hydrophobic functionalization of cellulosic substrate by tetrafluoroethane dielectric barrier discharge plasma at atmospheric pressure. *Carbohydrate Polymers* 253: 117272.

Samanta, K.K., M. Jassal, and A.K. Agrawal. 2008. Formation of nano-sized channels on polymeric substrates using atmospheric pressure glow discharge cold plasma. *Nano Trends-A Journal of Nano Technology & Its Applications* 4(1): 71–75.

Samanta, K.K., M. Jassal, and A.K. Agrawal. 2010. Atmospheric pressure plasma polymerization of 1, 3-butadiene for hydrophobic finishing of textile substrates. *Journal of Physics: Conference Series* 208: 12098.

Samanta, K.K., S. Basak, and S.K. Chattopadhyay. 2014. Environment-friendly textile processing using plasma and UV treatment. In *Roadmap to Sustainable Textiles and Clothing*, Springer, 161–201.

Samanta, K.K., S. Basak, and S.K. Chattopadhyay. 2015. Sustainable flame-retardant finishing of textiles advancement in technology. In *Handbook of Sustainable Apparel Production*, Routledge Handbooks Online, 18–53.

Samanta, K.K., S. Basak, and S.K. Chattopadhyay. 2017. Environmentally friendly denim processing using water-free technologies. In *Sustainability in Denim*, Elsevier, 319–48.

Samanta, K.K., T.N. Gayatri, A.H. Shaikh, S. Saxena, A. Arputharaj, S. Basak, and S.K. Chattopadhyay. 2014. Effect of helium-oxygen plasma treatment on physical and chemical properties of cotton textile. *International Journal of Bioresource Science* 1(1): 57–63.

Samanta, K.K., T.N. Gayatri, S. Saxena, S. Basak, S.K. Chattopadhyay, A. Arputharaj, and V. Prasad. 2016. Hydrophobic functionalization of cellulosic substrates using atmospheric pressure plasma. *Cellulose Chemistry and Technology* 50(7–8): 745–54.

Sengupta, S., S.N. Chattopadhyay, S. Samajpati, and A. Day. 2008. *Use of Jute Needle-Punched Nonwoven Fabric as Reinforcement in Composite*, CSIR.

Sengupta, S., L. Ammayappan, and K. K. Samanta. 2021. *Book on "Lignocellulosic Fibre based Sustainable Composite"*, Eds. D. B. Shakyawar and S. Sengupta. ICAR-NINFET publication, Kolkata, 1-100.

Teli, M.D., K.K. Samanta, P. Pandit, S. Basak, and S.K. Chattopadhyay. 2015a. Low-temperature dyeing of silk fabric using atmospheric pressure helium/nitrogen plasma. *Fibers and Polymers* 16(11): 2375–83.

Teli, M.D., K.K. Samanta, P. Pandit, S. Basak, and T.N. Gayatri. 2015b. Hydrophobic silk fabric using atmospheric pressure plasma. *International Journal of Bioresource Science* 2(1): 15–19.

Teli, M.D., P. Pandit, and K.K. Samanta. 2015. Application of atmospheric pressure plasma technology on textile. *Journal of the Textile Association*, 422–27.

Teli, M.D., P. Pandit, K.K. Samanta, S. Basak, and T.N. Gayatri. 2021. Salt-free and low temperature colouration of silk using He–N2 non-thermal plasma irradiation. *Journal of Cleaner Production* 296: 126576.

Vinogradov, I.P., and A. Lunk. 2005. Structure and chemical composition of polymer films deposited in a dielectric barrier discharge (DBD) in Ar/fluorocarbon mixtures. *Surface and Coatings Technology* 200(1–4): 660–63.

Zaman, H.U., M.A. Khan, R.A. Khan, M.Z.I. Mollah, S. Pervin, and M.D. Al-Mamun. 2010. A comparative study between gamma and uv radiation of jute fabrics/polypropylene composites: Effect of starch. *Journal of Reinforced Plastics and Composites* 29(13): 1930–39.

5 Particulate-Reinforced Composites

Subhajit Kundu and Debarati Mitra
University of Calcutta

Mahuya Das
Greater Kolkata College of Engineering and Management

CONTENTS

5.1 INTRODUCTION

A less powerful method of reinforcing than fibre reinforcement is particle reinforcement of composites. Particulate-reinforced composites mainly produce stiffness benefits, but can also achieve improvements in strength and power. The enhancements in both cases are smaller than can be done with a material lined with fibre. Applications where high levels of wear resistance are required, such as road surfaces, are found in particulate-strengthened composites. By using gravel as a reinforcing filler, the toughness of cement is greatly improved. The key benefit of composites reinforced by particles is their low cost and ease of processing and formation.

The particulate-reinforced composites can be sub-grouped into large-particle-reinforced composites and dispersion-strengthened composites. The distinction between these is based on the process of strengthening or reinforcement. The term large is used to suggest that it is not possible to treat particle-matrix interactions at the atomic or molecular level; instead, continuum mechanics is used. The particulate process in most of these composites is tougher and stiffer than the matrix.

DOI: 10.1201/9780429330575-5

These reinforcing particles in the vicinity of each particle appear to restrain the movement of the matrix process. The degree of mechanical activity reinforcement or enhancement relies on tight bonding at the interface of the particle matrix.

Particles are usually much smaller, with diameters between 0.01 and 0.1 µm (10 and 100 nm) for dispersion-strengthened composites. On the atomic or molecular level, particle-matrix interactions that contribute to strengthening take place. While the main part of an applied load is borne by the matrix, the small scattered particles obstruct or inhibit the motion of dislocations (Fu et al. 2008). Concrete, consisting of cement (the matrix) and sand and gravel (the particulates), is another familiar large-particle composite. The particles should be tiny and uniformly distributed in the matrix for efficient reinforcement. In addition, the volume fraction of the two phases affects the behaviour; with increasing particulate material, mechanical properties are increased (Callister et al. 2021).

By the uniform dispersion of several per cent volume of fine particles of a very hard and inert substance, metals and metal alloys can be reinforced and hardened. Metallic or nonmetallic can be the scattered phase; oxide products are also used. Again, as with precipitation hardening, the strengthening process requires interactions between the particles and dislocations within the matrix. The effect of dispersion strengthening is not as pronounced as with precipitation hardening; but at elevated temperatures and over longer periods of time, the strengthening is maintained because with the matrix process the dispersed particles are preferred to be unreactive. For precipitation-hardened alloys, as a consequence of precipitate growth or dissolution of the precipitate process, the increase in strength can disappear upon heat treatment (Egbo 2020).

5.2 FLY ASH-BASED PARTICULATE-REINFORCED COMPOSITES

One of the particulate wastes produced during the combustion in thermal power stations of coal is fly ash. The utilization will generate demand for this waste material and benefit the environment as well. Essentially, fly ash is a blend of metal oxides that can be used in metal and polymer composites as filler reinforcement. The quantity of metals and polymers is minimized by fly ash as a filler element, reducing integrated energy (Rohatgi, Menezes, and Lovell 2012).

In contrast with that of the matrix content, the low density of fly ash will help to reduce the structural weight without compromising mechanical properties. Lighter goods can help to minimize transportation life-cycle costs applications. Furthermore, opposed to costly engineered reinforcement materials, fly ash is a waste material that is usable for free (Desai, Paul, and Mallik 2020).

Considering synthesis method, Rohatgi P. applied cast method for Al-based composite and studied its applications in automotive fields (Singh and Chauhan 2019). The effect of silicon carbide particle size (0.5–2 wt%) on AA2219 stirred nanocomposites manufactured by the method based on ultrasonication was analysed by Murthy et al. (Babu et al. 2017). Alam et al. analysed the structural properties of A356 aluminium alloys reinforced with a very little wt% of nano-SiC particles formed by the 550 rpm speed, 10 minutes stirring time, and around 700°C molten slurry pouring, described an improvement of 41% in yield strength and 40%–45% of the final tensile strength (Alam, Arif, and Ansari 2017). Ravishankar et al. also observed that with the rise in particle size of hardened fly ash, compressive and tensile strength along with stiffness of the aluminium alloy (Al 6061) composites have reduced. The increase in fly ash particle weight fractions increases the ultimate tensile resistance, compressive strength and hardness and reduces the composite's ductility (Kumar, Hebbar, and Shankar 2011) (Table 5.1).

5.2.1 FLY ASH-BASED POLYMER COMPOSITES

Many polymers have a density in the range of 0.9–1.2 g/cm^3. The inclusion of cenospheres with a density of less than 0.7 g/cm^3 results in composites of lightweight polymer matrixes for applications sensitive to weight. However, owing to the low wettability of fly ash with polymers, achieving uniform distribution of cenospheres in lightweight polymer matrixes is a difficult task. Through

TABLE 5.1
Different Fly Ash-Based Metal Composites and Their Properties

Matrix	Fly Ash %	Properties (Compared with Respect to Unreinforced Matrix)	References
Ni	40 vol%	Hardness 135%; Coefficient of friction = −33%; Wear rate = −63%	Ramesh and Seshadri (2003)
A 356	35 vol% (enosphere)	Compressive yield strength= +19–41% (based on strain rate)	Rohatgi et al. (2006)
A 535	15 wt%	Tensile strength = −79%; Hardness = −40%	Ponnarengan et al. (2020)
Al-7Si	6 wt%	Hardness = 29%	Nagaraj, Mahendra, and Nagaral (2018)
AA 2024	10 wt%	Hardness = 73%	Rao et al. (2012)
AA 7075	12 wt%	Hardness = 20.89%; Ultimate tensile strength = 23%	Kasar et al. (2020)
AA Si10Mg	9 wt%	Hardness = 20%; Ultimate tensile strength = 26%	Khatkar, Suri, and Kant (2018)

TABLE 5.2
Different Fly Ash-Based Polymer Composites and Their Properties

Matrix	Fly Ash (wt%)	Properties (Compared with Respect to Unreinforced Matrix)	References
Epoxy	16	Compressive modulus = 68%	Kumar et al. (2016)
Epoxy	20	Tensile strength = 33.3%	Raju et al. (2018)
Epoxy	10	Tensile strength = −24%; Impact strength = −26.4%	Purohit et al. (2017)
Phenolic	70	Coefficient of friction = 0.12; Wear depth = 0.25 mm	Dadkar, Tomar, and Satapathy (2009)
Vinyl ester	40	Compressive strength = −25%; Compressive modulus = 45%	Ullas and Jaiswal (2020)
Vinyl ester	60	Compressive strength = −25%; Flexural strength = −73%; Flexural modulus = 47%	Labella et al. (2014)
Geopolymer	48.2	Compressive strength = 63%	Roviello et al. (2016)

treating the surface of the cenosphere with suitable agents, the cenosphere-polymer relationship can be enhanced.

Syntactic foam composites are produced by Shahapurkar et al. using hollow cenospheres inserted into the epoxy matrix as obtained and surface adjusted. From the experiments, they acquired that with a rise in volume fraction of cenosphere relative to the resin, the modulus of both untreated and treated syntactic foams rises (Shahapurkar et al. 2018). Goh et al. in another work immobilized some toxic metals such as Pb, Zn, Fe, Cr and Cu in polymer matrix, and their detail mechanical properties and leaching behaviour was observed (Nemati and Mahmoodabadi 2020). Gu et al. performed the damping tests of some epoxy-based fly ash composites by tension-compression mode with the variation of temperature and frequency. They concluded that for composites with approximate 30–50 vol.% of fly ash, the tan δ values reach their peak values at the glass transition temperatures, and the tan δ values slowly enervated with the frequency enhancement, which was the indication of the better damping properties of those composites compared with others (Gu, Wu, and Zhao 2008). There were some composites prepared by Bora et al. by solution-processing on fly ash composites based on polyvinyl butyral, which are suitable for microwave absorption applications, such as radar, robotics, military, electromagnetic gaskets and aircraft (Bora et al. 2018) (Table 5.2).

Some of the composites show improved properties, while due to the inclusion of fly ash cenospheres, some other composites show decreased mechanical properties. This may be possibly attributed to the weak interfacial adhesion between the fly ash particle and matrix. The composite may show a crossover point in mechanical properties at a certain volume fraction of the cenosphere.

5.2.2 Polymer (CFRP)

CFRP, a type of advanced composite, refers to a type of solid multi-phase plastic where a reinforcing phase is carbon fibres and a constant phase is the polymer as a resin matrix. This construction gives them lightweight, high strength, high rigidity, resistance to high temperatures and corrosion and other outstanding properties. They are therefore commonly used in high-end sports, entertainment, and the automotive and aerospace fields and are still rapidly growing into new applications (Keith et al. 2019). Many countries are likely to adopt regulations controlling CFRP waste over the coming years and are getting tighter. The disposal of CFRPs is still a problem, however. The key explanation is that the vast majority of polymer matrices used are very chemical matrices. Cross-linked thermosetting resins that offer excellent properties to CFRPs make it difficult to reclaim CFs by matrix degradation. At present, mechanical means (mainly grinding), pyrolysis and solvolysis are the commonly used recycling processes.

5.2.3 Recyclability of CFRP Composites

Recyclability of CFRP composites depends on the recycling nature of the matrix itself, and the matrix containing following chemical bonds is showing recyclability as per their bonding. To impart recyclability in CFRP, the most common matrix for CFRP, epoxy resin is usually modified by incorporation of different additives containing degradation-sensitive functional groups. Different bonds responsible for recyclability of CFRP are given in the following Table 5.3:

A closed-loop near 100% recycling model of CFRPs using epoxy vitrimers was developed by Qi et al. in which the triggering catalyst for dynamic transesterification was inserted into the cross-linked network of the resin matrix (Kuang et al. 2018) (Figure 5.1). In order to prepare CFRPs, they

TABLE 5.3
Degradability of CFRPs with Different Conditions

Name of the Bond	Degradation Temperature and Mechanism	Name of the Polymer with Which It Is Associated	Significance
Ester	Below 200°C	Epoxy, polyester	Easy degradation of thermoset waste and end-of-life control (Montarnal et al. 2011)
Acetal	Through hydrolysis at room temperature	Epoxy	Excellent degradability and high efficiency equal to or even better than traditional thermosets (Ma and Webster 2018)
Imine bond (C=N)	Hydrolysed under acidic conditions	Epoxy	Completely reversible bond, Successful closed loop recyclability (Wang et al. 2018a,b)
Disulphide bonds (–S–S–)	Cleaved by reducing reducers, the heat, UV or catalyst	Epoxy	Degradability, self-healing, disulphide metathesis (Johnson et al. 2015)
Hexahydro-triazine	Hydrolysed under conditions of acid	Epoxy	High thermal degradation temperature and high mechanical properties (Cromwell, Chung, and Guan 2015)
Boronic ester	Hydrolysis	Epoxy	Excellent recyclability, quick execution, zero emission (Yu et al. 2016)
Di-N-benzylaniline	Under acidic condition	Epoxy	Yet to explore (Cromwell, Chung, and Guan 2015)

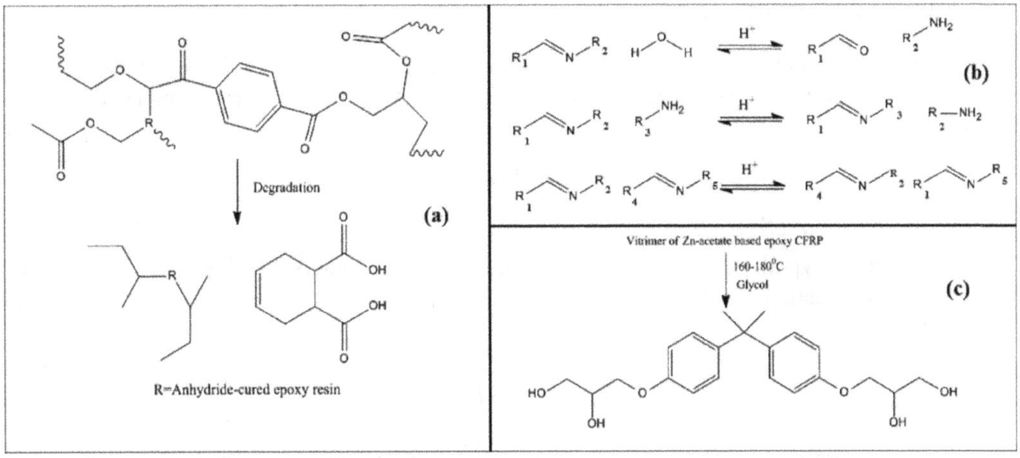

FIGURE 5.1 Degradation of (a) ester linkages; (b) acetal linkages; (c) Schiff base linkages.

used a carboxyl acid-epoxy covalent dynamic network comprising zinc acetate as the matrix. The depolymerization of the anhydride-epoxy network allowed by the selective ester bond cleavage process was significantly improved by Kuang et al. using a strong transesterification response assisted by alcohol and solvent (Hashimoto et al. 2012).

After being compounded with CFs, Hashimoto et al. introduced acetal linkage into the main chain of epoxy resins and investigated their properties and degradability using an additional reaction with a phenolic hydroxyl group with vinyl ether (Einstein 1956). Excellent thermosets degradability under moderate acidic conditions was monitored by Songqi et al. by spiro diacetal structure based on vanillin (Ma et al. 2019).

Zhang and his colleagues documented that malleable, mechanically compatible resilient base thermosets from Schiff have been added to CFRPs, which can contribute to successful closed-loop recycling (Jin et al. 2019). From a synthesised formyl group-containing vanillin-based monoepoxide and a diamine through in situ formulation of the Schiff base structure and epoxy network, a Schiff base epoxy thermoset combining excellent recyclability and high performance was readily prepared by Wang et al. (2019).

If the free radical is kindled rather than interacting with another free radical, the disulphide bond produces free radicals by heat, which can cause harm to the cross-linked network, so it is worth revisiting the thermal stability of the disulphide thermosetting resin. This was explained by Johnson et al. (2015) for oil discovery and processing, disulphide thermosetting resins.

The hexahydrotriazine structure was added by Yuan et al. to achieve a totally recyclable CFRP multiplication where they used propane-based polymer stability to react to p-formaldehyde to synthesise a poly-hexahydrotriazine that demonstrated excellent fracture resistance, thermal stability and high tensile strength (Yuan et al. 2017). Several hyperbranched epoxy resins containing hexahydrotriazine structure have been prepared by Ma et al. (2020) that have demonstrated flexibility in diglycidyl ether of bisphenol A modification, including sequential strengthening and toughening operation, adverse effects and recycling, and strong compatibility.

In order to generate monomer-recovery thermosets and closed-loop recyclable CFRPs, Wang et al. used boronic ester degradability. Novolac resin and phenylboronic acid have been selected as the raw materials for the resin matrix (Wang et al. 2018a,b). Via the copolymerization of an industrial bis-benzoxazine and alicyclic epoxy monomer to create compostable CFRPs, Lo et al. produced a di-N-benzylaniline linkage containing thermoset (Lo, Nutt, and Williams 2018).

5.3 PARTICULATE REINFORCEMENT THEORIES

The efficiency of particle reinforcement has been investigated by many theories and equations proposed are discussed below. Two types of matrices are considered here: non-rigid and rigid.

5.3.1 Rigid Inclusion Theory for a Non-Rigid Matrix

5.3.1.1 Einstein Equation

One of the earliest hypotheses for a composite structure for elastomers was developed and is based on the Einstein equation for the viscosity of a solid spherical inclusion suspension (Ahmed and Jones 1990; Einstein 1956).

$$\eta_c = \eta_m(1 + K_E V_P) \tag{5.1}$$

where η_c and η_m are the viscosity of the suspension and the matrix, respectively; K_E is termed as Einstein coefficient, which has a value of 2.5 for spheres; V_P is the volume fraction of the particulate inclusions. Eq. (5.1) is also applicable for change in modulus i.e. $\frac{\eta_c}{\eta_m} = \frac{G_c}{G_m}$

$$G_c = G_m(1 + 2.5V_P) \tag{5.2}$$

where the symbol G represents the shear modulus. The suffixes p, m and c are for particle, matrix and composite, respectively.

This equation has been useful for low filler concentrations only because the flow or strain fields surrounding particles combine to increase the volume fraction of the filler. Investigating the interactions, there have been some problems in evaluation considering Eq. (5.2). Thus, this equation was modified to:

$$G_c = G_m \exp\left(\frac{2.5V_P}{1 - SV_P}\right) \tag{5.3}$$

Here, S is the crowding factor which is the ratio of the volume occupied by the filler to the true volume of the filler. It has a value of 1.35 for close-packed spheres. Equation (5.3) known as Mooney equation can be used for both low-volume fraction as well as high-volume fractions. For non-spherical particles, Mooney equation was modified to Eq. (5.4), which is known as Brodnyan equation (Lee et al. 2019).

$$G_c = G_m \exp\left(\frac{2.5V_P + 0.407(p-1)^{1.508} V_P}{1 - SV_P}\right) \tag{5.4}$$

where aspect ratio, p, has a value in the range from 1 to 15.

Guth modified the Einstein equation by introducing some interaction parameters (Wang and Guth 1952).

$$G_c = G_m\left(1 + K_E V_P + 14.1{V_P}^2\right) \tag{5.5}$$

$$E_c = E_m\left(1 + K_E V_P + 14.1{V_P}^2\right) \tag{5.6}$$

$$E_c = E_m\left(1 + 0.67pV_P + 1.62p^2V_P{}^2\right) \quad (5.7)$$

Equation (5.7) is for non-spherical particles. The shape factor, *p*, has been defined as the ratio of particle length to width for non-spherical particles.

5.3.1.2 Kerner Equation

The Kerner equation is one of the most flexible and sophisticated equations for a composite substance composed of spherical particles in a matrix (Kerner 1956).

$$G_c = G_m\left[1 + \frac{V_P}{V_m}\frac{15(1-v_m)}{(8-10v_m)}\right] \quad (5.8)$$

where v_m is the Poisson ratio of the matrix.

It was further generalized by Nelson to give a generalized form of the equation (Wu et al. 2013).

$$M = M_m\frac{1 + ABV_P}{1 - B\delta V_P} \quad (5.9)$$

where *M* is the bulk modulus. Also, sometimes it is used for shear or elastic modulus. The term $B\delta$ depends upon the packing fraction of the particle. The constant *A* refers some specific factors like matrix Poisson ratio and filler geometry, and the constant *B* indicates the relative moduli of the filler and the matrix phase. The Mooney equation expects slightly more reinforcing operation than the Kerner equation, and a module with a high-volume fraction of filler that tends to infinity. The equation implies that $v_m = 0.5$ and that the filler module is exponentially larger than the matrix, both of which are not right for a rigid matrix, thus reducing the applicability of such models to polymeric matrices filled with rigid thermosetting.

5.3.2 Rigid Inclusion Theory for Rigid Matrix

A statistical methodology is necessary for the random distribution of the constituent phases in a completed scheme, but this involves knowledge of the distribution of the individual phases. As a result, the problem was generalized to a two-phase model in which in each of the stages, average stresses and strains are known to occur. A uniform strain field is induced in the composite when exposed to a gross uniform stress or strain and can be used to approximate the elastic constant. The other methods consist of specifying limits for the moduli using energy requirements in the theory of elasticity.

5.3.2.1 Series and Parallel Models

In the case of parallel arrangement, the uniform strain is assumed in the two stages, and Eq. (5.10) gives the upper limit (Tan, Jia, and Li 2016).

$$E_c = E_pV_P + E_mV_m \quad (5.10)$$

Whereas in case of series arrangement, the lower bound is given by Eq. (5.11), assuming the stress to be uniform in two phases.

$$E_c = \frac{E_pE_m}{E_pV_m + E_mV_P} \quad (5.11)$$

$$v_c = \left(\frac{v_c V_P E_m + v_m V_m E_p}{E_p V_m + E_m V_P} \right) \tag{5.12}$$

5.3.2.2 The Hashin–Shtrikman Model

This model highlights the contraction of Poisson ratio in the constituent phase. Eqs. (5.13) and (5.14) are for lower and upper bound, respectively (Bigoni et al. 1998).

$$E_c = \frac{9\left(K_m + \dfrac{V_P}{\left[\dfrac{1}{K_P - K_m}\right] + \left[\dfrac{3V_m}{3K_m + 4G_m}\right]}\right)\left(G_m + \dfrac{V_P}{\left[\dfrac{1}{G_P - G_m}\right] + \left[\dfrac{6(K_m + 2G_m)V_m}{5(3K_m + 4G_m)G_m}\right]}\right)}{3\left(K_m + \dfrac{V_P}{\left[\dfrac{1}{K_P - K_m}\right] + \left[\dfrac{3V_m}{3K_m + 4G_m}\right]}\right)\left(G_m + \dfrac{V_P}{\left[\dfrac{1}{G_P - G_m}\right] + \left[\dfrac{6(K_m + 2G_m)V_m}{5(3K_m + 4G_m)G_m}\right]}\right)} \tag{5.13}$$

$$E_c = \frac{9\left(K_p + \dfrac{V_m}{\left[\dfrac{1}{K_m - K_p}\right] + \left[\dfrac{3V_p}{3K_p + 4G_p}\right]}\right)\left(G_p + \dfrac{V_m}{\left[\dfrac{1}{G_m - G_p}\right] + \left[\dfrac{6(K_p + 2G_p)V_p}{5(3K_p + 4G_p)G_p}\right]}\right)}{3\left(K_p + \dfrac{V_m}{\left[\dfrac{1}{K_m - K_p}\right] + \left[\dfrac{3V_p}{3K_p + 4G_p}\right]}\right)\left(G_p + \dfrac{V_m}{\left[\dfrac{1}{G_m - G_p}\right] + \left[\dfrac{6(K_p + 2G_p)V_p}{5(3K_p + 4G_p)G_p}\right]}\right)} \tag{5.14}$$

where K_m and K_p are the bulk moduli of the matrix and particle, respectively, while G_m and G_p are the shear moduli of the matrix and particle, respectively.

The Poisson ratio is defined by: (Yin and Sun 2005)

$$v_c = \frac{3K_c - 2G_c}{2(3K_c + G_c)} \tag{5.15}$$

The ratio of E_p/E_m is known as the modular ratio of particle to matrix and is represented by m. The upper and lower bound separations depend on the modular ratio of the particle to the matrix. The bounds forecast values within 10% when the moduli of the constituent phases are closely balanced. In the case of a polymeric-filled rigid framework m is around 20. There are so many other models in this category which are mentioned in Table 5.4:

5.4 STRENGTH OF PARTICULATE-REINFORCED COMPOSITES

Other than the theories, many models, laws and equations have been proposed by different workers to explain the strength and failure of particulate composites and some are explained below:

TABLE 5.4
Different Models Regarding Particulate-Reinforced Theory

Name of the Model	Equation	Significance
Hirsch model (Xiong and Mashiur 2016)	$E_c = x(E_pV_P + E_mV_m) + (1-x)\dfrac{E_pE_m}{E_pV_m + E_mV_P}$	For concrete structures, this model was recommended to take account of the dynamic stress distribution in the human process
Takayanagi model (Zare and Garmabi 2017)	$E_c = \left(\dfrac{\alpha}{[1-\beta]E_m + \beta E_p} + \dfrac{[1-\alpha]}{E_p}\right)^{-1}$ α and β indicate the state of parallel and series coupling in the composite, respectively	For explaining the efficiency of an interpenetrating network, this model is especially suitable
Counto model (Goral 1956)	$\dfrac{1}{E_c} = \dfrac{1-\sqrt{V_P}}{E_m} + \dfrac{1}{\dfrac{(1-\sqrt{V_P})}{\sqrt{V_P}E_m} + E_p}$	The simpler model assumes absolute bonding between the particle and the matrix for a two-phase framework
Paul model (Zhou et al. 2020)	$E_c = E_m\left[\dfrac{1+(m-1)V_P^{\frac{2}{3}}}{1+(m-1)(V_P^{\frac{2}{3}} - V_P)}\right]$ modified to by Ishai and Cohen $E_c = E_m\left[1 + \dfrac{V_P}{\dfrac{m}{(m-1)} - V_P^{\frac{1}{3}}}\right]$	The elastic modulus of the composite is given when a uniform stress is applied at the boundary modified for uniform displacement at boundary
Chow model (Chow 1978)	$E_c = E_m(1 + \dfrac{\left(\dfrac{K_P}{K_m-1}\right)A_1 + \left(\dfrac{G_P}{G_m-1}\right)B_1}{2B_1A_3 + A_1B_3}$ where $A_i = 1 + \left(\dfrac{G_P}{G_m-1}\right)(1-V_P)\beta_i$ $B_i = 1 + \left(\dfrac{K_P}{K_m-1}\right)(1-V_P)\alpha_i;\ i = 1,3$	The anisotropy of the particles in the shape of the aspect ratio p
Cox model (Liu and Wei 2021)	$E_c = E_m(1-V_f) + E_fV_f\left(1 - \dfrac{\tanh z}{z}\right)$ where $z = \dfrac{l}{2r}\sqrt{\left(\dfrac{2G_m}{G_m \ln\dfrac{R}{r}}\right)}$	The load was moved from the surrounding matrix via a shear process to the fibre with an aligned short fibre composite ignoring any tensile stresses in the matrix

5.4.1 The Sahu–Broutman Model

In Sahu and Broutman's viewpoint, they believed that when one element is split as a result of a concentration of tension around the filler particle, the composite fails. It follows that with the inclusion of small quantities of filler, the power decreases easily. To model the composite, they used a finite element analysis with this assumption and compared the effects with the experimental strengths of a thermosetting resin filled with a glass sphere (Epaarachchi and Clausen 2005).

5.4.2 The Power Law

In the Power law, the strength is defined by Eq. (5.23), when there no transfer of stress which in turn the formation of a weak bond between the matrix and the filler. Also, there is the absence of a concentration of stress at the interface of the particle-matrix (Lee, Batt, and Hwang 1996).

$$\sigma_{cu} = \sigma_{mu}\left(1 - \alpha V_P{}^n\right) \tag{5.23}$$

where σ_{cu} and σ_{mu} are the ultimate tensile strengths of the composite and the matrix, respectively. The constants α and n depend on the shape and arrangement of the particle in the model composite.

For cubic particles inserted in cubic matrix, the Eq. (5.23) becomes

$$\sigma_{cu} = \sigma_{mu}\left(1 - V_P{}^{\frac{2}{3}}\right)K \tag{5.24}$$

[where, K (Stress concentration factor) = 0.5]

For uniformly dispersed spherical particles considering minimum cross-sectional area of the continuous phase, the following equation can be found:

$$\sigma_{cu} = \sigma_{mu}\left(1 - 1.21V_P{}^{\frac{2}{3}}\right) \tag{5.25}$$

Piggott and Leidner concluded that in reality the uniform filler structure expected in most models was impossible and implied an observational relationship (Landon, Lewis, and Boden 1977):

$$\sigma_{cu} = K\sigma_{mu} - bV_P \tag{5.26}$$

The constant b depends on the particle-matrix adhesion.

Landon et al. suggested another equation which is similar to Eq. (5.26) (Berka 1982).

$$\sigma_{cu} = \sigma_{mu}\left(1 - V_P\right) - k\left(V_P\right)d \tag{5.27}$$

where d is the average particle diameter and k the slope of the plot of tensile strength against mean particle diameter.

5.4.3 The Leidner–Woodhams Equation

It was made up of spherical particles, which were trapped in an elastic matrix. The particle size was approximated to a cylinder for the implementation of reinforcement theory. The tension distribution in the bead, at the breaking point, may be measured in this manner. The stress transfer between the particle and the matrix was believed to take place in the case of non-bonded particles as a function of the combination of particle-matrix friction and residual compressive stresses acting on the interface of the particle matrix. The stress is transmitted through the shear mechanism in the case of well-bonded particles. Therefore, the maximal stress in the particle is contingent on the shear strength of the matrix and on the particle-matrix bond strength. The composite's overall tensile strength was taken simply as the composite's ultimate strength (Peng et al. 2019).

$$\sigma_{cu} = \left(\sigma_a + 0.83\tau_m\right) + \sigma_a K\left(1 - V_P\right) \tag{5.28}$$

where σ_a indicates the strength of the interfacial bond and τ_m is the shear strength of the matrix.

$$\sigma_{cu} = 0.83\sigma_{th}\alpha V_P + k\sigma_{mu}(1 - V_P) \tag{5.29}$$

σ_{th} is the thermal compressive stress acting on the boundary of the particle, and α is the coefficient of friction.

Equation (5.28) holds for good interfacial adhesion, and Eq. (5.29) is for no interfacial adhesion.

A variety of attempts have been made to compare the power of particulate-filled structures to the particle diameter, *d*. Hojo et al., for example, have observed that as the particle size grows, the intensity of silica-filled epoxy decreases (Kushvaha and Tippur 2014).

$$\sigma_{cu} = \sigma_{mu} + kd^{-\frac{1}{2}} \tag{5.30}$$

There are models like micromechanical model and finite element model (FEA) but in a nutshell, the whole idea of the mechanism and modelling can be summarized as follows:

- If the secant modulus numerical approximation is implemented by the matrix alloy experimental curve, the new Eshelby type analytical model can predict stress–strain curve of both hard and soft matrix particulate enhanced composites very well.
- The FEA unit cell model with a rough matrix can only estimate the stress–strain curve of the composite. Bad prediction for the soft matrix composite means that for soft matrix composite, the use of calculated stress–strain curve of matrix monolithic alloy to replace the constitution law of the matrix would yield considerable error.
- In reinforcing the particulate-strengthened aluminium matrix composites with a precipitated rough matrix, micromechanical models only play a lesser role.
- The modelling work shows that load transfer from matrix to particle reinforcement plays an important role in strengthening composites of metal matrix, although it is the dominant process only for composites of hard matrix.
- The load sharing mechanism sheds new light on the need to monitor the geometry of particulate reinforcement, which has been ignored in practice because it has little impact on intensity in micromechanical models.
- Observation of the number of fragmented reinforcement fragments in tensile-fractured composite samples provides proof that the Eshelby model is also better served than the FEM model.
- The Eshelby model shows that during straining, the load transition is realised by the mismatch strain between the particles and the matrix, and the mismatch strain also induces a substantial increase in the matrix power.

5.5 LIMITATIONS OF THE PROPOSED MODELS/LAWS/THEORIES/EQUATIONS

All these proposals are not suitable to explain completely the reinforcing action of particulate composite but to some extent, and the limitations are as follows:

- The solutions of the lower and upper bound given by Eqs. (5.10) and (5.11) presume that, respectively, the individual stages are under uniform strain or stress. In fact, however, the filler particles may not be isolated from each other completely and the reinforcement factor may essentially be an aggregation of smaller particles at the microlevel. The stress would therefore be unevenly distributed in response to the applied load.
- The theories dealing with filled structures suggest that the elastic module for a given particle and matrix depends only on the filler volume fraction and not the particle size, so as the particle size decreases, the module usually increases. It was posited by Lewis and

Nielsen as the particle size decreases, the surface area increases and the interfacial bond is more powerful.
- Changes in particle form can also influence the properties of the composites. The effect was more pronounced with larger or non-spherically formed particles where the deformation behaviour could be changed by a desired orientation.

5.6 CONCLUSION

The particulate-reinforced composites are economically viable as well as eco-friendly due to utilization of waste material lower cost and density of particulate composites as well as the utilization of some waste material like fly ash as the reinforcing agent. The property of the composites depends strongly upon the uniform dispersion of the particulates. Fly ash composite gains interest due to its light hollow structure whereas the carbon fibre-based composite are famous for its recyclability. Different models are there to explain the properties of particulate composites but the success of proposed model depends of the structure and properties of the reinforcing particulate.

REFERENCES

Ahmed, SaFRJ, and FR Jones. 1990. A review of particulate reinforcement theories for polymer composites. *Journal of Materials Science* 25(12): 4933–42.

Alam, MT, S Arif, and AH Ansari. 2017. Mechanical behaviour and morphology of A356/SiC nanocomposites using stir casting. In *IOP Conference Series: Materials Science and Engineering*, 225:12293. IOP Publishing.

Babu, KK, K Panneerselvam, P Sathiya, A Noorul Haq, S Sundarrajan, P Mastanaiah, and CVS Murthy. 2017. Effects of mechanical, metallurgical and corrosion properties of cryorolled AA2219-T87 aluminium alloy. *Materials Today: Proceedings* 4(2): 285–93.

Berka, L. 1982. On stress distribution in a structure of polycrystals. *Journal of Materials Science* 17(5): 1508–12.

Bigoni, D, SK Serkov, M Valentini, and AB Movchan. 1998. Asymptotic models of dilute composites with imperfectly bonded inclusions. *International Journal of Solids and Structures* 35(24): 3239–58.

Bora, PJ, M Porwal, KJ Vinoy, PC Ramamurthy, and G Madras. 2018. Industrial waste fly ash cenosphere composites based broad band microwave absorber. *Composites Part B: Engineering* 134: 151–63.

Callister, WD, DG Rethwisch, A Blicblau, K Bruggeman, M Cortie, J Long, J Hart, R Marceau, M Ryan, and R Parvizi. 2021. *Materials Science and Engineering: An Introduction*. Wiley.

Chow, TS. 1978. Effect of particle shape at finite concentration on the elastic moduli of filled polymers. *Journal of Polymer Science: Polymer Physics Edition* 16(6): 959–65.

Cromwell, OR, J Chung, and Z Guan. 2015. Malleable and self-healing covalent polymer networks through tunable dynamic boronic ester bonds. *Journal of the American Chemical Society* 137(20): 6492–95.

Dadkar, N, BS Tomar, and BK Satapathy. 2009. Evaluation of flyash-filled and aramid fibre reinforced hybrid polymer matrix composites (PMC) for friction braking applications. *Materials & Design* 30(10): 4369–76.

Desai, AM, TR Paul, and M Mallik. 2020. Mechanical properties and wear behavior of fly ash particle reinforced al matrix composites. *Materials Research Express* 7(1): 16595.

Egbo, MK. 2020. *A Fundamental Review on Composite Materials and Some of Their Applications in Biomedical Engineering*. King Saud University.

Einstein, A. 1956. *Investigations on the Theory of the Brownian Movement*. Courier Corporation.

Epaarachchi, JA, and PD Clausen. 2005. A new cumulative fatigue damage model for glass fibre reinforced plastic composites under step/discrete loading. *Composites Part A: Applied Science and Manufacturing* 36(9): 1236–45.

Fu, SY, XQ Feng, B Lauke, and YW Mai. 2008. Effects of particle size, particle/matrix interface adhesion and particle loading on mechanical properties of particulate–polymer composites. *Composites Part B: Engineering* 39(6): 933–61.

Goral, ML. 1956. Empirical time-strength relations of concrete. *Journal Proceedings*, 53: 215–24.

Gu, J, G Wu, and X Zhao. 2008. Damping properties of fly ash/epoxy composites. *Journal of University of Science and Technology Beijing, Mineral, Metallurgy, Material* 15(4): 509–13.

Hashimoto, T, H Meiji, M Urushisaki, T Sakaguchi, K Kawabe, C Tsuchida, and K Kondo. 2012. Degradable and chemically recyclable epoxy resins containing acetal linkages: Synthesis, properties, and application for carbon fiber-reinforced plastics. *Journal of Polymer Science Part A: Polymer Chemistry* 50(17): 3674–81.

Jin, Y, Z Lei, P Taynton, S Huang, and W Zhang. 2019. Malleable and recyclable thermosets: The next generation of plastics. *Matter* 1(6): 1456–93.

Johnson, LM, E Ledet, ND Huffman, SL Swarner, SD Shepherd, PG Durham, and GD Rothrock. 2015. Controlled degradation of disulfide-based epoxy thermosets for extreme environments. *Polymer* 64: 84–92.

Kasar, AK, N Gupta, PK Rohatgi, and PL Menezes. 2020. A brief review of fly ash as reinforcement for composites with improved mechanical and tribological properties. *JOM* 72(6): 2340–51.

Keith, MJ, GA Leeke, P Khan, and A Ingram. 2019. Catalytic degradation of a carbon fibre reinforced polymer for recycling applications. *Polymer Degradation and Stability* 166: 188–201.

Kerner, EH. 1956. The elastic and thermo-elastic properties of composite media. *Proceedings of the Physical Society. Section B* 69(8): 808.

Khatkar, SK, NM Suri, and S Kant. 2018. A review on mechanical and tribological properties of graphite reinforced self lubricating hybrid metal matrix composites. *Reviews on Advanced Materials Science* 56(1): 1–20.

Kuang, X, Q Shi, Y Zhou, Z Zhao, T Wang, and HJ Qi. 2018. Dissolution of epoxy thermosets via mild alcoholysis: The mechanism and kinetics study. *RSC Advances* 8(3): 1493–1502.

Kumar, BRB, M Doddamani, SE Zeltmann, N Gupta, S Gurupadu, and RRN Sailaja. 2016. Effect of particle surface treatment and blending method on flexural properties of injection-molded cenosphere/HDPE syntactic foams. *Journal of Materials Science* 51(8): 3793–3805.

Kumar, HCA, HS Hebbar, and KSR Shankar. 2011. Mechanical properties of fly ash reinforced aluminium alloy (Al6061) composite. *International Journal of Mechanical and Minerals Engineering* 6: 41–45.

Kushvaha, V, and H Tippur. 2014. Effect of filler shape, volume fraction and loading rate on dynamic fracture behavior of glass-filled epoxy. *Composites Part B: Engineering* 64: 126–37.

Labella, M, SE Zeltmann, VC Shunmugasamy, N Gupta, and PK Rohatgi. 2014. Mechanical and thermal properties of fly ash/vinyl ester syntactic foams. *Fuel* 121: 240–49.

Landon, G, G Lewis, and GF Boden. 1977. The influence of particle size on the tensile strength of particulate—filled polymers. *Journal of Materials Science* 12(8): 1605–13.

Lee, JC-W, KM Weigandt, EG Kelley, and SA Rogers. 2019. Structure-property relationships via recovery rheology in viscoelastic materials. *Physical Review Letters* 122(24): 248003.

Lee, YS, TJ Batt, and SC Hwang. 1996. Strengthening caused by power law creep deformation of dispersed particles in an elastic matrix. *International Journal of Mechanical Sciences* 38(2): 203–18.

Liu, J, and X Wei. 2021. A universal fracture analysis framework for staggered composites composed of tablets with different wavy topologies. *Journal of the Mechanics and Physics of Solids* 151: 104387.

Lo, JN, SR Nutt, and TJ Williams. 2018. Recycling benzoxazine–epoxy composites via catalytic oxidation. *ACS Sustainable Chemistry & Engineering* 6(6): 7227–31.

Ma, S, and DC Webster. 2018. Degradable thermosets based on labile bonds or linkages: A review. *Progress in Polymer Science* 76: 65–110.

Ma, S, J Wei, Z Jia, T Yu, W Yuan, Q Li, S Wang, S You Liu, and J Zhu. 2019. Readily recyclable, high-performance thermosetting materials based on a lignin-derived spiro diacetal trigger. *Journal of Materials Chemistry A* 7(3): 1233–43.

Ma, X, W Guo, Z Xu, S Chen, J Cheng, J Zhang, M Miao, and D Zhang. 2020. Synthesis of degradable hyperbranched epoxy resins with high tensile, elongation, modulus and low-temperature resistance. *Composites Part B: Engineering* 192: 108005.

Montarnal, D, M Capelot, F Tournilhac, and L Leibler. 2011. Silica-like malleable materials from permanent organic networks. *Science* 334(6058): 965–68.

Nagaraj, N, KV Mahendra, and M Nagaral. 2018. Microstructure and evaluation of mechanical properties of Al-7si-fly ash composites. *Materials Today: Proceedings* 5(1): 3109–16.

Nemati, AR, and MJ Mahmoodabadi. 2020. Effect of micromechanical models on stability of functionally graded conical panels resting on winkler–pasternak foundation in various thermal environments. *Archive of Applied Mechanics* 90(5): 883–915.

Peng, W, S Rhim, Y Zare, and KY Rhee. 2019. Effect of 'Z' factor for strength of interphase layers on the tensile strength of polymer nanocomposites. *Polymer Composites* 40(3): 1117–22.

Ponnarengan, H, L Kamaraj, SR Balachandran, and SK Basha. 2020. Reusing exhausted alkaline battery powder as reinforcement in AA6061 composites and mechanical characterization. *Energy Sources, Part A: Recovery, Utilization, and Environmental Effects* 1–14.

Purohit, R, P Sahu, RS Rana, V Parashar, and S Sharma. 2017. Analysis of mechanical properties of fiber glass-epoxy-fly ash composites. *Materials Today: Proceedings* 4(2): 3102–9.

Raju, GKM, GM Madhu, M Ameen Khan, and PDS Reddy. 2018. Characterizing and modeling of mechanical properties of epoxy polymer composites reinforced with fly ash. *Materials Today: Proceedings* 5(14): 27998–7.

Ramesh, CS, and SK Seshadri. 2003. Tribological characteristics of nickel based composite coatings. *Wear* 255(7–12): 893–902.

Rao, JB, D Venkata Rao, I Narasimha Murthy, and NRMR Bhargava. 2012. Mechanical properties and corrosion behaviour of fly ash particles reinforced AA 2024 composites. *Journal of Composite Materials* 46(12): 1393–404.

Rohatgi, PK, JK Kim, N Gupta, S Alaraj, and A Daoud. 2006. Compressive characteristics of A356/fly ash cenosphere composites synthesized by pressure infiltration technique. *Composites Part A: Applied Science and Manufacturing* 37(3): 430–37.

Rohatgi, PK, PL Menezes, and MR Lovell. 2012. *Green Tribology*, Eds. M. Nosonovsky and B. Bhushan. Springer Berlin Heidelberg.

Roviello, G, L Ricciotti, O Tarallo, C Ferone, F Colangelo, V Roviello, and R Cioffi. 2016. Innovative fly ash geopolymer-epoxy composites: Preparation, microstructure and mechanical properties. *Materials* 9(6): 461.

Shahapurkar, K, CD Garcia, M Doddamani, GC Mohan Kumar, and P Prabhakar. 2018. Compressive behavior of cenosphere/epoxy syntactic foams in arctic conditions. *Composites Part B: Engineering* 135: 253–62.

Singh, J, and A Chauhan. 2019. A review of microstructure, mechanical properties and wear behavior of hybrid aluminium matrix composites fabricated via stir casting route. *Sādhanā* 44(1): 1–18.

Tan, JF, YJ Jia, and LX Li. 2016. A series–parallel mixture model to predict the overall property of particle reinforced composites. *Composite Structures* 150: 219–25.

Ullas, AV, and B Jaiswal. 2020. Halloysite nanotubes reinforced epoxy-glass microballoons syntactic foams. *Composites Communications* 21: 100407.

Wang, MC, and E Guth. 1952. Statistical theory of networks of non-gaussian flexible chains. *The Journal of Chemical Physics* 20(7): 1144–57.

Wang, S, S Ma, Q Li, W Yuan, B Wang, and J Zhu. 2018a. Robust, fire-safe, monomer-recovery, highly malleable thermosets from renewable bioresources. *Macromolecules* 51(20): 8001–12.

Wang, S, S Ma, Q Li, X Xu, B Wang, W Yuan, S Zhou, S You, and J Zhu. 2019. Facile in situ preparation of high-performance epoxy vitrimer from renewable resources and its application in nondestructive recyclable carbon fiber composite. *Green Chemistry* 21(6): 1484–97.

Wang, S, X Xing, X Zhang, X Wang, and X Jing. 2018b. Room-temperature fully recyclable carbon fibre reinforced phenolic composites through dynamic covalent boronic ester bonds. *Journal of Materials Chemistry A* 6(23): 10868–78.

Wu, J, G Huang, H Li, S Wu, Y Liu, and J Zheng. 2013. Enhanced mechanical and gas barrier properties of rubber nanocomposites with surface functionalized graphene oxide at low content. *Polymer* 54(7): 1930–37.

Xiong, J, and R Mashiur. 2016. Mechanical property measurement and prediction using hirsch's model for glass yarn reinforced polyethylene composite fabric formwork. *Journal of Textile Science and Engineering* 6: 241–50.

Yin, HM, and LZ Sun. 2005. Elastic modelling of periodic composites with particle interactions. *Philosophical Magazine Letters* 85(4): 163–73.

Yu, K, Q Shi, ML Dunn, T Wang, and HJ Qi. 2016. Carbon fiber reinforced thermoset composite with near 100% recyclability. *Advanced Functional Materials* 26(33): 6098–106.

Yuan, Y, Y Sun, S Yan, J Zhao, S Liu, M Zhang, X Zheng, and L Jia. 2017. Multiply fully recyclable carbon fibre reinforced heat-resistant covalent thermosetting advanced composites. *Nature Communications* 8(1): 1–11.

Zare, Y, and H Garmabi. 2017. Predictions of takayanagi model for tensile modulus of polymer/CNT nanocomposites by properties of nanoparticles and filler network. *Colloid and Polymer Science* 295(6): 1039–47.

Zhou, Z, Z-z Liu, H Yang, W-y Gao, and C-c Zhang. 2020. Freeze-thaw damage mechanism of elastic modulus of soil-rock mixtures at different confining pressures. *Journal of Central South University* 27(2): 554–65.

6 Graphene-Reinforced Polymer Composites

Bidita Salahuddin
University of Wollongong

Shazed Aziz
The University of Queensland

Md. Mokarrom Hossain
The University of New South Wales

Aziz Ahmed
University of Wollongong

Md. Arifur Rahim and Mohammad S. Islam
The University of New South Wales

Shaikh N. Faisal
University of Wollongong
University of Technology Sydney

CONTENTS

6.1 INTRODUCTION

Tough and lightweight composites are the keystones of the evolution of next-generation aerospace, automotive and sports manufacturing (Mirabedini et al. 2020; Rajak et al. 2019). To develop the metal-free lightweight structural composite materials, the combination of polymer matrix and reinforcing fillers provides more than 40% weight reduction than traditional metallic structures. Among the reinforcing fillers, carbon fibres (CF) have been proven to be the best material as the primary load element and for high strength and modulus properties of composites (Ma et al. 2016). To induce the efficient mechanical property of the composites, the high volume fractions of fibres (>50%) are necessary as the mechanical property is extensively dependent on the interfacial characteristics of the filler and matrix material. In addition, the strong surface bonding like hydrogen bonding

DOI: 10.1201/9780429330575-6

and Van der Waals forces between the CF and polymer matrix are crucial throughout composite processing for strong interfacial adhesion. The fibre/matrix interfacial adhesion energy need to be higher than the cohesive energy of the matrix (Dvir, Jopp, and Gottlieb, 2006; Li et al. 2018; Petersen, Lemons, and McCracken 2006; Zhang et al. 2011). The surface alteration of CF can influence the fibre/matrix interface to improve the mechanical properties (Dvir, Jopp, and Gottlieb 2016; Jones 1991). Typically, CF has a non-polar surface with crystallized graphitic basal planes formed during the high temperature carbonization process in synthesis. In parallel with this, the surface lipophobicity, smoothness and low absorption properties of the CF surface create inadequate bonding with polymer matrix during reinforced composite preparation (Paiva, Bernardo, and Nardin, 2000; Park and Kim 2005; Sharma et al. 2014).

The confinement of low-dimensional (0D, 1D and 2D) graphitic carbon materials, such as zero-dimensional graphene quantum dots (GQD), one-dimensional carbon nanotubes (CNT), two-dimensional graphene (Gr) and graphitic nanoplatelets (GNP) as well as transition metal dichalcogenides (TMDs) and hexagonal boron nitride (hBN) on the CF surface *via* adhesion, coating, grafting or chemical bonding, can highly manipulate the fibre/matrix interfacial adhesion. This process enhances the interlocking mechanism of the fibres with the matrix as well as retardation of the cracking and fractures (Hu et al. 2019; Kumar, Kumar, and Bhadauria 2020; Tang et al. 2019; Wazalwar, Sahu, and Raichur 2021). Among these low-dimensional materials, nanocarbons, especially CNT and Gr, contribute majorly to introducing electrical, optical, electromagnetic interferences and energy storage properties in the composites along with enhanced mechanical properties (Singh et al., 2013, 2015; Islam et al. 2016, 2017a, b, 2020, 2021). The insertion of nanocarbons minimized the complications related to the matrix predominating properties for developing advanced tough fibre-reinforced polymer composites. The addition of nanocarbons to attach on CF surfaces either by adhesion, carbon-carbon bonding, hydrogen bonding, Van der Wales bond and chemical bonding like ester linkages process can be achieved by several experimental processes such as chemical vapour deposition, spray up methods, dip coating, electrophoretic deposition and solvothermal (Islam et al. 2016, 2017a, b, 2020, 2021; Lubineau and Rahaman 2012; Bekyarova et al. 2007; Sharma and Lakkad 2011; Dong et al. 2017; Park et al. 2008; Li et al. 2017). The functionalization of graphitic basal planes of the nanocarbons enhances the surface energy and increases the roughness of the CF surface that improves the mechanical interlocking and interfacial strength between the fibre and polymer matrix. In parallel with nanocarbons, transition metal dichalcogenides such as MoS_2 and 2D layered hBN have recently gained attention as promising nanofiller in reinforced polymer composites, which open up the future opportunities for other 2D TMDs to develop next-generation tough composites (Wazalwar, Sahu, and Raichur 2021; Riaz and Park 2021; Xia et al. 2016).

This chapter focuses on the general concepts of low-dimensional graphitic materials, especially nanocarbons confined to CF-reinforced polymer composites. It first summarizes the different types of low-dimensional nanocarbons and their confinement process on CF composites. It then focuses on the different polymer matrix materials, which are essential to fabricate the tough fibre-reinforced polymer composites.

6.2 GRAPHENE QUANTUM DOTS

Graphene quantum dots (GQDs) are a subset of graphene-based nanomaterials having a diameter in the range of 10–50 nm (Tian et al. 2018; Facure et al. 2020; Hassan et al. 2014, 2018). GQDs are different from the carbon dots as it contains the graphene lattice inside the tiny dot-like structure. The unique structure-related properties with quantum confinement, edge effects and edge-functionalization aid to infuse in the polymer composites with high attachment and chemical bonding (Tian et al. 2018; Seibert et al. 2019; Gobi et al. 2017). The infusion of GQDs in epoxy improved the thermal conductivity by 144%, glass transition temperature by 10% and storage modulus by 9% (Seibert et al. 2019). In parallel with the modulation of thermal conductivity and glass transition temperature, the infusion of GQDs in the epoxy matrix from 1 to 10 wt.% via in situ polymerization was demonstrated. The enhancement of tensile strength, Young modulus's and nominal strain at break by 125%, 153.4% and 18.1% were observed by adding 2.5 wt.% of GQDs in epoxy matrix (Gobi et al. 2017). Figure 6.1 shows the outstanding tensile strength and modulus of elasticity of GQD-reinforced composites compared with other reinforcing agents. To obtain a tough polymer composite, the proper dispersion and nano-inclusion of GQDs are crucial. The surface polarities and functionalities on the edge of the GQDs play an important role to avoid agglomeration and

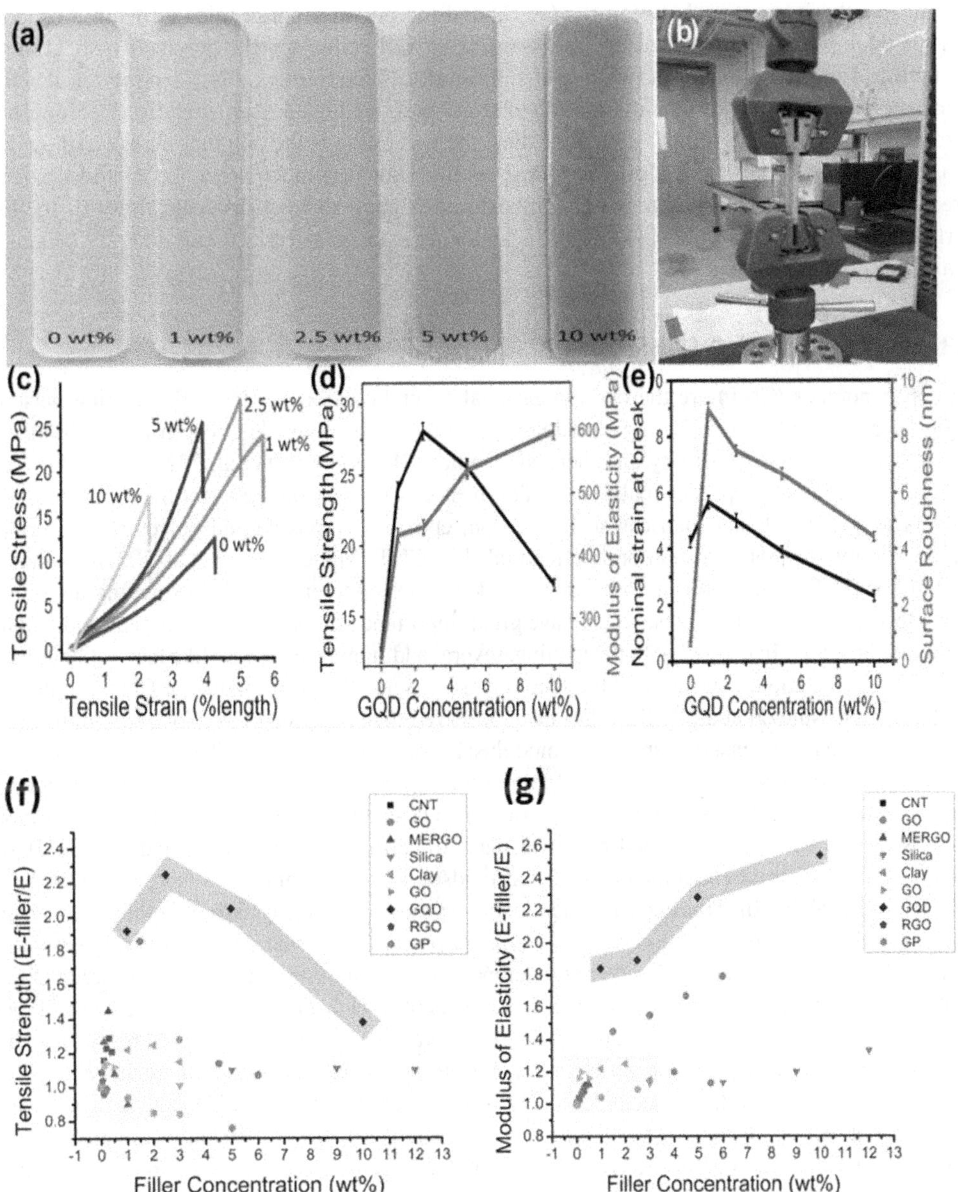

FIGURE 6.1 (a) E–GQD nanocomposites with different weight loadings of GQDs synthesized at 85°C. (b) Instron setup for tensile test. (c) Stress–strain curves of different nanocomposites. (d) Tensile strength and modulus of elasticity against GQD concentrations, showing that the addition of GQDs made the epoxy stiffer and stronger. (e) Tensile strain and surface roughness of composites. (f) Comparison of the ultimate tensile strength of different nanoparticles relative to neat epoxy, demonstrating that the GQDs provided better enhancement. (g) Comparison of moduli of elasticity of composites of different nanoparticles relative to those of neat epoxy reveals that GQDs made the epoxy stiffer compared to that of all other nanoparticles. The GQDs are highlighted in the purple-shaded area of both graphs (Gobi et al. 2017).

non-uniform distribution throughout the polymer matrix. In addition, it improves physisorption-assisted anchoring of polymer chains onto GQDs surfaces (Ganguly et al. 2019). The formation of interpenetrating network of amino-functionalized GQDs with carboxylated nitrile latex via hydrogenation demonstrated comparably better composites in physical properties and ageing resistances (Xie et al. 2020). The oxygen functionalized GQDs not only boost the quality dispersion but also improve the covalent attachment between the surfaces and host the polymer's pendant groups. It indicates the interaction between the GQDs and the polymer phase initiated from their edges and the edge polarity enhances easy load transfer. It well connected the polymer-to-sheet during stretching when the nanocomposites were subjected to stretch uniaxially (Ganguly et al. 2019). During stressing a polymer, the dissipation of stress was conducted from a relatively soft phase to hard phase. In the GQD-reinforced composites, the soft phase is the polymer matrix and the hard phase is the GQDs. The typical polymers without filler are very much prone to fail under external stress but for GQD-reinforced polymers, the externally applied stress dissipates towards the filler particles and the whole composites can withstand more external stresses (Gobi et al. 2017; Ganguly et al. 2019).

6.3 CARBON NANOTUBES

Carbon nanotubes (CNT) are the one-dimensional graphitic carbon material that exhibit outstanding properties of ultra-high Young's modulus (1 TPa), excellent tensile strength (11–63 GPa), with exceptional thermal conductivity and thermal stability (Mohd Nurazzi et al. 2021; Salahuddin et al. 2021). These unique properties make the CNTs most suitable reinforcing agent for composite materials. A range of polymer matrices such as nylon, epoxy, polyepoxide and polyetherimide exhibit higher affinity towards nano-reinforcing materials like CNT (Mohd Nurazzi et al. 2021). An optimal loading for CNT in the polymer composites is a key criterion to employ its outstanding mechanical characteristics in the composites, which have great importance in the area of aerospace and automotive manufacturing. In parallel with the loading, several additional parameters like fabrication process, dispersion, orientation, chirality and the length can influence the performance of CNTs in the composites (Mohd Nurazzi et al. 2021). It has been observed that increasing CNT loading can primarily lead to an increase in tensile strength and modulus; however, increasing CNT loading has an adverse effect on the modulus and strength of the composite beyond a critical weight fraction. For example, the deterioration on tensile strength, failure strain and elastic modulus of CNT reinforced polystyrene composites has been found beyond a critical mass fraction (Agnihotri, Basu, and Kar 2011). The dispersion of CNT throughout the polymer matrix also plays a crucial role on mechanical properties (Mirabedini et al. 2020). This can cause a reduction in the load carrying capacity of the composite as well. Therefore, it is necessary to grow CNT directly on the CFs surface. The processing parameters such as the growth time, catalyst and growth temperature play an important role on the grade of CNT confinement and the fibre surface coverage. The effects of CNT loading on the CNT-coated CF/polyester composites' properties have been studied (Agnihotri, Basu, and Kar 2011). The researchers have found that the optimization of the multiscale composites are achievable by changing the reactor duration for chemical vapour deposition (CVD). Dispersion, degradation of the CNTs and matrix viscosity are some challenging issues for CNT insertion. In some cases, the dispersion techniques of CNTs coating in the polymer composites resulted in low CNT graphitization, poor nanotube alignment, CNT agglomeration, inadequacy to small weight percentage accumulations, deficiency of morphology control and reduced matrix infusion capability. Sharma and Lakkad (2011) investigated the CF before and after the growth of CNTs/CNFs (Sharma and Lakkad 2011) It is clear that the CNT covered the fibre surface uniformly and grew long enough with various diameter and length. Typically, the long CNTs may orient itself with the direction of drawing of the fibre when it is being pulled out through the very small orifice of the wire drawing die. The cause of the alignment is due to the viscous force of the polymer matrix and also for the frictional force due to the compactness of the opening of the die. Throughout the fibre pulling across the die, the exertion of the forces took place. The increase in the

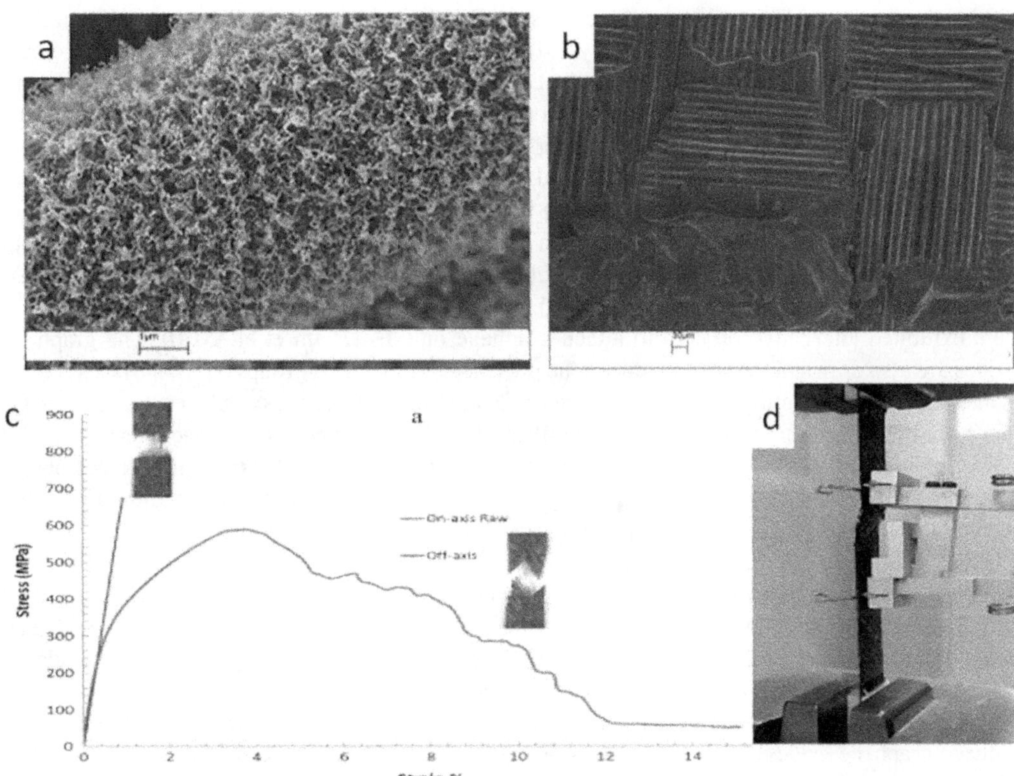

FIGURE 6.2 (a) SEM micrographs of GSD surface grown MWCNTs on SiO_2 precoated and nickel-sputtered carbon fibres. (b) SEM micrographs of the impregnation of the fibres with epoxy matrix in composites based on carbon fibres with surface grown CNTs. (c) Sample stress–strain behaviour of on- and off-axis samples based on raw carbon fibres, including fracture surfaces. Note that the stress values for the off-axis curve are magnified by a factor of 5.0 to allow better visibility. (d) Off-axis tensile test of composite sample beyond the yield point (Boroujeni et al. 2014).

composites' tensile strength is due to this partial alignment. The grafting of CNTs on carbon fibre by epoxy bonds through solvothermal process demonstrated the possibilities of CNTs bonded chemically with carbon fibre that can tolerate the failure stress up to 31 GPa (Islam et al. 2016). It indicates the addition of CNTs with CF before fabricating polymer matrix has a strong influence on the properties of the composites. Further studies on adding CNTs to the polymeric matrix directly before fabricating the fibre-based polymer composite showed noteworthy enhancements for vibration attenuation (25.8%), impact energy absorption (21.3%) and axial strain to failure (12%) by adding multi-walled CNTs of 2.0 wt% to an epoxy matrix of a fibre-reinforced polymer (Boroujeni et al. 2014) Although CNT-reinforced polymer composites have exhibited great promise for various structural applications, some technological challenges such as reproducibility in large-scale production, safety, reliability and durability of these composites are still subject to exploration (Figure 6.2).

6.4 GRAPHENE AND GRAPHENE OXIDE

The discovery of graphene and its outstanding electrical and mechanical properties have drawn significant attention as an alternative reinforcing agent in developing tough polymer composites (Geim and Novoselov 2010; Kumar and Xavior 2014; Sreenivasulu, Ramji, and Nagaral 2018; Pourazadi et al. 2018; Faisal et al. 2021). The two-dimensional sp2-hybridized planar structure naturally formed of hexagonal lattices resists variable in-plane deformations during chemical and mechanical interactions in polymers (Sreenivasulu, Ramji, and Nagaral 2018). In the polytetrafluoroethylene composites, the graphene/CF hybrid as a multi-functional interfacial nano-reinforcement played a crucial role to enhance the mechanical and electrical properties (Li et al. 2016). The growth of functionalized graphene onto CFs as reinforcing interface for epoxy composites demonstrated the improvement of tensile strength of the composites (Karakassides et al. 2020). GNFs were grown via microwave plasma-enhanced CVD method. 800 W microwave power and 600°C reaction temperature were found most suitable for making GNFs/CF fibres with best interfacial adhesion with the surrounding epoxy matrix. Compared with bare CF, a 28% improvement in the tensile strength was seen for the hybrid fibres through single-fibre tensile strength tests, while the interfacial shear strength augmented by 101.5% (Figure 6.3a–c). The grafting of graphene on functionalized carbon fibre exhibited alternative process to attach graphene on CFs (Islam et al. 2020). The graphene oxides are also drawn high attention due to the presence of functional groups on the surfaces (Faisal et al. 2021) The availability of numerous oxygen functional groups, for example, epoxide, hydroxyl and carbonyl groups on the surfaces, can significantly enhance the adhesion at the interface of CF/polyethersulfone composites owing to the hydrophilic oxygen functional groups in the basal planes of graphene oxide (Li et al. 2016). The sizing was conducted in a method similar to the Hummers method through acid oxidation of graphite powders. The preparation of pure polyethersulfone was conducted using the injection moulding technique, which helps to observe the effect on the mechanical properties of the polyethersulfone matrix when extrusion compounding process is used. Dried graphene oxide-coated short CFs were re-dispersed in deionized water and intensively stirred. The coating of short CFs was conducted in order to assess the coating efficiency of graphene oxide on short CFs in the physical adsorption process. The optimal graphene oxide content of 0.5 wt% can efficiently improve the overall mechanical performance of the composite. Moreover, the usage of reduced graphene oxide has the ability to improve the interfacial and mechanical properties of CF. The reduced graphene oxide-based CF has been found to be more successful than CF to enhance

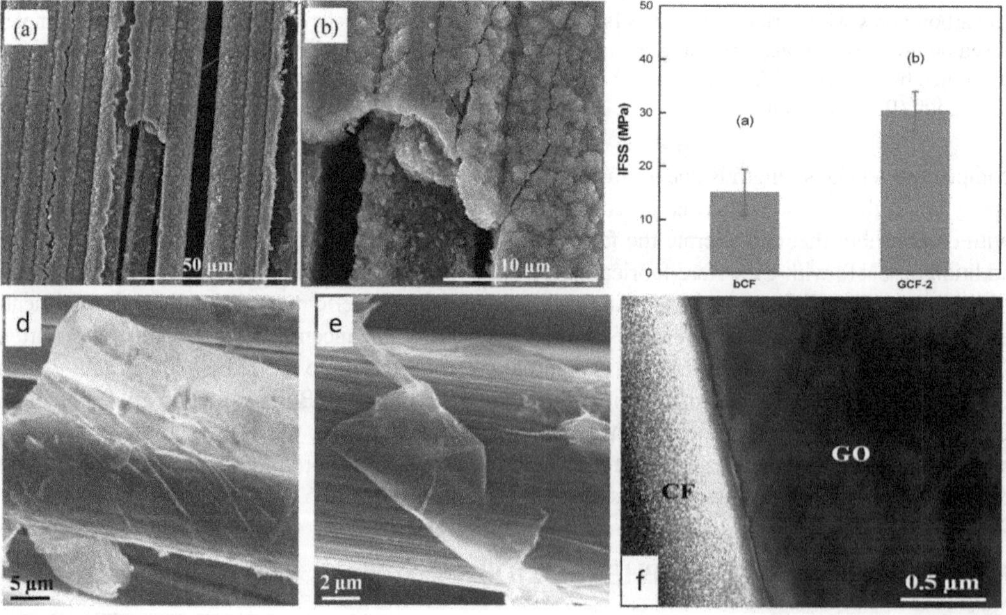

FIGURE 6.3 (a) SEM images of Adhesion tests on the grown GNFs. (b) Removal of the GFNs with a blade. (c) Interfacial shear strength (IFSS) results for (a) bare (bCF) and (b) GCF-2 (800 W at 600°C) hybrid fibre conducted via single-fibre fragmentation test (Karakassides et al. 2020). (d) SEM images of graphene oxide grafted on CF at 5μ and (e) 2μ scale. (f) Thickness map of graphene oxide grafted on CF (Islam et al. 2020).

the unsaturated polyester-based composites in terms of electromagnetic interference shielding property. Directly grafting graphene on CF can contribute to enhance the mechanical properties of the polymer composites. Figure 6.3d–f shows the SEM image of grafted GO on CF by ester linkage in a low temperature (>100°C) solvothermal process (Islam et al. 2021). Zhang et al. (2016) has studied the novel hierarchical reinforcement of CF on which graphene oxide is directly grafted. The interphase strength between the CF and resin matrix was boosted due to the grafting of graphene oxide onto CF. In fact, surface modification can effectively increase the polarity and wettability of the surface of CF without compromising the tensile strength. The efficient graphene oxide reinforcement between CF and matrix resin results in these improvements. Viscous aqueous dispersion of ultra-large sheets of liquid crystalline graphene oxide makes it possible to draw fibres without any additive and could be an alternative agent for reinforced composites (Seyedin et al. 2015; Roy et al. 2021). The addition of graphene as reinforcing agent showed exceptional improvement in thermoset polymers specially epoxy, which has a large commercial value. Typically, epoxy has excellent chemical and mechanical properties, low shrinkage, high temperature stability (>200°C) and good adhesion with fibres. In addition, epoxy have outstanding corrosion resistance and less shrinkage once cured (Li et al. 2015; Xu et al. 2021; Eyckens et al. 2021; Zhang et al. 2013; Siddiqui et al. 2011). The GO-modified CF-reinforced composites showed the improvement of interfacial strength. In the report, it was found that the neat CF-epoxy interface on the fractured surface of the composite has low mechanical properties, whereas the interface of the fractured graphene oxide modified CF-reinforced composites samples was comparatively stronger (Altin et al. 2020). Due to the presence of hydroxyl groups in epoxy similar to graphene oxide, stronger chemical bonds through functional groups of epoxies with graphene oxide surface created. Thus, fibre-thermoset matrix resin has the potential to guarantee more effective properties such as delamination resistance, interlaminar shear strength, fatigue and corrosion resistance in composites due to their strong interfacial interactions. In parallel with the thermoset polymers, graphene can also manipulate the mechanical properties in thermoplastic polymers like polyethersulfone (PES), which is an amorphous, amber-coloured, polymer with relatively large mechanical strength, high glass transition temperature (Tg, ~225°C–230°C) mostly used in automobile, aerospace and sports industries (Gabrion et al. 2016; Yao et al. 2018). The GO-coated CF-reinforced PES composites prepared by melt blending and injection blending methods showed improvement on interfacial adhesion with CF and PES and thus enhance the mechanical performances (Zhang et al. 2014).

6.5 GRAPHENE NANOPLATELETS

Graphene nanoplatelets (GnP) are the few to multilayered graphene flakes that produced from the liquid exfoliation of graphite by shear mixture or by electrochemical exfoliation of graphite (Liu et al. 2019; Li et al. 2020). The exfoliation processes are scalable, and it is possible to produce graphene flakes at kilogram to ton scale commercially. The high surface area and aspect ratio of GnP contributed to percolate conducting network with the addition of 2 vol.% and enhanced electrical, thermal and mechanical properties in GnP/Cu coated CF epoxy composite (Park et al. 2008) This is very crucial to manufacture composite in large scale. The major advantages of using GnP as a reinforcing agent are that it improves the electrical and thermal conductivity along with the mechanical performance. It has been demonstrated that the addition of 15 wt.% GnP can enhance 48% in thermal conductivity for CF-reinforced polymer composite (Kostagiannakopoulou et al. 2016). In another study, it has been shown that the addition of 3 wt.% GnP can increase 19% interlaminar shear strength (Li et al. 2017). Although numerous studies have exemplified performance enhancement of polymer materials through the incorporation of GnP into polymer matrix, only a handful of graphene-based composite products have made the transition to the commercial application (Mirabedini et al. 2020). One critical barrier is transition from laboratory scale to industrial scale, which would require the capability to produce large-scale composite materials without significantly changing in the existing manufacturing methods and infrastructure (Mirabedini et al. 2020).

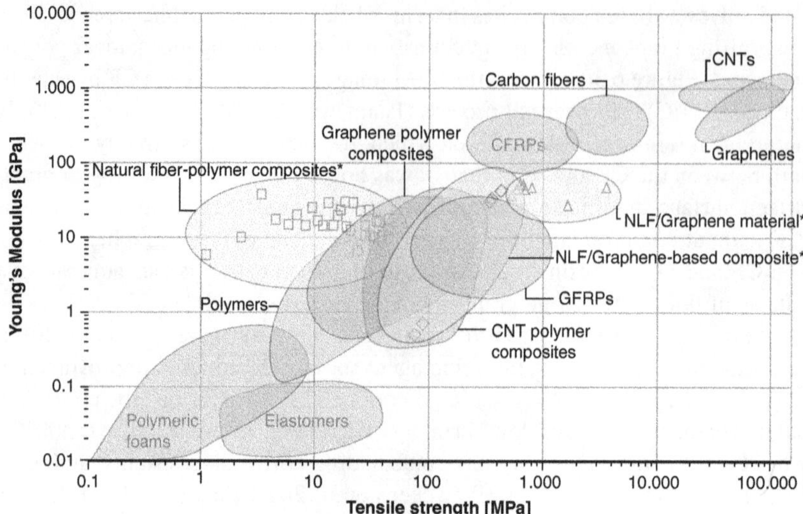

FIGURE 6.4 Ashby plot of young's modulus vs. tensile strength of CNT or graphene-reinforced polymer composite. (Adopted from Kinloch et al. 2018.)

Despite these limitations, commercial development of graphene-based composite has made significant progress in recent years. For example, Elmarakbi et al. developed prototype of graphene-based composite for automotive application (car bumper) (Mirabedini et al. 2020; Milberg 2018). Briggs automotive company (BAC), in collaboration with Haydale and Pentaxia, has recently developed lightweight BAC Mono R body using graphene-composite materials (Keighley 2019). Figure 6.4 shows the outstanding performance of graphene-reinforced polymer composites in a plot of Young's modulus vs. tensile strength comparing with other commercial composites (Kinloch et al. 2018).

6.6 SUMMARIES AND CONCLUSIONS

Graphenic materials with different nano-dimension are playing a crucial role to develop next-generation composites. The advancement of scalable and cost-effective production of graphene takes it from the laboratory scale to research translation for commercialization opportunities. The automative and aerospace industries are already employing graphene-reinforced composites for limited applications; however, further development is required for widespread applications. Innovation in new polymers, coating material and processing techniques are also important. In this chapter, we have discussed the variations in graphene with different nano-dimensions and their role in fibre-reinforced composites. For future development, the researchers need to work on the current limitation of the graphene like scalable production of GQDs and few layer graphene, reduce the cost of production, agglomeration of CNTs as well as direct growth of CNTs on CF. For graphene, the functionalization and proper dispersion are the key challenges. The homogeneity and size of the GnP for proper insertion in polymer matrix are also necessary. In parallel with the material development, the development in process, fabrication mixing and curing are also crucial for developing advanced graphene-reinforced tough polymer composites.

ACKNOWLEDGEMENTS

The authors gratefully acknowledge the technical and intellectual support from ARC Centre of Excellence for Electromaterials Science, The University of Technology Sydney, The University of Wollongong, University of New South Wales, and the University of Queensland.

REFERENCES

Agnihotri, P., S. Basu, and K.K. Kar, 2011. Effect of carbon nanotube length and density on the properties of carbon nanotube-coated carbon fiber/polyester composites. *Carbon* 49(9): 3098–106.

Altin, Y., et al., 2020. Graphene oxide modified carbon fiber reinforced epoxy composites. *Journal of Polymer Engineering* 40(5): 415–20.

Bekyarova, E., et al., 2007. Multiscale carbon nanotube–carbon fiber reinforcement for advanced epoxy composites. *Langmuir* 23(7): 3970–4.

Boroujeni, A.Y., et al., 2014. Hybrid carbon nanotube–carbon fiber composites with improved in-plane mechanical properties. *Composites Part B: Engineering* 66: 475–83.

Dong, J., et al., 2017. Improved mechanical properties of carbon fiber-reinforced epoxy composites by growing carbon black on carbon fiber surface. *Composites Science and Technology* 149: 75–80.

Dvir, H., J. Jopp, and M. Gottlieb, 2006. Estimation of polymer–surface interfacial interaction strength by a contact AFM technique. *Journal of colloid and interface science* 304(1): 58–66.

Eyckens, D.J., et al., 2021. Covalent sizing surface modification as a route to improved interfacial adhesion in carbon fibre-epoxy composites. *Composites Part A: Applied Science and Manufacturing* 140: 106147.

Facure, M.H.M., R. Schneider, L.A. Mercante, and D.S. Correa, 2020. A review on graphene quantum dots and their nanocomposites: from laboratory synthesis towards agricultural and environmental applications. *Environmental Science: Nano* 7: 3710–34.

Faisal, S.N., C.M. Subramaniyam, M.M. Islam, A.I. Chowdhury, S.X. Dou, A.K. Roy, A.T. Harris, and A.I. Minett, 2021. 3D copper-confined N-doped graphene/carbon nanotube network as high-performing lithium-ion battery anode. *Journal of Alloys and Compounds* 850: 156701.

Gabrion, X., et al., 2016. About the thermomechanical behaviour of a carbon fibre reinforced high-temperature thermoplastic composite. *Composites Part B: Engineering* 95: 386–94.

Ganguly, S., P. Das, S. Banarjee, and N.C. Das, 2019. Advancement in science and technology of carbon dot-polymer hybrid composites: a review. *Functional Composites and Structures* 1: 022001.

Geim, A.K., and K.S. Novoselov, 2010. The rise of graphene. *Nanoscience and Technology: A Collection of Reviews from Nature Journals* 11–19.

Gobi, N., D. Vijayakumar, O. Keles, and F. Erogbogbo. 2017. Infusion of graphene quantum dots to create stronger, tougher, and brighter polymer composites. *ACS Omega* 2(8): 4356–62.

Hassan, M., E. Haque, K.R. Reddy, A.I. Minett, J. Chen, and V.G. Gomes, 2014. Edge-enriched graphene dots for enhanced photo-luminescence and supercapacitance. *Nanoscale* 6: 11988–94.

Hassan, M., V.G. Gomes, A. Dehghani, and S.M. Aedekani, 2018. Engineering carbon quantum dots for photomediated theranostics. *Nano Research* 11: 1–41.

Hu, C., et al., 2019. The fabrication and characterization of high density polyethylene composites reinforced by carbon nanotube coated carbon fibers. *Composites Part A: Applied Science and Manufacturing* 121: 149–56.

Islam, M.M., Faisal, S.N., Akhter, T., Roy, A.K., Minett, A.I., Konstantinov, K., and Dou, S.X. 2017a. Liquid-crystal-mediated 3D macrostructured composite of Co/Co3O4 embedded in graphene: Free-standing electrode for efficient water splitting. *Particle & Particle Systems Characterization* 34: 1600386.

Islam, M.S., S.N. Faisal, L. Tong, A.K. Roy, J. Zhang, E. Haque, A.I. Minett, and C.H. Wang, 2021. N-doped reduced graphene oxide (rGO) wrapped carbon microfibers as binder-free electrodes for flexible fibre supercapacitors and sodium-ion batteries. *Journal of Energy Storage* 37: 102453.

Islam, M.S., Y. Deng, L. Tong, A.K. Roy, S.N. Faisal, M. Hassan, A.I. Minett, and V.G. Gomes, 2017b. In-situ direct grafting of graphene quantum dots onto carbon fibre by low temperature chemical synthesis for high performance flexible fabric supercapacitor. *Material Today Communications* 10: 112–9.

Islam, M.S., Y. Deng, L. Tong, S.N. Faisal, A.K. Roy, and A.I. Minett, 2020. High grafting strength from chemically bonded 2D layered material onto carbon microfibres for reinforced composites and ultra-long flexible cable electronic devices. *Material Today Communication* 24: 100994.

Islam, M.S., Y. Deng, L. Tong, S.N. Faisal, A.K. Roy, A.I. Minett, and V.G. Gomes, 2016. Grafting carbon nanotubes directly onto carbon fibers for superior mechanical stability: Towards next generation aerospace composites and energy storage applications. *Carbon* 96: 701.

Jones, C., 1991. The chemistry of carbon fibre surfaces and its effect on interfacial phenomena in fibre/epoxy composites. *Composites Science and Technology* 42(1): 275–98.

Karakassides, A., et al., 2020. Radially grown graphene nanoflakes on carbon fibers as reinforcing interface for polymer composites. *ACS Applied Nano Materials* 3(3): 2402–13.

Keighley, S. 2019. *Haydale Graphene-Enhanced Composite Tooling and Automotive Body Panels Showcased on New BAC Mono R.*

Kinloch, I.A., J. Suhr, J. Lou, R.J. Young, and P.M. Ajayan, 2018. Composites with carbon nanotubes and graphene: An outlook. *Science* 362(6414): 547–53.

Kostagiannakopoulou, C., et al., 2016. Thermal conductivity of carbon nanoreinforced epoxy composites. *Journal of Nanomaterials* 2016, Article ID 1847325.

Kumar, H.G.P., and M. A. Xavior 2014. Graphene reinforced metal matrix composite (GRMMC): A review. *Procedia Engineering* 97: 1033–40.

Kumar, M., P. Kumar, and S.S. Bhadauria, 2020. Interlaminar fracture toughness and fatigue fracture of continuous fiber-reinforced polymer composites with carbon-based nanoreinforcements: A review. *Polymer-Plastics Technology and Materials* 59(10): 1041–76.

Li, F., et al., 2015. Enhanced mechanical properties of short carbon fiber reinforced polyethersulfone composites by graphene oxide coating. *Polymer* 59: 155–65.

Li, F., et al., 2016. Greatly enhanced cryogenic mechanical properties of short carbon fiber/polyethersulfone composites by graphene oxide coating. *Composites Part A: Applied Science and Manufacturing* 89: 47–55.

Li, Q., A.L. Woodhead, J.S. Church, and M. Naebe, 2018. On the detection of carbon fibre storage contamination and its effect on the fibre–matrix interface. *Scientific Reports* 8(1): 1–10.

Li, Y., et al., 2017. Graphite nanoplatelet modified epoxy resin for carbon fibre reinforced plastics with enhanced properties. *Journal of Nanomaterials* DOI:10.1155/2017/5194872.

Li, Z., R.J. Young, C. Backes, W. Zhao, X. Zhang, A.A. Zhukov, E. Tillotson, A.P. Conlan, F. Ding, S.J. Haigh, K.S. Novoselov, and J.N. Coleman, 2020. Mechanisms of liquid-phase exfoliation for the production of graphene. *ACS Nano* 14: 10976–85.

Liu, F., C. Wang, X. Sui, M.A. Riaz, M. Xu, L. Wei, and Y. Chen, 2019. Synthesis of graphene materials by electrochemical exfoliation: Recent progress and future potential. *Carbon Energy* 1: 173–99.

Lubineau, G. and A. Rahaman, 2012. A review of strategies for improving the degradation properties of laminated continuous-fiber/epoxy composites with carbon-based nanoreinforcements. *Carbon* 50(7): 2377–95.

Ma, Q., et al., 2016. Effects of surface treating methods of high-strength carbon fibers on interfacial properties of epoxy resin matrix composite. *Applied Surface Science* 379.

Milberg, E. 2018. *Composite Manufacturing.* Arlington, VA: American composites manufacturing association (ACMA).

Mirabedini, A., A. Ang, M. Nikzad, B. Fox, K.-T. Lau, and N. Hameed, 2020. Evolving strategies for producing multiscale graphene-enhanced fiber-reinforced polymer composites for smart structural applications. *Advanced Science* 7(11): 1903501.

Mirabedini, A., et al., 2020. Evolving strategies for producing multiscale graphene-enhanced fiber-reinforced polymer composites for smart structural applications. *Advanced Science* 7(11): 1903501.

Mohd Nurazzi, N., M.R. Muhammad Asyraf, A. Khalina, N. Abdullah, F. Athiyah Sabaruddin, S. Hasnah Kamarudin, S. Ahmad et al. 2021. Fabrication, functionalization, and application of carbon nanotube-reinforced polymer composite: An overview. *Polymers* 13(7): 1047.

Paiva, M.C., C.A. Bernardo, and M. Nardin, 2000. Mechanical, surface and interfacial characterisation of pitch and PAN-based carbon fibres. *Carbon* 38(9): 1323–37.

Park, J.K., et al., 2008. Electrodeposition of exfoliated graphite nanoplatelets onto carbon fibers and properties of their epoxy composites. *Composites Science and Technology* 68(7): 1734–41.

Park, S.-J. and B.-J. Kim, 2005. Roles of acidic functional groups of carbon fiber surfaces in enhancing interfacial adhesion behavior. *Materials Science and Engineering: A* 408(1): 269–73.

Petersen, R.C., J.E. Lemons, and M.S. McCracken, 2006. Stress-transfer micromechanics for fiber length with a photocure vinyl ester composite. *Polymer Composites* 27(2): 153–169.

Pourazadi, E., E. Haque, S.N. Faisal, and A.T. Harris, 2018. Identification of electrocatalytic oxygen reduction (ORR) activity of boron in graphene oxide; incorporated as a charge-adsorbate and/or substituinal p-type dopant. *Materials Chemistry and Physics* 207: 380–8.

Rajak, D., et al., 2019. Fiber-reinforced polymer composites: Manufacturing, properties, and applications. *Polymers* 11: 1667.

Riaz, S., and S.-J. Park, 2021. A comparative study on nanoinclusion effect of MoS2 nanosheets and MoS2 quantum dots on fracture toughness and interfacial properties of epoxy composites. *Composites Part A: Applied Science and Manufacturing* 146: 106419.

Roy, A.K., S.N. Faisal, A. Spickenheuer, C. Scheffler, J. Wang, A.T. Harris, A.I. Minett, and M.S. Islam, 2021. Loading dependency of 2D MoS2 nanosheets in the capacitance of 3D hybrid microfibre-based energy storage devices. *Carbon Trends* 2021: 100097.

Salahuddin, B., S.N. Faisal, T.A. Baigh, M.N. Alghamdi, M.S. Islam, B. Song, X. Zhang, S. Gao, and S. Aziz, 2021. Carbonaceous materials coated carbon fibre reinforced polymer matrix composites. *Polymers* 13(16): 2771.

Seibert, J.R., Keles, O., Wang, J., and Erogbogbo, F. 2019. Infusion of graphene quantum dots to modulate thermal conductivity and dynamic mechanical properties of polymers. *Polymer* 185: 121988.

Seyedin, S., M.S. Romano, A.I. Minett, and J.M. Razal, 2015. Towards the knittability of graphene oxide fibres. *Scientific Reports* 5(1): 1–12.

Sharma, M., et al., 2014. Carbon fiber surfaces and composite interphases. *Composites Science and Technology* 102: 35–50.

Sharma, S.P. and S.C. Lakkad 2011. Effect of CNTs growth on carbon fibers on the tensile strength of CNTs grown carbon fiber-reinforced polymer matrix composites. *Composites Part A: Applied Science and Manufacturing* 42(1): 8–15.

Siddiqui, N.A., et al., 2011. Manufacturing and characterization of carbon fibre/epoxy composite prepregs containing carbon nanotubes. *Composites Part A: Applied Science and Manufacturing* 42(10): 1412–20.

Singh, A.P., et al., 2013. Multiwalled carbon nanotube/cement composites with exceptional electromagnetic interference shielding properties. *Carbon* 56: 86–96.

Singh, A.P., et al., 2015. Probing the engineered sandwich network of vertically aligned carbon nanotube–reduced graphene oxide composites for high performance electromagnetic interference shielding applications. *Carbon* 85: 79–88.

Sreenivasulu, B., B.R. Ramji, and M. Nagaral, 2018. A review on graphene reinforced polymer matrix composites. *Materials Today Proceedings* 5: 2419–28.

Tang, B. et al., 2019. Preparation and properties of lightweight carbon/carbon fiber composite thermal field insulation materials for high-temperature furnace. *Journal of Engineered Fibers and Fabrics* 14: 1558925019884691.

Tian, P., L. Tang, K.S. Teng, and S.P. Lau, 2018. Graphene quantum dots from chemistry to applications. *Materials Today Chemistry* 10: 221–58.

Wazalwar, R., M. Sahu, and A.M. Raichur, 2021. Mechanical properties of aerospace epoxy composites reinfoced with 2D nano-fillers: Current status and road to industrialization. *Nanoscale Advances* 3: 2741–2776.

Xia, C., A.C. Garcia, S.Q. Shi, Y. Qiu, N. Warner, Y. Wu, L. Cai, H.R. Rizvi, N.A. D'Souza, and X. Nie, 2016. Hybrid boron nitride-natural fiber composites for enhanced conductivity. *Scientific Reports* 6: 34726.

Xie, F., Z. Yang, E. Xu, L. Zhang, and D. Yue, 2020. Preparation of graphene quantum dots modified hydrogenated carboxylated nitrile rubber interpenetrating cross-linked film. *Colloid and Polymer Science* 298: 1361–8.

Xu, N., et al., 2021. Enhanced mechanical properties of carbon fibre/epoxy composites via in situ coating-carbonisation of micron-sized sucrose particles on the fibre surface. *Materials & Design* 200: 109458.

Yao, S.-S., et al., 2018. Recent advances in carbon-fiber-reinforced thermoplastic composites: A review. *Composites Part B: Engineering* 142: 241–50.

Zhang, J., et al., 2013. Effect of nanoparticles on interfacial properties of carbon fibre–epoxy composites. *Composites Part A: Applied Science and Manufacturing*, 55: 35–44.

Zhang, R.L., et al., 2016. Directly grafting graphene oxide onto carbon fiber and the effect on the mechanical properties of carbon fiber composites. *Materials & Design* 93: 364–9.

Zhang, Y., J. Stringer, R. Grainger, P.J. Smith, and A. Hodzic, 2014. Improvements in carbon fibre reinforced composites by inkjet printing of thermoplastic polymer patterns. *Physica Status Solidi (RRL)–Rapid Research Letters* 8(1): 56–60.

Zhang, Z.Q., et al., 2011. Interfacial characteristics of carbon nanotube-polyethylene composites using molecular dynamics simulations. *ISRN Materials Science* 145042.

7 Matrices for Composite Materials

Polymers, Metals, Ceramics, and Cements

G.S. Mukherjee
DRDO

CONTENTS

DOI: 10.1201/9780429330575-7

7.1 INTRODUCTION

Engineers have to design materials to meet the requirements prescribed by the specifications and deliver the products in a time schedule without violating the rules of the contemporary environmental protocols. To meet such requirements, engineers have evolved the concept of composite materials.

A formal definition of composite materials given by the ASM Handbook is a macroscopic combination of two or more distinct materials having a recognizable interface between them. The IUPAC defines a composite material as a multicomponent material consisting of multiple different (non-gaseous) phase domains in which at least one type of phase domain is a continuous phase; and nanocomposite is a composite in which at least one of the phase domains has at least one dimension of the order of nanometers (Banerjee, Sachdeva, and Mukherjee 2012). In a composite material, the discontinuous phase(s) is/are embedded in a continuous phase, where the continuous phase is known as matrix. The matrix is homogeneous monolithic in nature and is the continuous phase in which the reinforcements are embedded and/or dispersed therein. The discontinuous phase is known as reinforcement.

7.1.1 Role of Matrices

A matrix is a relatively ductile material that holds the reinforcements of different shapes and dimensions (fiber, particles, platelets, whiskers, etc.) in such a way they function together (Agarwal and Broutman 1980). A matrix provides a continuous medium for binding the reinforcements together and consolidates them into a solid composite material, and it protects the reinforcements from abrasion, and environmental and physical damage. Matrices must act in harmony with the reinforcement to avoid failure as much as possible. Matrices keep the reinforcements separate and decrease the chance of cracking by redistributing the load equally among all fibers, even though only a small proportion of an applied load is sustained by the matrix phase as the elastic modulus of the matrix is less than that of reinforcements. The ability of engineering composite to withstand heat and/or mechanical load, or to conduct heat or electricity depends largely on the matrix properties since this is a continuous phase (Agarwal and Broutman 1980; Chawla 2012).

7.1.2 Matrix and their Influence on the Properties

The matrix separates the fibers and, by virtue of its relative softness and plasticity, prevents the propagation of brittle cracks from fiber to fiber, otherwise they could lead to catastrophic failure; in other words, the matrix phase serves as a barrier to crack propagation; in case a fiber is broken or a fiber is discontinuous, then the matrix material helps redistribute the load in the vicinity of the broken site. The matrix material enhances the properties—particularly the transverse strength of a lamina—and, for that matter, impacts resistance of the final structural product (Agarwal and Broutman 1980; Chawla 2012). Adequate bonding between the matrix and the reinforcement is achieved through proper coupling agents to maximize the stress transmittance from the weak matrix to the strong fibers. Coupling agents play an important role in such bonding at the interface between the matrix resin and reinforcement (Figure 7.1).

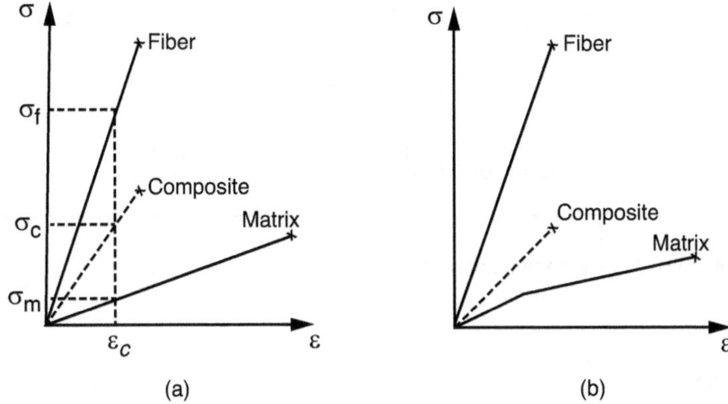

FIGURE 7.1 Stress pattern of the composite material.

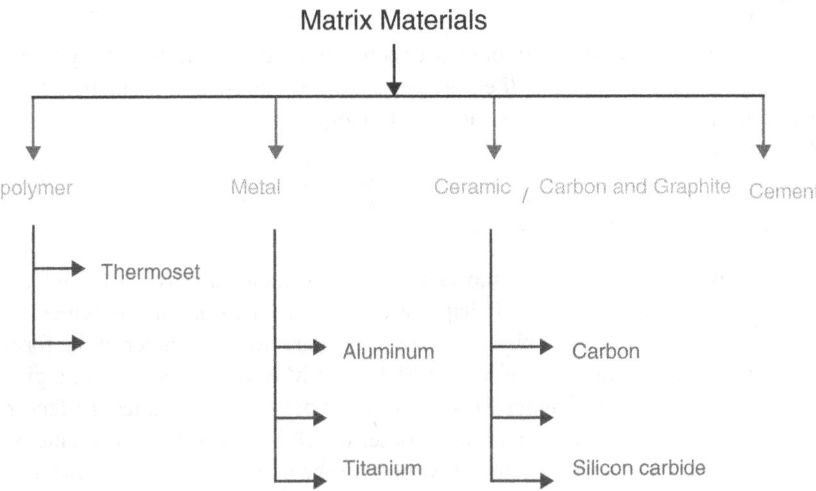

FIGURE 7.2 Different classes of matrix materials.

7.1.3 Classification of Matrices

Matrices are a subclass of materials (https://nptel.ac.in/courses/112/104/112104229/), and they are also generally of four kinds: (i) polymer, (ii) metal, (iii) ceramic, and (iv) cement. In this chapter, all these four types of matrices are summarized. Metal or ceramic matrices are inorganic in nature (Figure 7.2).

7.1.4 Service Temperature for Matrices

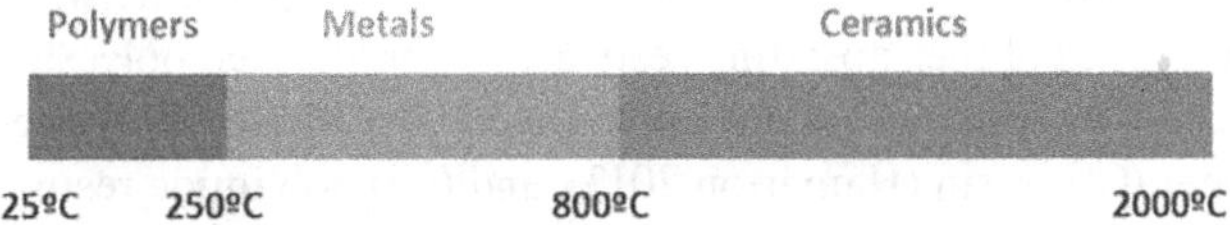

FIGURE 7.3 Temperature range for application of different matrices (https://nptel.ac.in/courses/112/104/112104229/).

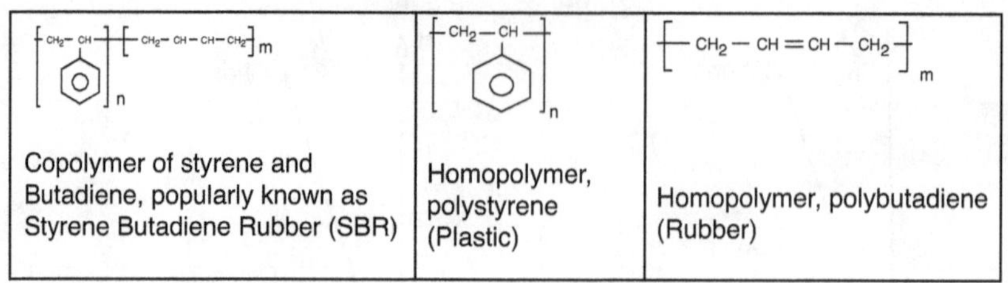

Copolymer of styrene and Butadiene, popularly known as Styrene Butadiene Rubber (SBR)
Homopolymer, polystyrene (Plastic)
Homopolymer, polybutadiene (Rubber)

FIGURE 7.4 Examples of homopolymer and copolymer.

7.2 POLYMERS

Polymers are often interchangeably referred to as resins in the composite industries. Polymers are composed of long chain-like molecules derived from the combination of many "*mers,*" where *mer* is nothing but the simple chemical repeating units known as "monomer." Polymeric matrices are used because they are lightweight and easy to fabricate (Ghosh 1990; Brydson 1999; Ghosh and Mukherjee 1999) (Figure 7.3).

Characteristics of the polymer or copolymer depend on the composition of polymerizing monomers; homopolymers are derived from the polymerization of one type of monomer, whereas copolymers are produced from the copolymerization of monomers with more than one type of monomer (Ghosh 1990; Brydson 1999) (Figure 7.4).

7.2.1 Characteristics of Polymers

Characteristics of polymers depend on the factors like its chemical structure, molecular weight (MW), configuration, conformation, and shape of molecular chain. Even a polymer with a given chemical structure can differ in their physical structural characteristics either in the form of rubber or plastic or fiber. Generally, the order of MW is $MW_{rubber} > MW_{Plastic} > MW_{fiber}$. For a given molecular structure, the polymer with a higher MW exhibits rubbery characteristics. In fact, rubber is a class of polymer whose MW is very high on the order of 10^6 D. The rubber-like state is characterized by vibrational motion of units to effect flexibility in the polymer chain at ambient temperature (Ghosh 1990; Brydson 1999).

Polymer matrices are lighter than other category of matrix materials like metals and ceramics so afford high specific strength. Mechanical characteristics of polymers, for the most part, are highly sensitive to the strain rate and chemical nature of environment. Metals rarely elongate plastically beyond 100%, but highly elastic polymers may elongate more than 1,000% (Balasubramaniam 2009) (Figure 7.5).

7.2.2 Types of Polymer

Polymers are broadly classified into two categories: (i) "thermosetting" polymers and (ii) "thermoplastic" (Ghosh 1990; Brydson 1999; Ghosh and Mukherjee 1999).

7.2.2.1 Thermosets

Thermosets are formed from *in situ* non-reversible reactions to form network structure infusible mass. The important classes of thermosetting resin are (i) unsaturated polyesters (UPEs), (ii) epoxy resin, (iii) vinyl ester (VE) resin, (iv) diallyl phthalate (DAP)-based polyester resin, (v) phenolic resin, (vi) cyanate ester (CE) resin (Hamerton 2013), and (vii) polyimide resin.

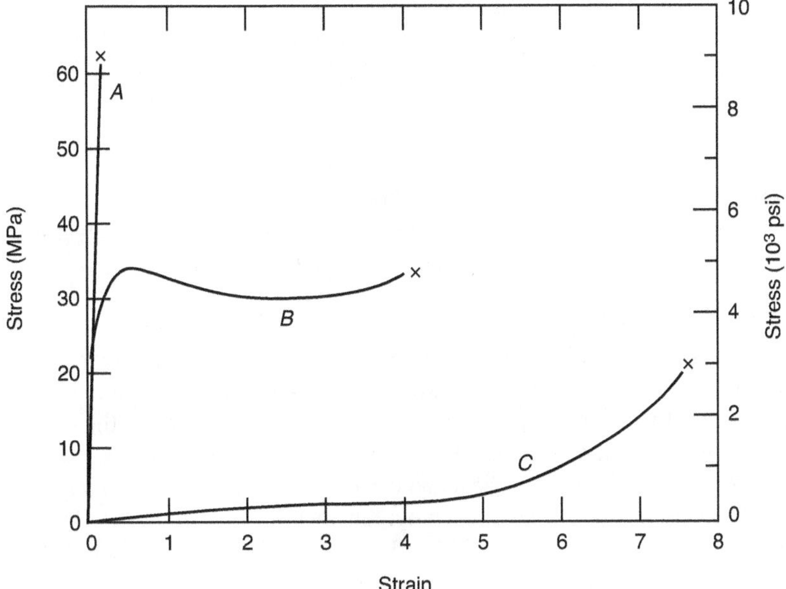

FIGURE 7.5 Typical stress–strain pattern for different types of polymer system (Balasubramaniam 2009).

Thermosets are prevalently used in industry as they are generally easier to process using compression molding, autoclave, extrusion, etc.; and are more chemically resistant; and generally less expensive than thermoplastics, but they are toxic in their uncured state and are susceptible to cracking if over-cured.

7.2.2.2 Thermoplastics

Thermoplastics are like metals; they soften on application of heat and eventually melt reversibly and thus can be processed repeatedly using different molding techniques, such as injection molding, extrusion, blow molding, and rotational molding. Major types of thermoplastics include (i) vinyl polymers and (ii) condensation polymers.

7.2.2.2.1 Major Types of Thermosetting Resin

7.2.2.2.1.1 Polyester Resin UPE matrix resins are also simply referred to as "polyesters" in the industry. Polyester matrix resins are formed by the reaction of a glycol or polyol with a dibasic carboxylic acid. Polyester resins are the most widely used general-purpose resin, particularly suitable for glass fiber, and they can afford to provide a balance of properties in respect of mechanical, chemical, electrical, and dimensional stability; ease of processing; and lower cost. Polyesters can withstand many chemicals, but not alkaline substrates. They can be used in some moderate temperatures (~80°C), but not at high temperatures (Ghosh 1990; Brydson 1999).

7.2.2.2.1.2 Epoxy Resin Epoxy resins are synthesized from the reaction a dihydric phenol like bisphenol-A with excess epichlorohydrin in an alkaline medium (Clayton May 1987; May 1979). Epoxy resins are used primarily for the fabrication of high-performance composites with superior mechanical properties. Characteristically, they exhibit low-cure shrinkage on curing, have low internal stress, and maintain better dimensional stability and adhesion properties to substrates. Epoxies are resistant to many corrosive liquids and environments. High-performance epoxy matrices are generally achieved by multifunctional epoxies cured with multifunctional amines or organic acids or anhydrides. Epoxies are compatible with most composite manufacturing processes, particularly vacuum bag molding, autoclave molding, pressure bag molding, compression molding, and hand layup. Epoxy resins are widely used in filament-wound composites and are suitable for molding prepregs (Mukherjee 2012).

The large family of epoxy resins represents some of the highest performance resins to ensure balanced properties in the composite product. Prepregs of epoxies have limited shelf life (Clayton May 1987; Mukherjee 2012). The cure time of epoxy systems is often relatively longer depending on curing temperature. Epoxies do not show excellence in UV resistance.

7.2.2.2.1.3 DAP Resin: An Improved Version of UPEs Most UPEs contain styrene or vinyl monomer, causing environmental problems, and their curing leads to very high shrinkage, which may act adversely on the adhesion between resin and reinforcements with allied defects. Thus, an improved version of polyesters is evolved by using DAP in place of styrene in the UPE formulations, which enormously enhances the mechanical properties of the composite (May 1979).

Its products are used in critical electrical and electronic applications with high reliability under long-term adverse environmental conditions, providing an excellent balance of physical and chemical properties. It causes less pollution than UPEs. Prepreg handleability is excellent, and it has good tack and drapability.

7.2.2.2.1.4 VE Resins VE is derived from the esterification of an epoxy resin and an unsaturated monocarboxylic acid like acrylic and/or methacrylic acids, as shown in Figure 7.8 (Benmokrane et al. 2017). It may be considered as a polyester resin strengthened with epoxy molecules in the backbone of the molecular chain; its cure is accomplished by vinyl polymerization in the presence or absence of other vinyl monomers like styrene as in UPEs. Characteristically, VE resins stand between polyester and epoxy resins (Figure 7.6).

7.2.2.2.1.5 CE Resin CE resins are relatively a new generation of thermosetting resin, in which its cyanate group is the reactive group. It has a wide scope of applications right from the domain of microelectronics to aerospace composites (Hamerton 2013).

The cyanate monomer is prepared in the industry from the reaction of halogenated cyanide with phenolic compounds in the alkaline condition (Hamerton 2013).

$$\mathrm{ArOH + XCN = ArOCN + HX}$$

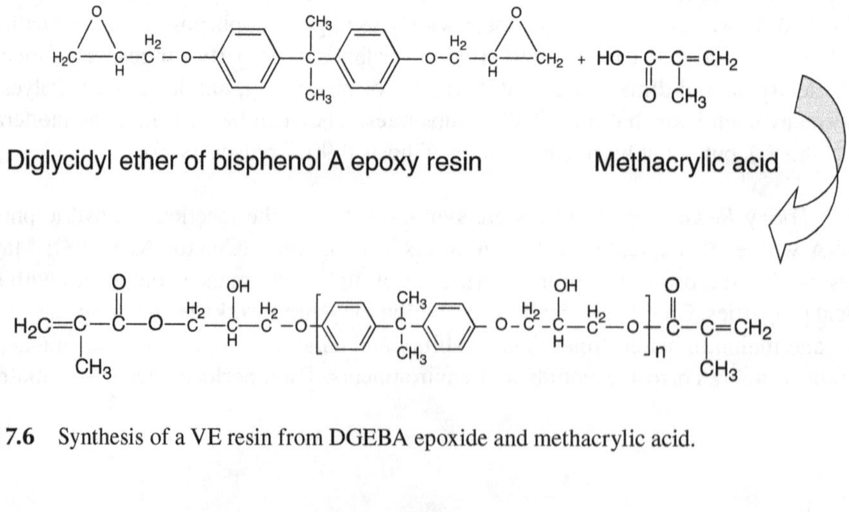

Diglycidyl ether of bisphenol A epoxy resin

Methacrylic acid

FIGURE 7.6 Synthesis of a VE resin from DGEBA epoxide and methacrylic acid.

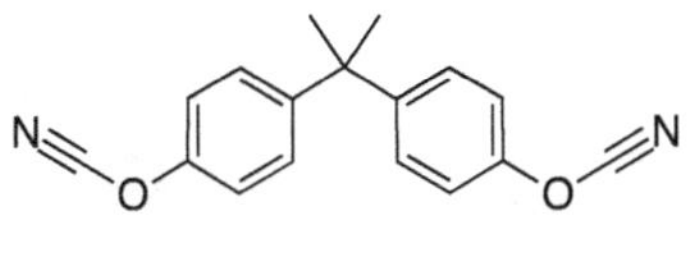

FIGURE 7.7 Typical CE monomer. CE, cyanate ester.

where "X" can be Cl, Br, or I, and "ArOH" can be single phenols, polyphenols, or aliphatic hydroxy compounds (Figure 7.7).

CEs constitute a high-temperature resin family traditionally associated with space applications because of their good thermal stability, low dielectric constant, and extremely low moisture uptake compared with other resins of their class. CE is known for its built-in toughness, micro-crack resistance, and ease of processing.

7.2.2.2.1.6 Polyimide Resins Polyimide resins are a class of high-temperature resins and are often known as *polymerization of monomer reactant* (PMR) resin, which is preferred over epoxy resin for the fabrication of composite materials (May 1979).

In the PMR approach, the reinforcing fibers are impregnated with a solution containing a mixture of monomers dissolved in a low boiling point alkyl alcohol solvent. Such monomers remain dormant at room temperature, but become reactive *in situ* at elevated temperatures to form a thermo-oxidatively stable polyimide matrix. These addition-type polyimides are highly processable and can be processed by compression or by autoclave (May 1979; Ghosh 1996) molding technique to realize high-temperature polymer matrix composites.

During composite fabrication, volatilization of the solvent and condensation by-products results in high void content in the composites to cause inferior mechanical properties and thermo-oxidative stability. To overcome the problem of volatilization, polyimides were developed through addition reaction where low-MW amide–acid prepolymers are synthesized whose chain terminals were end-capped with norbornenyl groups. Such norbornenyl groups undergo addition polymerization at elevated temperatures (275°C–350°C) without the evolution of volatiles, making it possible to reduce the void content in the composites.

7.2.2.2.1.6.1 Polybismaleimides or Bismaleimides

Polybismaleimides are high-performance thermosetting addition-type polyimides, and they are known for their high strength and high temperature performance, which lie between lower temperature-resistant epoxy systems and very high temperature-resistant polyimides (Ghosh 1996).

7.2.2.2.1.7 Phenolic Resin This family of resin is produced from the condensation reaction between phenol(s) and formaldehydes in an acidic or basic condition depending on the type of resin to be prepared. Broadly, there are two types of phenolic resins, namely, Resol and novolac, as generated from basic or acidic composition, respectively (Brydson 1999; Ghosh and Mukherjee 1999). An important use of phenolic resin is for the manufacture of carbon–carbon composite material (https://nptel.ac.in/courses/112/104/112104229/).

7.2.2.2.2 Major Types of Thermoplastic Resin

Thermoplastic resins are generally recognized as ductile, tough polymer systems compared to thermosets but have only recently been utilized with graphite fibers for advanced composite applications. The mechanical behaviors of high-performance thermoplastic resin (e.g., Polyether ether ketone (PEEK), polysulfone, Polyphenylene sulphide (PPS) and polyphenylene sulfide) and their composites can manifest nonlinear curves with very high strain to failure. Such characteristic translates into higher transverse strain to failure with improved delamination resistance, improved impact strength, and toughness (Gardner et al. 1994).

Polymer	Repeat unit	K/E*	T_g (°C)	T_m (°C)
Poly(ether ether ketone)	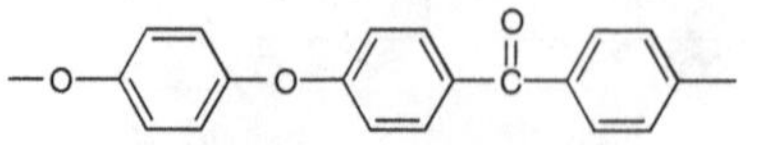	0.5	142[b]	340[b]
Poly(ether ketone)	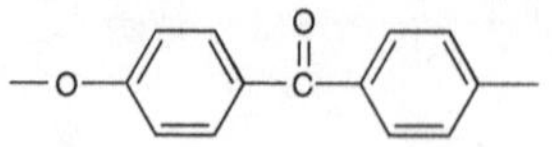	1.0	156[b]	370[b]
Poly(ether ketone ketone)	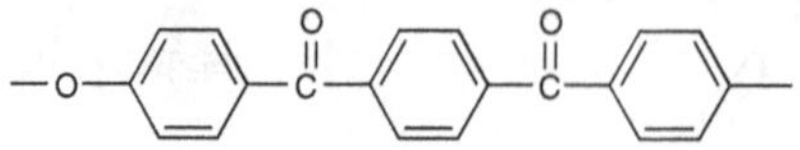	2.0	155	358

FIGURE 7.8 Different ketone-containing repeat units.

7.2.2.2.2.1 Polyaryl Ether Ketone Polyarylether ketone is an important class of thermoplastic material that came to the market in the early 1980s. The three most common of this family of resin are (i) polyether ketone and (ii) common polyaryl ether ketones (Figure 7.8).

K/E*, that is, ketone/ether ratio (Gardner et al. 1994).

PEEK is a biocompatible, semicrystalline thermoplastic with excellent mechanical and chemical resistance properties. PEEK is a high-temperature (up to 260°C) engineering thermoplastic having m.p. of 335°C and T_g of 143°C, which are excellent for applications where thermal, chemical, and combustion properties are critical to performance. PEEK emits little smoke or toxic gas when exposed to flame. This material is tough, strong, and rigid and has superior creep resistance and also resists radiation. With its resistance to hydrolysis, PEEK can withstand boiling water and superheated steam used with an autoclave and sterilization equipment at temperatures higher than 250°C. Carbon-reinforced PEEK provides excellent wear capabilities. PEEK HPV grade offers outstanding bearing performance. Typical applications are in the areas of marine, nuclear, oil well, automotive, and aerospace industries.

7.2.2.2.3 Other Major Types of Thermoplastics

Major types of engineering thermoplastics include (i) addition polymers: polyolefins, polyvinyl chloride, polymethyl methacrylate (PMMA) etc.; (ii) condensation polymers: nylon, PPS, Poly(p-phenylene oxide) (PPO) and (v) polycarbonate, etc. (Brydson 1999).

7.2.3 Rubber as Matrix Material

Elastomer is generally a subclass of polymer that displays rubber-like elasticity. Rubber or elastomers are traditionally used in industrial components because of their special characteristics like flexibility toughness, high resilience, impact resistant, shock absorbing ability, relatively lighter, amenable to processing, and moldability (Ghosh 1990; Sothern 1983).

Reinforced rubber products are one of the largest classes of composite materials to provide certain important essential engineering properties for aviation and automobile pneumatic tires of various kinds for vehicular transport items, marine components, hose pipes, conveyor and power transmission belts, and skirts for air cushion vehicle or hovercraft (Figure 7.9).

Rubber-based engineering items are made of various natural and synthetic rubbers reinforced with different reinforcing materials, such as carbon black, fibers, and other reinforcing fillers. Raw rubber is not useful unless it is compounded and cured or crosslinked appropriately in the presence of different additives and curing agents (Ghosh 1990; Sothern 1983).

7.2.3.1 Types of Rubber

There are two types of rubber, namely, (i) natural rubber (NR) and (ii) synthetic rubber (Sothern 1983). The most commonly used rubber is NR, which is chemically cis-polyisoprene derived from

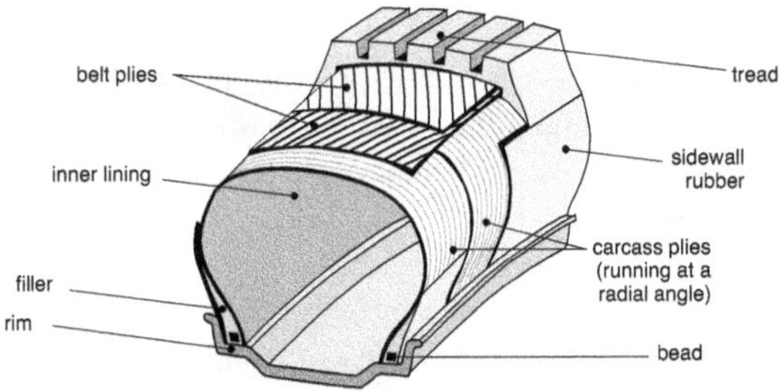

FIGURE 7.9 Components of a typical tire. (Courtesy: Winspear 1968.)

TABLE 7.1
Typical Chemicals Used for the Synthesis PMR Resins

Chemical Structure	Name of Chemicals	Acronym
MeO-C(=O), CF_3, C(=O)-OMe; HO-C(=O), CF_3, C(=O)-OH	Dimethyl ester of 3,3′, 4,4′- (benzphenonetetracarboxylic acid)	BTDE
H_2N–⟨benzene ring⟩–NH_2	4,4′-Methylenedianiline	MDA

PMR, polymerization of monomer reactant.

the natural resource, *Hevea brasiliensis*. All other rubbers are synthetic rubber. A common example of the compatible blend matrix material is the blend of a plastic like polyvinyl chloride and a rubber like acrylonitrile butadiene rubber (nitrile butadiene rubber) (Sen and Mukherjee 1993). Features of some important rubbers are given below:

7.2.3.1.1 Natural Rubber

NR is a very useful matrix material because of its high resilience and a low hysteresis, less heat buildup during dynamic conditions.

7.2.3.1.2 Styrene Butadiene Rubber

Styrene butadiene rubber (SBR) is a synthetic rubber obtained from the copolymerization of styrene and butadiene (Table 7.1). Its properties are similar to those of NR. SBR has better heat and abrasion resistance, and hardens on aging, whereas NR softens on aging.

7.2.3.1.3 Ethylene Propylene and Ethylene Propylene Diene Rubber

These synthetic elastomers are synthesized by the copolymerization of ethylene and propylene in the absence or presence of a diene monomer like ethylidene norbornene to produce ethylene propylene or ethylene propylene diene rubber, respectively. Its dynamic and mechanical properties are generally in between NR and SBR. It is resistant to steam, oxygenated solvents, and weather, however relatively less resistant to petroleum, hydrocarbon, and oils (Sothern 1983).

7.2.3.1.4 Nitrile Rubber

Nitrile butadiene rubber is the copolymer of acrylonitrile and butadiene (Sothern 1983). It is resistant to oils and fuels so suitable for use in sealants and hose pipes in the oil field, and in the automotive and aeronautical industry. The stability of nitrile butadiene rubbers in the temperature range of −40°C to 108°C makes them an ideal material for aeronautical applications.

7.2.3.1.5 *Polychloroprene or Neoprene Rubber*

Polychloroprene or neoprene rubber is a synthetic polymer produced by polymerization of chloroprene, and its products are very pliable, more resistant to oil and chemicals than NR or other synthetic rubbers, retain flexibility over a wide temperature range, and have good electrical insulation characteristics (Sothern G.R 1983).

7.2.3.1.6 *Butyl Rubber*

Butyl rubber is produced by copolymerization of 98% of isobutylene with a smaller amount of about 2% of isoprene such that the isobutylene part can provide for a highly saturated backbone, whereas the isoprene component provides a crosslinking site. It has very low permeability to air and gases (Sothern 1983) and is used for design of automobile tire liner and inner tubes.

7.2.3.1.7 *Polybutadiene Rubber*

Polybutadiene (PB) rubber is synthesized from the polymerization of 1,3 butadiene (Gardner et al. 1994). Characteristically, PB is relatively resistant to wear and is used especially in the construction of tire; the other major use is in the manufacture of toughened plastics like acrylonitrile butadiene styrene plastics and toughened polystyrene.

7.2.4 Toughened Matrix Resin

Incorporation of rubber particles can improve the toughness in the matrix and can enhance the matrix-dominated properties including the transverse tensile strength and interlaminar shear strength. A brittle matrix can be toughened by incorporating rubber particles like hydroxy-terminated PB, carboxy-terminated PB (Sothern 1983; Brydson 1999).

7.3 METAL MATRIX MATERIALS

Metal matrix composites (MMCs) are a class of materials like metals, alloys, or intermetallic compounds which are incorporated with various reinforcing phases, such as particulates, whiskers, or continuous fibers. Matrices in MMCs are ductile metals; superalloys and alloys of lighter metals like Al, Mg, Ti, and Cu are commonly used matrix materials (Balasubramaniam 2009; Kainer 2006; Malaki et al. 2019). The matrix is usually a low-density metal alloy. One of the most important parameters of a metal for the development of MMCs is the density and is so important; oftentimes MMC is taken as an equivalent to *light MMCs*. In structural applications, the matrix is normally a lighter metal such as Al, Mg, or Ti and provides a compliant support for the reinforcement, and in high-temperature applications, Co and Co/Ni alloy matrices are common.

7.3.1 Mechanical Features

For most metallic materials, elastic deformation persists only to strains of about 0.005. As the material is deformed beyond this point, the stress is no longer proportional to strain (Hook's law ceases to be valid), but permanent, nonrecoverable, or plastic deformation occurs (Figure 7.10).

The transition from elastic to plastic is a gradual one for most metals. From the atomic perspective, plastic deformation corresponds to the breaking of bonds with original atom neighbors and then reforming bonds with new neighbors as a large number of atoms or molecules move relative to one another; upon the removal of the stress, they do not return to their original positions. The mechanism of the deformation is different for crystalline and amorphous materials. For crystalline solids, deformation is accomplished by means of a process called slip, which involves the motion of dislocations. Plastic deformation in noncrystalline solids occurs by a viscous flow mechanism.

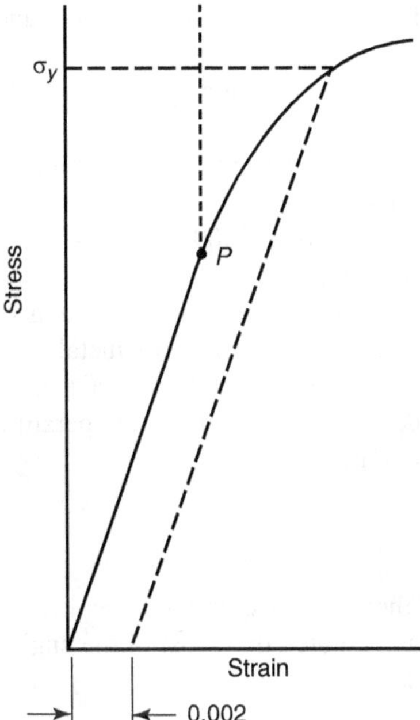

FIGURE 7.10 Schematic of tensile stress–strain behavior in the plastic region for a typical metal (Balasubramaniam 2009).

7.3.2 Importance of MMCs

MMCs have great potential to replace monolithic metals in many engineering applications because of their enhanced properties, for example, higher strength and stiffness, better wear resistance, or higher operating temperature. However, the application of MMCs has been limited primarily due to their high cost, complexity involved in the synthesis process, relative low fracture toughness, and reliability despite their attractive mechanical properties.

However, the synthesis of discontinuous reinforcement-incorporated MMCs was successfully attempted using a range of processing techniques (Kainer 2006; Malaki et al. 2019). Incorporation of nano-scale reinforcements primarily selected from oxide, carbide, boride, and nitride families led to promising results in improving a wide range of properties pertaining to metallic matrices. With the advent of carbon nanotubes, graphene, buckyballs, and carbon nanoplatelets were also critically investigated in both MMCs and polymer composites. Some of the properties that can be enhanced by using nano-scale reinforcements in the development of MMCs are tensile and compressive behavior, ductility, or elongation to failure, a must for "bend than break" design philosophy; high-temperature mechanical properties; creep; dynamic mechanical properties; wear resistance, including scratch resistance; coefficient of thermal expansion; damping; machining; ignition resistance; dry/wet corrosion resistance.

MMCs are lightweight structural materials used in a specific number of aircraft, helicopters, and spacecraft. The MMC material comprises hard reinforcing particles embedded within a metal matrix phase.

7.3.3 Microstructure

Development of high-performance MMCs requires a careful microstructure design, which can improve the fracture toughness of the material while maintaining high strength. The microstructure influences the fracture toughness of MMCs (Kainer 2006; Malaki et al. 2019), and their effects on the interplay between plastic deformation and crack formation, and their effects on the competing failure mechanisms are known.

Hybrid MMCs are advanced materials used for lightweight high strength purpose in automobile and aerospace sectors.

The attractive feature of the MMCs is their higher temperature use; the aluminum matrix composite can be used in the temperature range above 300°C, while the titanium matrix composites can be used above 800°C (Balasubramaniam 2009; Kainer 2006; Malaki et al. 2019; Li, Cao, and Williams 2019). Metal matrices for composite materials have an intermediate working temperature range between 200°C and 800°C. Metal matrices provide high mechanical properties like strength, stiffness, ductility, and electrical conductivity. Alloys are metallic substances made by fusing two or more metals, or a metal and sometimes a nonmetal, to obtain desirable qualities such as hardness, lightness, and strength and for designing an extended temperature range for a given application. Brass, bronze, and steel are some of the examples of alloys.

7.3.4 General Information about Metals

Among the metals, only a few of them have acquired practical value so far; their utility as engineering materials is concerned. It may be noted that most of the engineering materials are alloys, and only a few are used as pure metals.

7.3.4.1 Types of Metals

Metals may be categorized into ferrous materials and non-ferrous materials (Kainer et al. 2006). The largest use in the industry perhaps is of ferrous metals and alloys.

Non-ferrous materials include all other metals; among them, the important practical materials are copper (Cu), nickel (Ni), aluminum (Al), titanium (Ti), zinc (Zn), tin (Sn), and lead (Pb) and their alloys.

There is another category of metals known as special metals which include chromium (Cr), manganese (Mn), tungsten (W), antimony (Sb), bismuth (Bi), and boron (B) and are mostly used as alloying elements for matrix development. However, other metals like zirconium (Zr), (Ta), silver (Ag), and gold (Au) and their alloys are used for very specific applications as they are too expensive (Balasubramaniam 2009).

7.3.4.1.1 Non-Ferrous Materials

7.3.4.1.1.1 Aluminum and Aluminum Alloys The density of aluminum is 2.70 g/cm^3 and Poisson's ratio is 0.35. Among the non-ferrous metals, it is soft and has relatively low melting point (660°C) and low strength. The major characteristic features of aluminum are that it is much lighter than iron, about one-third of the density of that of iron, good resistance to corrosion, good electrical conductivity, relatively less costly, and is used as a matrix material for production of MMC. Aluminum and its alloys are extensively used in the aerospace industry because of its ability to develop considerable strength by suitable alloying and in some cases by subsequent heat treatment for annealing (Balasubramaniam 2009; Kainer 2006; Malaki et al. 2019; Li, Cao, and Williams 2019).

7.3.4.1.1.2 Copper and Copper Alloys The density of copper is 8.96 g/cm^{3}and Poisson's ratio is 0.34. Copper is the second most important non-ferrous metal after aluminum. Copper and its alloys are highly ductile and can be easily cast and processed to produce various engineering components; the largest use of copper alloys in process industries is in tubing and piping, particularly where resistance to certain specific corrosive media is required; the preferred method of using copper and its alloys is to use as lining over the steel base for fabrication of pressure vessels, tanks, etc.

7.3.4.1.1.3 Nickel and Nickel Alloys The density of nickel is 8.908 kg/m^3 and Poisson's ratio is 0.31. Nickel is costlier than aluminum and copper. Selection of nickel and its alloys is specially made where resistance to the highly corrosive environment is required (Balasubramaniam 2009).

7.3.4.1.1.4 Lead and Lead Alloys The density of lead is 11.34 kg/m^3 and Poisson's ratio is 0.44. Lead and its alloys as engineering material have limited but important uses; its alloys are used for soldering (Pb–Sn, Pb–Sn–Sb) and bearings (Pb–Sn–Sb, Cu–Pb, Cu–Sn–Pb). Lead-based matrices are used in the service condition of the corrosive environment, specifically when it has to face an environment involving dilute sulfuric acid (Balasubramaniam 2009). Lead has a relatively low melting point (327°C), but by alloying, its normal strength and creep can be increased and are often utilized as a lining material.

7.3.4.1.1.5 Titanium and Titanium Alloys The density of Ti is 4.506 g/cm^3 and Poisson's ratio is 0.32. Titanium is approximately 50% lighter metal than iron and thus has the advantage to provide higher specific strength (Balasubramaniam 2009). Ti6 Al–4V is a widely used alloy, where strength and toughness properties are important requirement. In the process industry, unalloyed titanium (i.e., commercially or chemically pure titanium) is commonly used. Titanium has excellent corrosion resistant properties against a large variety of environments; however, they are selected where high strength is not required. It is a costly metal; thus, the use is limited to exchanger tubes involving sea water as coolant and some specific corrosive chemicals. The variations in mechanical properties are dependent on interstitial solid solution (oxygen, hydrogen, nitrogen) and impurity (iron) levels.

7.3.4.1.1.6 Other Non-Ferrous Metals The other non-ferrous metals have either limited use under special conditions (magnesium, tantalum, zirconium, cobalt, etc.) or no utility in process industries (silver, zinc, tin, etc.).

7.4 CERAMIC MATRICES FOR CERAMIC MATRIX COMPOSITES

7.4.1 What Are the Ceramic Matrix Materials?

Ceramics are inorganic and non-metallic materials. Their main constituents are generally silica (SiO_2), alumina (Al_2O_3), and other inorganic non-metallic substances (Balasubramaniam 2009). Ceramic material comprises all kinds of constituents in the sense that they are a combination of a metal (e.g., aluminum) or intermediate metal (e.g., silicon) with a non-metal (i.e., oxygen, nitrogen, or carbon). Ceramics are also used as a matrix material to develop ceramic matrix composite materials (CMCs) to obtain desired engineering properties. Although ceramics are generally insulators, they can also be formed to serve as electrically conductive materials; however, some ceramics, like superconductors, also display magnetic properties (Sciti et al. 2018; Trento 2019).

Ceramics and CMCs are often used interchangeably because ceramics generally need modification by incorporating another ceramic of its own or different to develop CMCs. Generally, CMC comprises ceramic fibers embedded in a ceramic matrix.

7.4.2 Preparation of Ceramics

Ceramic materials if processed properly can create useful products of high-tech and low-tech areas and find different uses for ceramic products in everyday life. Ceramics are generally made by taking mixtures of earthen elements, clay, powders, and water and shaping them into desired forms; once it is shaped, it is fired in a high-temperature oven kiln. Depending on their processing procedural method of formation; ceramics can provide a wide range of products with controlled density from lightweight to dense products (Balasubramaniam 2009).

7.4.3 Characteristics of Ceramics

Ceramics exhibit excellent strength and hardness properties and provide a high elastic modulus yet low density; however, the disadvantages of ceramic materials are their relatively lower tensile strength and brittleness, causing susceptibility to failure. Ceramic materials are somewhat limited in applicability by their mechanical properties and, however, in many respects, are inferior to those of metals. The principal drawback is a disposition to catastrophic fracture with very little energy absorption. Both crystalline and noncrystalline ceramics almost always fracture without any plastic deformation in response to a tensile load (Balasubramaniam 2009; Sciti et al. 2018; Trento 2019).

7.4.4 Toughened Ceramics

The brittleness of ceramics can be reduced by using a reinforcing material. For example, Al_2O_3/Al is a ceramic composite, where a ductile material like aluminum serves as the reinforcing phase. Carbon, silicon carbide, and silicon nitride are ceramics and used as matrix materials.

By taking advantage of the inherent high strength and Young's modulus of the ceramic matrix, it has been possible by incorporating reinforcements in the matrix to enhance the fracture toughness of the CMC.

The measure of a ceramic material's ability to resist fracture is specified in terms of fracture toughness (Balasubramaniam 2009). The plane strain fracture toughness K_{Ic} is defined according to the expression

$$K_{Ic} = Ys\sqrt{\pi a} \tag{7.1}$$

where Y is a dimensionless parameter or function that depends on both specimen and crack geometries, σ is the applied stress, and a is the length of a surface crack or half of the length of an internal crack (Balasubramaniam 2009). Crack propagation will not occur as long as the right-hand side of Eq. (7.1) is less than the plane strain fracture toughness of the material. Plane strain fracture toughness values for ceramic materials are smaller than those for metals: typically, they are below 10 MPa $\sqrt{m}$. However, under some circumstances, the fracture of ceramic materials will occur by the slow propagation of cracks, when stresses are static in nature, and the right-hand side of Eq. (7.1) is less than K_{Ic}. This phenomenon is called *static fatigue* or *delayed fracture*. Fracture toughness values for ceramic materials are low and typically lie between 1 and 5 MPa $\sqrt{m}$. By contrast, K_{Ic} values for most metals are much higher (15 to greater than 150 MPa $\sqrt{m}$).

7.4.4.1 Features of Fracture

The brittle fracture process consists of the formation and propagation of cracks through the cross section of the material in a direction perpendicular to the applied load. Crack growth in crystalline ceramics may be either transgranular (i.e., through grains) or intergranular (i.e., along grain boundaries); for transgranular fracture, cracks propagate along specific crystallographic (or cleavage) planes, planes of high atomic density (Balasubramaniam 2009).

The fracture toughness of ceramics have been improved significantly by the development of a new generation of ceramic-matrix composites (CMCs)—particulates, fibers, or whiskers of one ceramic material that have been embedded into a matrix of another ceramic. CMCs have extended fracture toughness between 6 and 20 MPa $\sqrt{m}$. Crack initiation normally occurs with the matrix phase, whereas crack propagation is impeded by the reinforcement phase.

7.4.5 Thermal Features

The bonding in ceramics may be either ionic or covalent, and generally, they are hard, strong with high compressive strength, oxidation-resistant, and corrosion-resistant and can withstand even at very high temperature ranges even above 2,000°C. The thermal conductivity of ceramic materials plays an important role in desired applications as it can be enhanced by some suitable methods to facilitate heat conduction, convection, and heat radiation so as to further expand its scope of applications in desired fields. Ceramic materials with high thermal conductivity are mainly composed of nitrides, carbides, oxides, and borides, such as polycrystalline diamond ceramics, aluminum nitride, silicon nitride, silicon carbide, and beryllium oxide (Balasubramaniam 2009; Trento 2019).

7.4.6 Different Types of Ceramics

In general, the nomenclature of a CMC is made in accordance with the type of reinforcement and type of the matrix selected (i.e., *type of fiber/type of matrix*). Illustratively, *C/C* stands for carbon fiber-reinforced carbon (CFRC) ceramic composite, or for that matter, carbon fiber-reinforced silicon carbide (*C/SiC*) ceramic composite.

Ceramic fiber reinforcement increases the initial resistance to crack propagation and ensures the CMC to avoid catastrophic failure as generally observed characteristically in the case of monolithic ceramics. More importantly, such behavior is distinct from that of ceramic fibers in MMCs and in polymer matrix composites, where the fibers typically fracture first before the matrix starts to fail as a result of higher failure strain capabilities of these matrices (Balasubramaniam 2009).

One interesting point for the ceramic system is that by incorporation of a ceramic reinforcement material to a ceramic matrix, a ceramic composite material can be generated. For example, special silicon carbide (SiC), alumina (Al_2O_3), mullite (Al_2O_3–SiO_2), and carbon fibers are the most commonly used ceramic material for CMCs, where the matrix materials are usually the same that is, SiC, Al_2O_3, Al_2O_3–SiO_2, and carbon. A separate class of ceramics is ultrahigh-temperature ceramics; and this ceramic particle or fiber reinforces the ceramic matrix to produce ultrahigh-temperature ceramic composites (Sciti et al. 2018; Trento 2019; Mungiguerra et al. 2018; Krenkel Walter 2021; Kochendorfer Richard 2009).

Silicon carbide has a dual role depending on how it is applied as a matrix or as a reinforcing fiber, for example, carbon fiber-reinforced silicon carbide *matrix* composites (C/Si) and silicon carbide fiber-reinforced silicon carbide *matrix* composites (SiC/SiC).Interestingly, silicon carbide is a hard material with a low erosion, and it forms a silica glass layer during oxidation, which prevents further oxidation of the inner material (Figure 7.11).

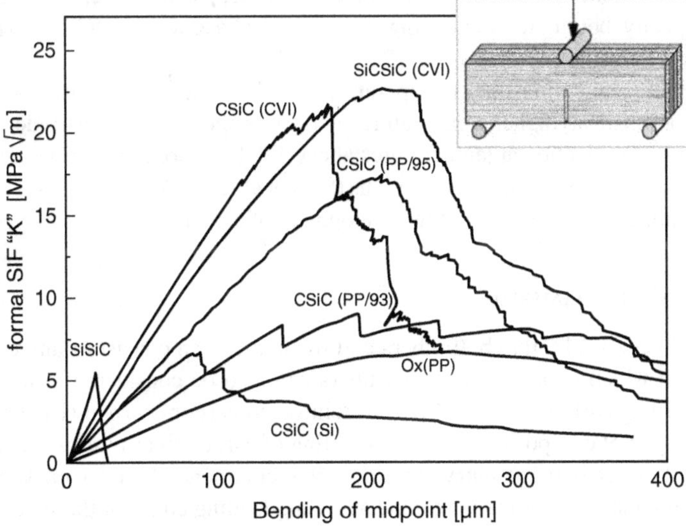

FIGURE 7.11 Single-edge notch bend tests for different CMCs. (Courtesy: Trefilov 1995.)

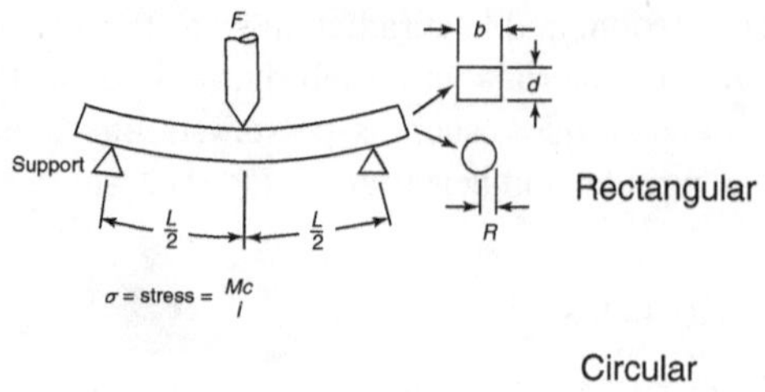

FIGURE 7.12 Three-point loading testing scheme.

7.4.7 Flexural Load in Ceramics

Ceramics are rather brittle, so it is difficult to grip such materials without being fractured. In fact, ceramics fail only after about 0.1% strain; thus, their tensile specimens should be perfectly aligned to avoid the bending stresses. Thus, a more suitable transverse bending test is generally used, in which a circular or rectangular cross-sectional specimen is bent until fracture using a three- or four-point loading method (Figure 7.12).

It represents a typical three-point bending for ceramic materials, where M = maximum bending moment; c = the distance from the center of the specimen to outer fibers; I = moment of inertia; F = applied load; stress $\sigma = 3FL/2bd^2$ (rectangular), and $\sigma = FL/\pi R^3$ (circular).

At the point of loading, the top surface of the specimen experiences compression, while the bottom surface remains in tension. Since the tensile strengths of ceramics are about one-tenth of their compressive strengths, the fracture has to occur on the tensile specimen face; thus, the flexure test is a reasonable substitute for the tensile test.

7.4.8 Applications

Ceramic materials are inherently resilient to oxidation and deterioration at elevated temperatures, especially for components in automobile and aircraft gas turbine engines. CMCs are innovated and widely used in the aerospace sector for the development of gas turbines and structural re-entry thermal protection, and in the energy sector for heat exchangers and fusion reactor walls. Development of CMC-based heat shield systems promises the advantages; for instance, reduced weight, higher load-carrying capacity, better steering during the re-entry phase with CMC flap systems, and reusability for several re-entries.

However, for such applications, high temperatures preclude the use of oxide fiber CMCs because of the creeping problem at higher temperature, and amorphous silicon carbide fibers lose their strength due to re-crystallization at temperatures above 1,250°C. So, carbon fibers in the silicon carbide matrix (C/SiC) are used for these applications (Sciti et al. 2018; Mungiguerra et al. 2018; Krenkel Walter 2021). Ceramics are also used in the development of armor against high -speed projectiles.

7.4.9 Carbon Matrix Material

It can be seen in Section 7.4.6 that both fibers and the matrix can consist of any ceramic material, in accordance with which carbon and carbon fibers can also be considered as a ceramic material of special kind. CFRC, carbon–carbon (C/C), or reinforced carbon–carbon (Kochendorfer Richard 2009) is a special type of composite material consisting of carbon fiber reinforcement in a matrix of graphite developed for use in the re-entry vehicles of intercontinental ballistic missiles, which is most widely known as the material for the nose cone and wing leading edges of the space shuttle orbiter.

C/C composites are employed in rocket motors, as friction materials in aircraft and high performance automobiles, for hot pressing molds, in components for advanced turbine engines, and as ablative shields for re-entry vehicles. However, C/C composites are very expensive due to their relatively complex processing techniques. They are prepared from the 2D or 3D carbon fiber structure impregnated with resins like phenolic resin, then molded, and pyrolyzed in an inert atmosphere. It is then heat-treated at high temperature for densification for strengthening.

The CFRC matrix (C/C) can be used up to 3,000°C because carbon is the element with the highest melting point; however, C/C is an ablative material that dissipates energy through self-sacrificing. C/C is well-suited to structural applications at high temperatures, or where thermal shock resistance and/or a low coefficient of thermal expansion is needed. Although it is less brittle than many other ceramics, it lacks impact resistance.

7.5 CEMENT AS MATRIX MATERIAL

7.5.1 What Is Cement?

Cement is a kind of binder used for construction that sets, hardens, and adheres to other materials like reinforcement to bind them together. As such, cement is seldom used on its own but rather to bind sand and gravel (aggregate) together (Balasubramaniam 2009). Concrete is a common large-particle composite in which both matrix and dispersed phases are ceramic materials. Since the terms "concrete" and "cement" are sometimes incorrectly interchanged, perhaps it is appropriate to make a distinction between them. Broadly speaking, concrete implies a composite material comprising aggregate particles that are bound together in a solid matrix binding medium as cement. The terms cement and concrete often are used interchangeably; however, in fact, cement is actually an ingredient of concrete. Concrete is a mixture of aggregates and paste. Generally, such aggregates are sand and gravel or crushed stone, and the paste is a combined mixture of water and Portland cement, where water takes part in the hydration process by which cement and water harden and bind the aggregates to a rock-like mass. Generally, cement constitutes around 10%–15% of the concrete mix by volume.

Organic polymers are sometimes used as cements in concrete. Polymer cements (Kumar and Monteiro 2014) are made from organic chemicals that polymerize and ensure water proofing with tensile strength.

The two most common concretes are those made with Portland cement and asphaltic cements, where one of the matrix binders is inorganic and the other is organic in nature, and the reinforcements are in the form of aggregates of gravel and sand. Asphaltic concrete is widely used primarily as a paving material; on the other hand, Portland cement concrete is employed mostly as a structural building material and, of late, also used to construct metal roads. Asphalt cement requires a petroleum-based binder (Balasubramaniam 2009).

7.5.2 Types of Cements

Cement materials are classified into two distinct categories: (i) hydraulic cements and (ii) nonhydraulic cements in accordance with their settings and hardening mechanisms (Figovsky and Dmitry 2013). The hydraulic cement setting and hardening involve hydration reactions and therefore require water, while non-hydraulic cements only react with gas and can directly set under air.

7.5.2.1 Hydraulic Cement

Cements are generally hydraulic cement, and they harden by hydration of the clinker minerals (limestone and alumina silicate clay-based materials during the cement kiln stage) on water addition. Hydraulic cements or Portland cement is made of a mixture of four major mineral phases of the clinker such as alite silica-based ($3CaO{\cdot}SiO_2$); belite silica-based ($2CaO{\cdot}SiO_2$), Celite- or alumina-based tricalcium aluminate ($3CaO{\cdot}Al_2O_3$), and alumina- and iron oxide-based brown millerite ($4CaO{\cdot}Al_2O_3{\cdot}Fe_2O_3$). Silicates are responsible for the mechanical properties of cement.

7.5.2.2 Nonhydraulic Cement

Another kind of cement, even though less known, is *non-hydraulic cement* such as slaked lime that is calcium oxide mixed with water, which gets hardened by carbonation in the presence of the atmospheric carbon dioxide.

7.5.3 Polymer Concretes

Polymer concretes are some kind of particulate polymer composite; they are a type of concrete in which polymers act as binders to replace the lime-type cements. However, if polymer is used in addition to Portland cement in the composition, it is known as polymer cement concrete or polymer-modified concrete. Polymers in concrete are actually overseen by the American Concrete Institute since 1971.

Polymer concrete may be used for new construction or repairing of old concrete (Do Suh 2008; Kumar and Monteiro 2014). The adhesive properties of polymer concrete allow repair of both polymer and conventional cement-based concretes.

7.5.4 Properties

Exact properties of the polymer concrete depend on the mixture, polymer, aggregate used, etc. But generally speaking, such materials have certain important characteristics as given below:

Such materials have significantly greater tensile strength than unreinforced Portland concrete because plastic is "stickier" than cement and has reasonable tensile strength (Kumar et al. 2014). They have low permeability to water and aggressive solutions. They have good chemical resistance and also corrosion resistance. Such polymer concretes exhibit compressive strength similar to or greater than Portland concrete. They undergo much faster curing yet good adhesion to most surfaces, including reinforcements. Because of the presence of polymer, they have good long-term durability against exposure to temperature fluctuations because of their ability to withstand freeze and thaw cycles. Obviously, their density is lower, and it can be controlled by adjusting the resin content of the mixture and thus lighter in weight than traditional concrete (Kumar and Monteiro 2014; Figovsky and Dmitry 2013).

7.6 SUMMARY

This chapter has provided a glimpse of different classes of matrix materials used for the fabrication of composite materials. Such matrix materials should be selected in accordance with the design of engineering composite materials for many purposes. Polymer matrices are used for the design of lighter composite materials with high specific strength. MMCs are tough and are used when the material has to face higher temperature. Ceramic matrices are used for composites in high-tech areas, where desired conductivity, rigidity, and high heat resistance properties are to be maintained. Carbon matrices are used for carbon/carbon composite materials, which are generally used in the critical areas of applications. Cement is also a kind of matrix material used for construction of reinforced concrete structures in civil engineering.

REFERENCES

Agarwal, B.D. and Broutman, L.J. 1980. *Analysis and Performance of Fibre Reinforced Composite*. Wiley, New York.

ASM Handbook. 1984. ASM International Society.

Balasubramaniam, R. 2009. *Callister's Materials Science and Engineering: Indian Adaptation (W/Cd)*. John Wiley & Sons.

Banerjee, M., Sachdeva, P., and Mukherjee, G.S. 2012. Preparation of PVA/Co/Ag film and evaluation of its magnetic and micro-structural properties. *Journal of Applied Physics* 111: 094302.

Benmokrane, B, Ali, A.H., Mohamed, H.M., El-Safty, A. and Manalo, A. 2017. Laboratory assessment and durability performance of vinyl-ester, polyester, and epoxy glass-FRP bars for concrete structures. *Composites Part B: Engineering* 114: 163–174.

Brydson, J.A. 1999. *Plastics Materials*, 7th ed. Butterworth-Henchmann, Oxford.

Chawla, K.K. 2012. *Composite Materials: Science and Engineering*. Springer Science & Business Media, Switzerland.

Clayton May, C. 1987. *Epoxy Resins: Chemistry and Technology*, 2nd ed. CRC Press, Taylor & Francis Group, New York.

Do Suh, J. 2008. Design and manufacture of hybrid polymer concrete bed for high-speed CNC milling machine. *International Journal of Mechanics and Materials in Design* 4(2): 113–121.

Figovsky, O. and Dmitry, B. 2013. *Advanced Polymer Concretes and Compounds*. Tailor & Francis Group, Boca Raton, FL: 26.

Gardner, K.H., Benjamin, S.H., and Katherine, L.F. 1994. Polymorphism in poly (aryl ether ketone) s. *Polymer* 35(11): 2290–2295.

Ghosh, M. 1996. *Polyimides: Fundamentals and Applications*. CRC Press, Taylor & Francis Group, Boca Raton, FL.

Ghosh, P. 1990. *Polymer Science and Technologies of Plastics and Rubbers*. Tata-McGraw-Hill, New Delhi.

Ghosh, P. and Mukherjee, G.S. 1999. *Polymers for Advanced Technologies* 10(12): 687–694.

Hamerton, I. 2013. *Chemistry and Technology of Cyanate Ester Resins*, Springer, Tunbridge Wells.

Kainer, K.U. 2006. Basics of metal matrix composites. *Metal Matrix Composites* 1–54.

Kochendorfer, R. 2009. Ceramic matrix composites—from space to earth: The move from prototype to serial production. In Mrityunjay singh and Todd Jessen (eds.), *25th Annual Conference on Composites, Advanced Ceramics, Materials, and Structures A*, 248, John Wiley & Sons, Westerville, OH: 11.

Krenkel, W. 2021. *Ceramic Matrix Composites: Fiber Reinforced Ceramics and Their Applications*. John Wiley & Sons, Weinheim.

Kumar, P.M. and Monteiro, PJM. 2014. *Concrete: Microstructure, Properties, and Materials*. McGraw-Hill Education: 505–510.

Li, Y., J. Cao, and C. Williams. 2019. Competing failure mechanisms in metal matrix composites and their effects on fracture toughness. *Materialia* 5: 100238.

Malaki, M., Xu, W., Ashish, K.K., Pradeep, L.M., Hajo D., Varma, R.S., and Manoj, G. 2019. Advanced metal matrix nanocomposites. *Metals* 9(3): 330.

May, C.A. 1979. Resins for aerospace. *ACS Symposium Series* 132.

Mukherjee, G.S. 2012. Evaluation of processing temperature in the production of fibre reinforced epoxy composites. *Journal of Thermal Analysis and Calorimetry* 108(3): 947–950.

Mungiguerra, S., Di Martino, G.D., Raffaele, S., Luca, Z., Diletta, S., and Miguel, A.L. 2018. Ultra-high-temperature ceramic matrix composites in hybrid rocket propulsion environment. *International Energy Conversion Engineering Conference*: 4694.

NPTEL, 2014. *Introduction to Composites, Matrix Materials, Modul*e 1. NPTEL, Kanpur.

Sciti, D., Laura, S., Frédéric, M., Antonio, V., and Luca, Z. 2018. Introduction to H2020 project C3HARME–next generation ceramic composites for combustion harsh environment and space. *Advances in Applied Ceramics* 117: 70–75.

Sen, A.K. and Mukherjee, G.S., 1993. Studies on the thermodynamic compatibility of blends of poly (vinyl chloride) and nitrile rubber. *Polymer* 34(11): 2386–2391.

Sothern, G.R. 1983. *Rubber Technology & Manufacture: CM Blow*. C. Hepburn Butterworths Elsevier, Buttenvorths: 947–948.

Trefilov, V.I. 1995. Ceramic-matrix composites. In Academician V.I. Trefilov (ed.), *Ceramic-and Carbon-Matrix Composites*, Springer, Kiev: 1–254.

Trento, C., 2019. What are the ceramic materials with high thermal conductivity? In *Stanford Advanced Materials*. https://www.samaterials.com/content/what-are-the-ceramic-materials-with-high-thermal-conductivity.html#:~:text=Ceramic%20materials%20with%20high%20thermal%20conductivity%20are%20mainly%20composed%20of,silicon%20nitride%2C%20and%20silicon%20carbide

Winspear, G.G. 1968. *The Vanderbilt Rubber Handbook*. RT Vanderbilt Company, Maryville, TN.

8 Metallurgical Failure Analysis of an Excavator Boom Body Comprised of Several Members Manufactured Out of Welded Composite Structures

Tapas Das
LMATS Brisbane Laboratory

CONTENTS

8.1 INTRODUCTION AND BACKGROUND

A boomerang-shaped excavator boom body comprises of several fabricated members: a boom front member, an intermediate member and a rear member. An arm-connection bracket is jointed to the boom front member at the shovel end, and a vehicle body-mounting bracket is jointed to the boom rear member at the cabin end, thereby forming a boom (see Figure 8.1).

This boomerang-shaped structure, as viewed from side in Figure 8.1, is formed into a hollow structure of rectangular cross section in which an upper lateral plate, a lower lateral plate, and left and right vertical/side plates are all welded at right angles to each other so as to manufacture a welded composite structure with the main aim to reduce the overall weight of the boom body. It is important to note that such a designed engineering composite structure of the boom is generally less

DOI: 10.1201/9780429330575-8

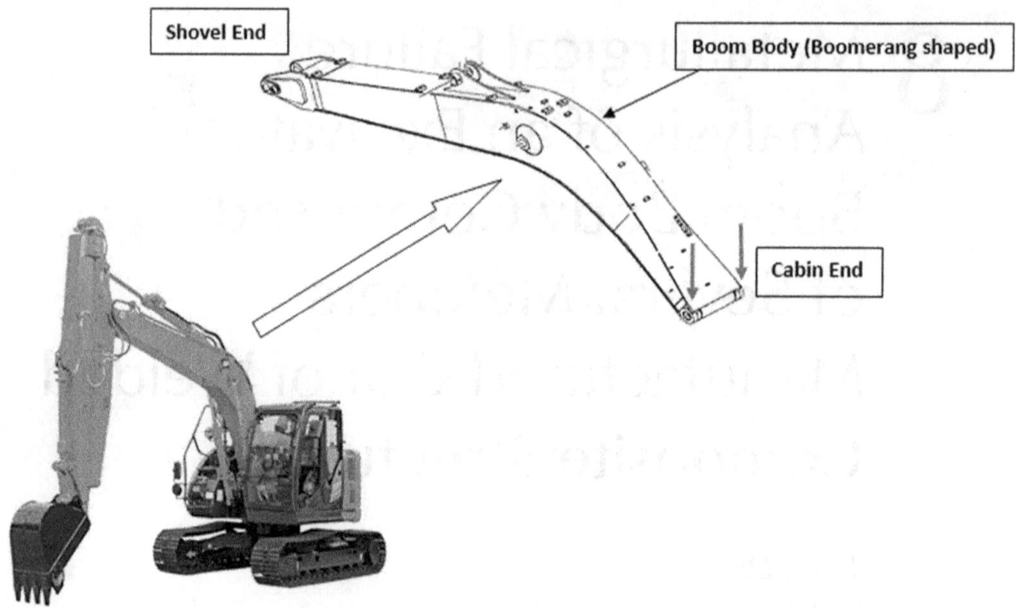

FIGURE 8.1 A line diagram of a boomerang-shaped boom body. Note that the failure had taken place at the cabin end as indicated by red arrows.

prone to deform which also allows the use of steel plates with reduced thickness without compromising the rigidity and load-carrying capacity of the structure.

A complete metallurgical failure analysis was carried out on a supplied failed excavator boom constructed out of welded structural steel plates as discussed above. The boom had failed in service well under the expected life of several years of operation. It was reported that the steel plates used for construction of the failed boom were of Grade SS400 structural steel as per a Japanese Industrial Standard specification, JIS G3106. These plates were welded by means of a manual metal arc welding route.

These steel plates were 12 mm thick, and they were both butt and fillet-welded as required following a weld procedure, the details of which and the welding consumable were unknown at the time while the failure analysis was carried out.

Following were the scopes of work undertaken to complete the failure analysis of the supplied composite boom structure:

- Visual observations and stereomicroscopy/digital photography of as-received failed boom and locations for samples collected.
- Chemical analyses of the steel plate materials.
- Vickers hardness testing/hardness traverse of samples of the two welded plates (butt and fillet types) at various regions, for example, parent materials, heat-affected zones (HAZs) and welds to determine their mechanical properties.
- Tensile testing of a plate sample obtained from the boom in order to verify the compliance/steel grade.
- Weld-macro assessments of the two welded plates and both macro- and microscopic observations of the weld with bottom plate and the cylindrical roll.
- Scanning electron microscopy (SEM) and energy-dispersive spectroscopy (EDS) of various regions on the fracture surface and any areas of interest on the fracture surface in order to determine mostly their chemical compositions.

- Optical microscopy of the samples extracted from various regions of the welded plates including the welds and HAZs.
- Evaluation of all the results.
- Conclusions and recommendations.

Samples obtained from the welded plates collected from the failed boom together with various tests conducted on them are all listed in Table 8.1.

8.2 VISUAL EXAMINATION

A number of photographs were taken showing the general condition of the failed boom in the as-received conditions. Photographs provided in Figures 8.2–8.6 revealed the location(s) of failure at the cabin end indicating that the failure of the boom had taken place along the weld between the top plate and the cylindrical base support of the boom body which then carried over to the bottom plate following the crack path mostly along the weld seam on the side plate. It can be further observed that the cylindrical base support of the boom body was almost separated from the boom top plate as revealed in Figures 8.2–8.4.

Various weld-macro samples, mating halves of the fracture surfaces and a tensile test piece were collected from the areas of interest as shown in Figures 8.5 and 8.6 for in-depth investigations. The most important area to observe was the mating halves that are identified as M3 and M4 in Figure 8.6 in addition to adjacent areas and areas underneath the top plate at the weld region.

TABLE 8.1
List of Various Samples with Identifications and Tests Carried Out on Them

Sample Id	Visual Observations	Chemical Analysis	Hardness Testing	Tensile Testing	Weld Macro	SEM/EDS	Optical Microscopy
Top plate	Yes	Yes	Yes	Yes	Yes	Yes	Yes
Bottom plate	Yes	No	No	No	Yes	Yes	Yes
Side plate	Yes	Yes	Yes	No	Yes	Yes	Yes

FIGURE 8.2 A photograph showing the failure of the boom had taken place along the weld between the top plate and the cylindrical roll.

FIGURE 8.3 A photograph showing a close side view of the failed boom.

FIGURE 8.4 A photograph showing a close side view of the failed boom. The arrow indicates the location with about 80% separation of the top plate from cylindrical roll which is the base support of the excavator boom body.

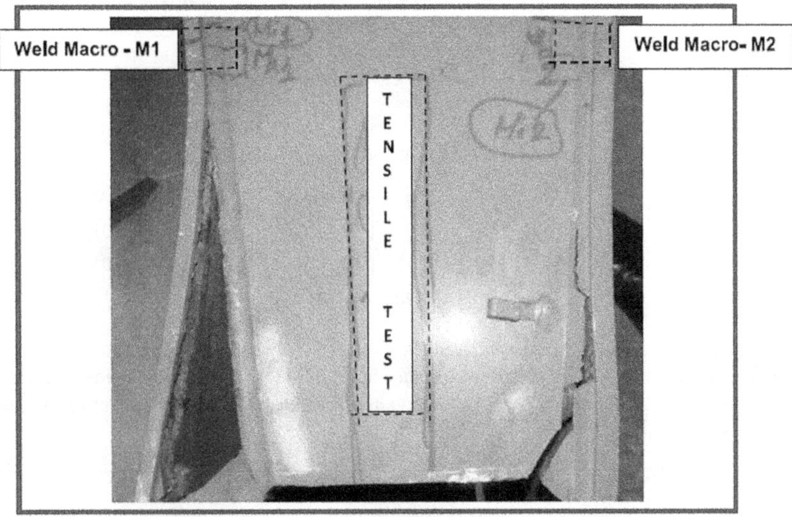

FIGURE 8.5 A photograph showing the locations from where the weld samples for macroscopic examination and tensile test piece were extracted.

TABLE 8.2
Macroscopic Examinations on the Welds of the Two Welded Plates

Sample ID	Observations	Comments
M1	Fillet weld between the top and side plates.	No discontinuities were evident.
M2	Fillet weld between the bottom and side plates.	No discontinuities were evident.

8.3 WELD-MACRO ASSESSMENTS

Weld-macro assessments were carried out on the two welded plates that were obtained from areas marked M1 and M2 on the failed boom as shown in Figure 8.5. A total of two samples containing fillet welds, both the top and bottom plates with a side plate close to the fracture path, have been selected. These two samples were extracted by cold-cutting and later ground from coarse to finer stages up to 1,200 grit SiC paper and further ground to the final stage to 2,000 grit SiC paper. They were then etched with 10% HNO_3 in ethyl alcohol solution to reveal macrostructures of the weld and HAZ.

Weld assessments were then carried out on the above two welded plate samples following AS 2205.5.1:2003 standard specification. Discontinuities observed in the two samples are listed in Table 8.2. The details with enlarged views at the welds of the two samples were taken by means of a stereomicroscope and have been included in Appendix 8.1.

8.4 CHEMICAL ANALYSIS

Two representative samples were removed by cutting sections across the length of the top plate and the side plate in order to determine chemical composition of the bulk plate materials that were used for manufacturing the composite structure. These samples were analysed by means of an optical emission spectroscopy to determine chemical composition of the steel plate samples. The nearest matching steel grade for individual plates was obtained from the available steel data base.

TABLE 8.3
Chemical Analysis Results (wt%) of the Bulk Plate Materials Used for Manufacturing the Composite Boom Structure

	Chemical Composition – Elements (Average wt%)													
Sample ID	**Fe**	**C**	**Si**	**Mn**	**P**	**S**	**Cr**	**Mo**	**Ni**	**Al**	**Cu**	**Nb**	**Ti**	**V**
Boom top plate	≈Bal	0.15	0.02	0.86	0.02	<0.01	0.01	0.01	0.01	0.029	0.01	<0.01	<0.01	<0.01
Boom side plate	≈Bal	0.14	0.02	0.74	0.01	0.01	0.01	<0.01	0.01	0.03	0.05	0.01	<0.01	<0.01
JIS G3101/SS400 grade steel	≈Bal	*		**	<0.05	<0.05								

Note: Both carbon and manganese (* and **) are not specified in JIS 3101 standard for SS400 grade structural steel. The weldability group number can be calculated from the carbon equivalent, 0.3 (as calculated from the above composition of plate materials).

Results of all chemical analyses of the two steel plates (both the boom top side plates) are given in Table 8.3. An average of three analyses at three different areas/spots on each plate sample has been considered for reporting the chemical composition of individual plates. The chemical compositions of both the boom top plate and the side plate would satisfy JIS G3101/SS400 grade Japanese Industrial Standard steel specification as can be seen in Table 8.3. This particular steel grade, SS400, would generally be suitable for the intended general structural applications such as requiring a tensile strength between 400 and 500 MPa. The minimum yield strength required for a plate thickness of <16 mm is 245 MPa. In Japan, this particular steel grade is most commonly used as hot-rolled structural steel plates for fabrication/manufacturing mechanical equipment structures.

For better understanding the weldability of group members, the carbon equivalent (which was ≈ 0.3) was calculated from the chemical composition of the steel plates to be welded together in order to determine the total heat input and preheat temperature requirements for a given plate thickness. This would generally ensure a good weld quality. Also, at the time of developing a weld procedure specification (WPS), the chemical compositions/carbon equivalent of the plate materials as given in Table 8.3 would be a great help.

8.5 HARDNESS TESTING

Two representative samples were removed from the failed boom keeping the fillet welds in between the top and bottom plates and the side plate of the boom at a place further away from the opened weld. These samples are identified as M1 and M2 as can be viewed in Figure 8.6.

These two samples were cold-mounted in epoxy resin and were then prepared using standard metallographic techniques up to a 2,000 grit SiC paper in accordance with ASTM E3-2004 standard specification. Vickers hardness testing was performed on these two prepared specimens in accordance with AS1817.1 standard specification with an applied load of 10 kgf (AS1817.1 2003).

Results of the hardness measurements of one sample with a fillet weld between the top weld plate and the side plate (identified as M2) are given in Table 8.4. As can be seen in Table 8.4, the average hardness of the weld, HAZ and the bulk plate material, in general, have complied with the requirements of AS/NZS 1554.1 standard specification (AS 1554.1 2014). However, in the absence of any given standard(s) from the manufacturer, it was not possible to verify the compliance.

In addition, the hardness of both the top and side plate materials were about 145 and 154 VHN, which are equivalent to tensile strengths of 564 and 485 MPa when converted from the above hardness levels following the AS5016 standard specification (a conversion table provided for low carbon steels) (AS 5016 2004). Both these tensile strength levels would comply with the tensile strength requirement for a SS400 grade steel. However, the difference of tensile strengths for the two plates

FIGURE 8.6 A photograph showing the locations from where the two mating halves of the fracture surface, identified as M3 and M4, were extracted.

TABLE 8.4
Vickers Hardness Test Results of the Welded Component (M2)

			Harness Measurements (VHN)	
Sample ID	**Zone**	**S. No.**	**HV10 (Load 10 kgf)**	**Average Hardness (Load 10 kgf)**
Top plate (T2)	Weld	1	195	192
		2	192	
		3	191	
	HAZ	1	169	166
		2	168	
		3	162	
	Parent	1	144	145
		2	145	
		3	146	
Side plate (T1)	Weld	1	199	202
		2	203	
		3	204	
	HAZ	1	176	172
		2	171	
		3	170	
	Parent	1	152	154
		2	153	
		3	158	

TABLE 8.5
Tensile Test Results of the Boom Plate

Plate ID	Size (mm)	Test Temp. (°C)	0.2% Proof Load (kN)	UTS Load (kN)	0.2% Proof Strength (MPa)	UTS (MPa)	Elongation (%)
Boom plate (side)	11.89 × 12.58	Ambient	50.5	66.6	337	445	38

could have been due to the rolling direction/manufacturing process employed. Also, the hardness levels of both the HAZ (i.e., 170 VHN) and the weld (i.e., 195 VHN) were within the acceptable levels and are commonly observed for sound welding of structural materials.

8.6 TENSILE TESTING

One representative sample from the top plate was removed from the failed boom. It was then further machined to obtain a rectangular test specimen. The tensile test was carried out as per AS1391 standard specification (AS1391 2020) at an ambient temperature of 23°C.

The ultimate tensile strength and 0.2% proof stress obtained from the above test were 445 and 337 MPa, respectively, while the percentage elongation was 38% (see results of the tensile test in Table 8.5). It is important to note that all these mechanical properties have complied with the requirements of Grade SS400 structural steel as per JIS G3101 Japanese Industrial Standard specification.

8.7 STEREOMICROSCOPY/FRACTOGRAPHY

Both mating halves of the fracture surface (at the location identified as M3 in Figure 8.6) were closely observed under a stereomicroscope for any specific features related to the failure. Figures 8.7 and 8.8, in general, revealed the mating halves of the fracture surface showing a number of ratchet marks adjacent to the inner surface of the top plate as marked by arrows (red). These ratchet marks

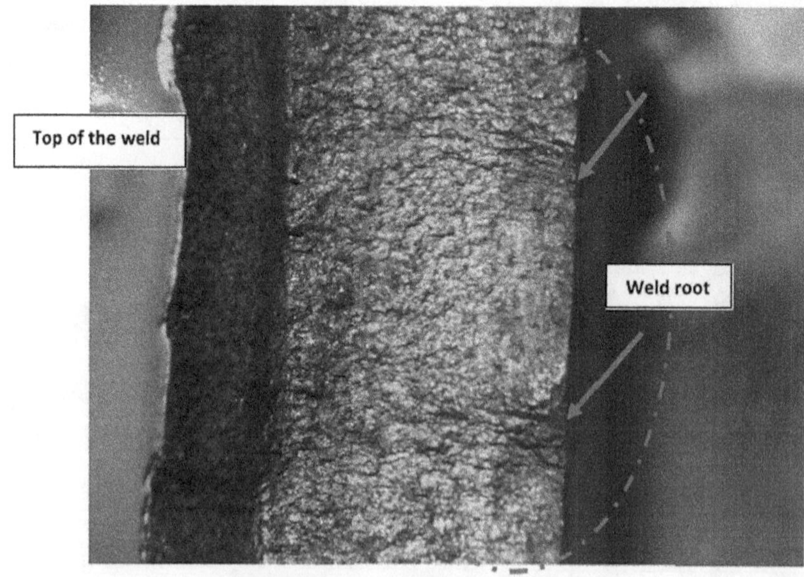

FIGURE 8.7 A photograph showing the fracture surface and a number of fatigue crack initiation sites as marked by arrows at the weld root.

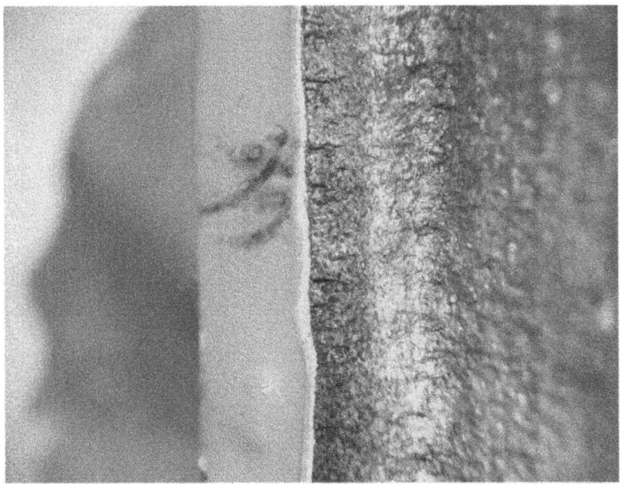

FIGURE 8.8 A photograph showing an area of the fracture surface at the top of the weld where it failed at the final stage.

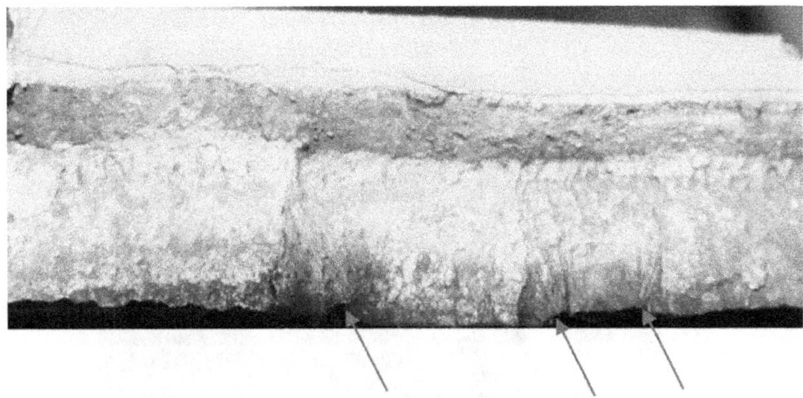

FIGURE 8.9 A photograph of another area adjacent to M3 showing the sites from where fatigue cracks were initiated.

are indicative of fatigue crack initiation sites where fatigue cracks were initiated under a repetitive/cyclic stress field and later progressed toward the other side of the weld/plate leading to the final catastrophic failure with little plastic deformation at the end as a result of a gross overload (see Figure 8.9).

Figures 8.9 and 8.10 show another fracture surface of an area adjacent to M3 which clearly indicated ratchet marks and faint fatigue arrest beach marks as the fatigue cracks were initiated and further progressed. Figures 8.11 and 8.12 revealed discontinuities resulting due to poor surface preparation as viewed underneath the top plate adjacent to the fracture surface and at the weld root (Figure 8.13).

In addition, a section across the fillet weld between the bottom plate and the cylindrical drum diametrically opposite to the fracture surface at the area close to M3 was also cold-cut, ground, polished and etched with 10% Nital to further observe any weld defects which was undisturbed as a result of the failure. Figure 8.12 had clearly indicated the lack of root fusion including the presence of slag inclusions. Figures 8.14 and 8.15 show the same areas at a slightly higher magnification

FIGURE 8.10 A photograph of the above area adjacent to M3 as shown in Figure 8.9 at a slightly higher magnification showing the same features as described earlier.

FIGURE 8.11 A photograph showing surface irregularities underneath the top plate adjacent to the fracture surface at the weld root which would have participated in the fracture process.

indicating the above features. It is important to note that such areas formed due to lack of fusion at the weld root are generally susceptible to initiate fatigue cracks in a cyclic stress field during normal operation of an excavator. Once initiated, these fatigue cracks generally progress through the weld over a time period leading to a final catastrophic failure.

It is important to note that both the top and bottom steel plates were fillet-welded to a thick cylindrical drum. Extra care is generally required in this type of fillet-welding where a thinner plate is fillet-welded with a much thicker section (of a cylindrical drum). Correct adjustments of weld travel

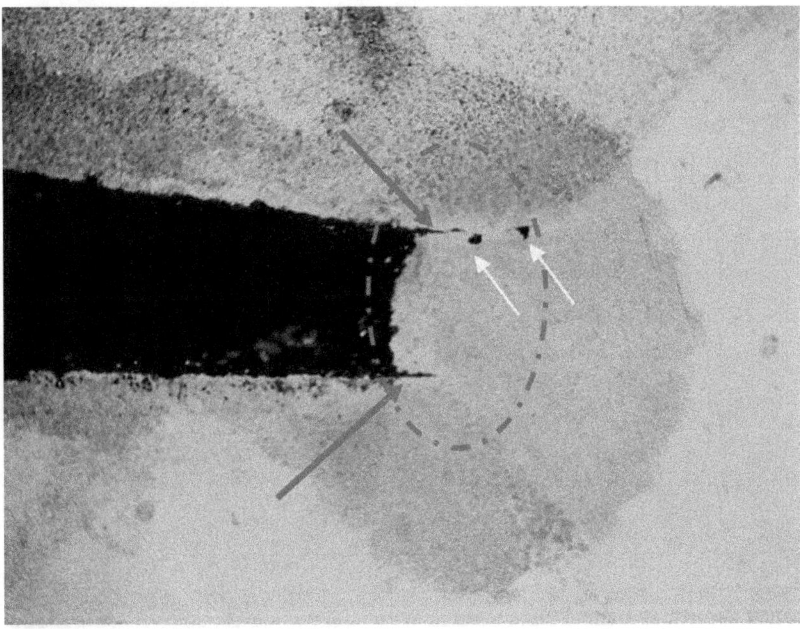

FIGURE 8.12 A photograph showing a section across the fillet weld between the bottom plate and the cylindric drum diametrically opposite to the fracture surface at M3 indicating lack of root fusion (as marked by red arrows) and presence of slag inclusions (as marked by yellow arrows) at the weld root.

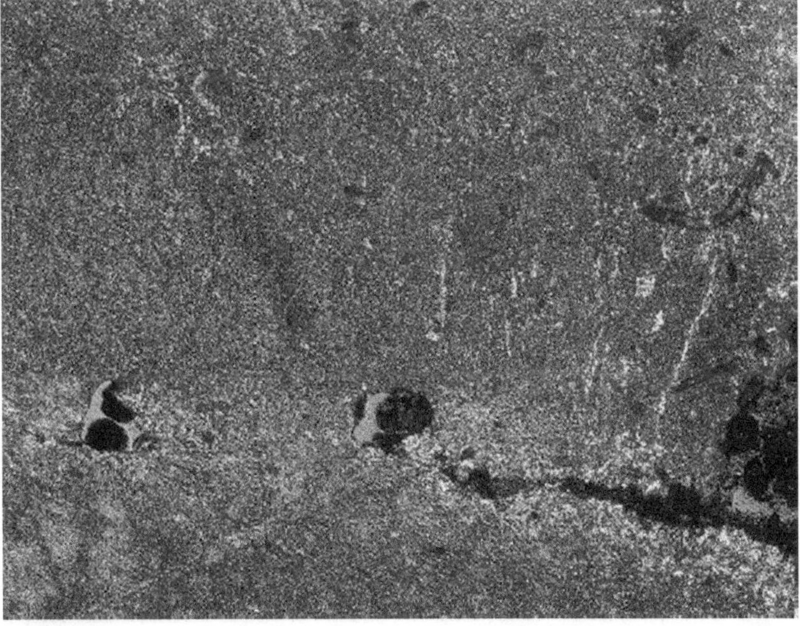

FIGURE 8.13 A photograph showing a close view of one side of the above section across the fillet weld (as shown in Figure 8.12) at a higher magnification.

FIGURE 8.14 A photograph showing a close view of the other side of the above section (as shown in Figure 8.13) across the fillet weld at a higher magnification.

FIGURE 8.15 A SEM micrograph (SE image) of the fracture surface of the M3 area of the weld at a higher magnification.

speed, electrode angle with the selection of right electrode diameter with respect to the plate thickness and thickness of the cylindrical drum, and the amount heat input (i.e., preheating the cylindrical drum) would improve the overall weld quality.

The other remedial measures are to prepare a qualified weld procedure for such a difficult-to-weld area, and sometimes a better weld design is required for tackling such a rigid-/difficult-to-access area/weld.

8.8 SCANNING ELECTRON MICROSCOPY/FRACTOGRAPHY AND ENERGY-DISPERSIVE SPECTROSCOPY (SEM/EDS)

Two mating fracture surfaces of areas M3 and M4, as can be seen in Figure 8.6, were further examined with a ZEISS Sigma VP field emission scanning electron microscope equipped with an Oxford XMax 50 Silicon Drift EDS detector/analyser for rapid X-ray mapping and accurate analysis of any particular areas of interest, for example, weld metal and slag inclusions, etc.

SEM micrographs or the secondary electron (SE) images of the fracture surfaces are shown in Figures 8.15–8.17. Figures 8.15 and 8.16 are the SE images showing general features of the fracture surfaces of areas M3 and M4 at the weld while Figure 8.17 shows the area on the fracture surface where the final fracture had taken place with a small amount of plastic deformation.

Elemental maps/spectra are obtained by means of an EDS analyser on selected areas that are marked on the back-scattered electron images of the fracture surface as shown in Figures 8.18, 8.20 and 8.23 indicating weld metal and/or possible presence of slag inclusions in the weld at M3 area. These elemental maps/spectra shown in Figures 8.19, 8.21, 8.22 and 8.24, in general, indicated the presence of major elements, for example, iron (Fe), carbon (C) and oxygen (O), which appeared from the weld metal and mild surface oxidation after the fracture had taken place. The peaks of other minor elements, for example, manganese (Mn), silicon (Si), etc., also appeared in the spectra arising from the constituents of the weld metal.

FIGURE 8.16 A SEM micrograph (SE image) of the fracture surface of the M4 area of the weld at a higher magnification.

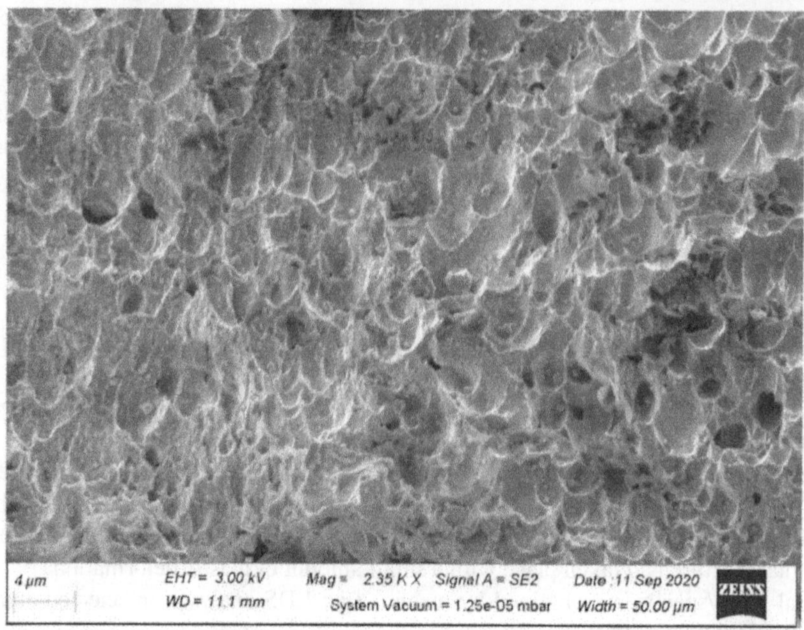

FIGURE 8.17 A SEM micrograph (SE image) of the fracture surface of the M4 area where the final fracture had taken place with small amount of plastic deformation with the formation of shallow dimples revealed at a much higher magnification.

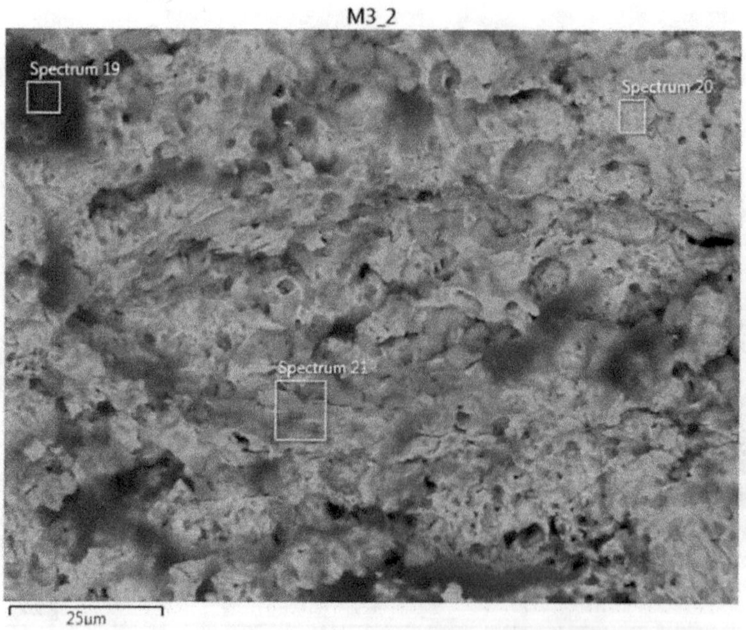

FIGURE 8.18 Elemental maps were obtained of the areas as marked on back-scattered electron (BSE) images of the fracture surface of M3 by means of an EDS analyser.

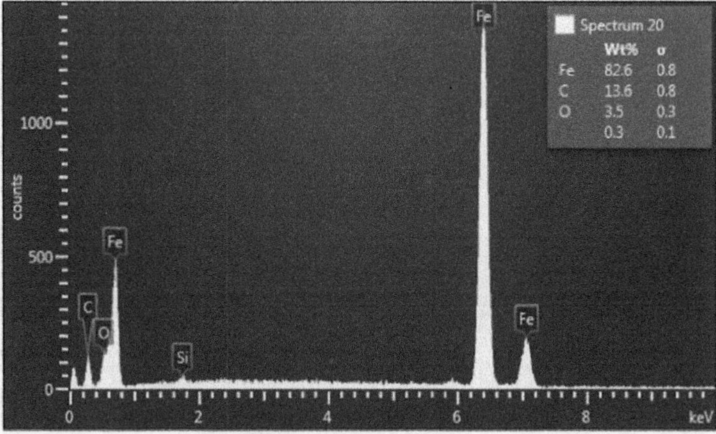

FIGURE 8.19 An elemental map obtained from an EDS analysis of an area no. 20 as marked in Figure 8.19 indicating the presence of major elements.

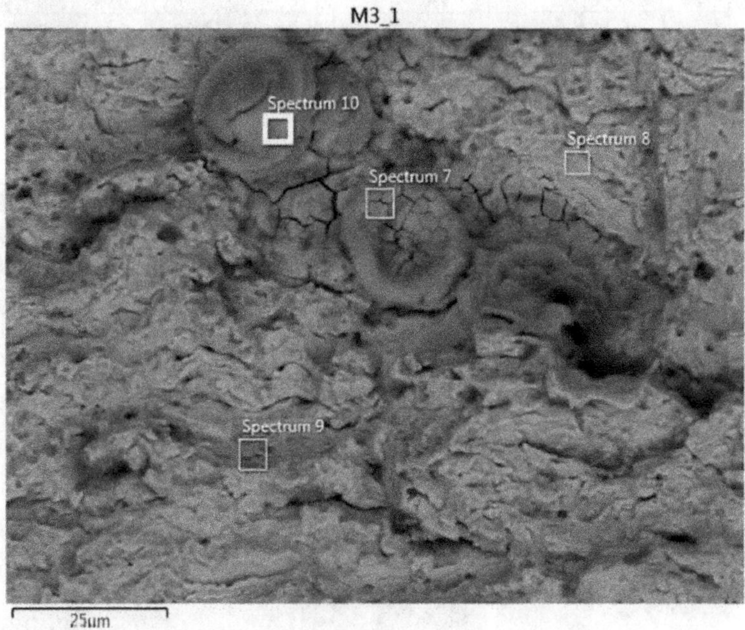

FIGURE 8.20 Elemental maps were obtained of the areas as marked on back-scattered electron (BSE) images of the mating fracture surface of M3 by means of an EDS analyser. The BSE image revealed areas of suspected slag inclusions which were also analysed by means of the EDS analyser.

Slag inclusions present in the fracture surface can be seen in Figure 8.20. The spectrum representing major constituents of the slag inclusion, as seen in Figure 8.20, appeared due to the presence of high levels of iron (Fe) and oxygen (O) which most likely have originated from iron oxide (FeO), major constituents of a slag inclusion.

It is important to note that these slag inclusions would generally appear in the weld due to incorrect manipulation of the electrode (at difficult-to-access areas) and/or improper cleaning between interrun passes or lack of cleanliness of the surfaces to be welded. Also, such slag inclusions present in the weld would generally be detrimental to weld properties and decrease overall joint strength.

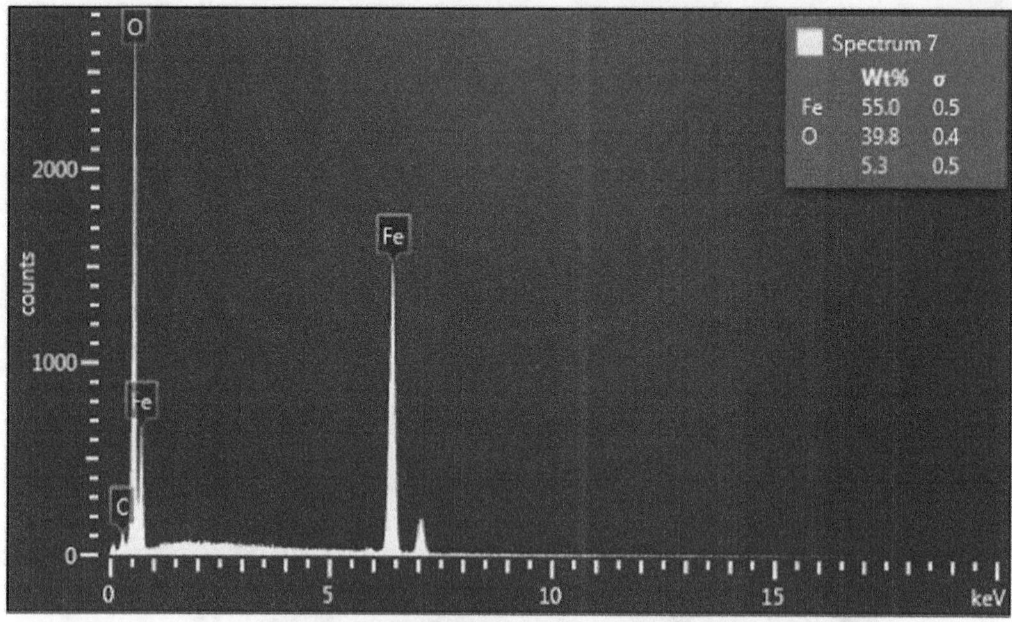

FIGURE 8.21 An elemental map obtained from an EDS analysis of an area no. 7 as marked in Figure 8.21.

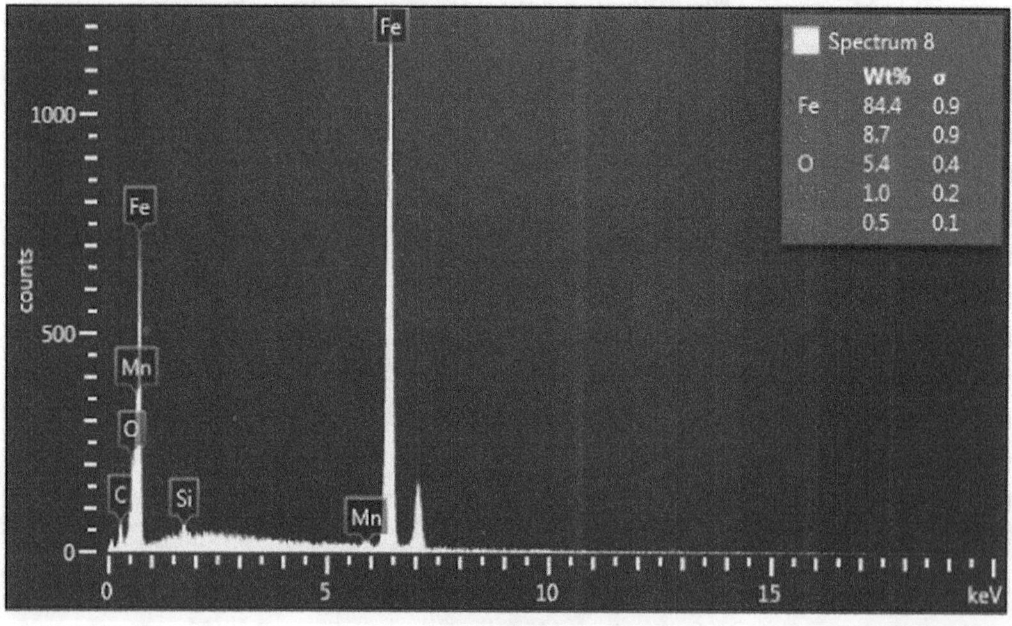

FIGURE 8.22 An elemental map obtained from an EDS analysis of an area no. 8 as marked in Figure 8.21.

Also, weight percentages (wt%) of various elements, as analysed by means of an EDS analyser attached to the SEM, are derived from the semiquantitative analysis of various areas of interest, for example, weld and slag inclusion, and are listed at the right top corner of maps, and their respective peaks can be viewed in spectra in Figures 8.19, 8.21, 8.22 and 8.24.

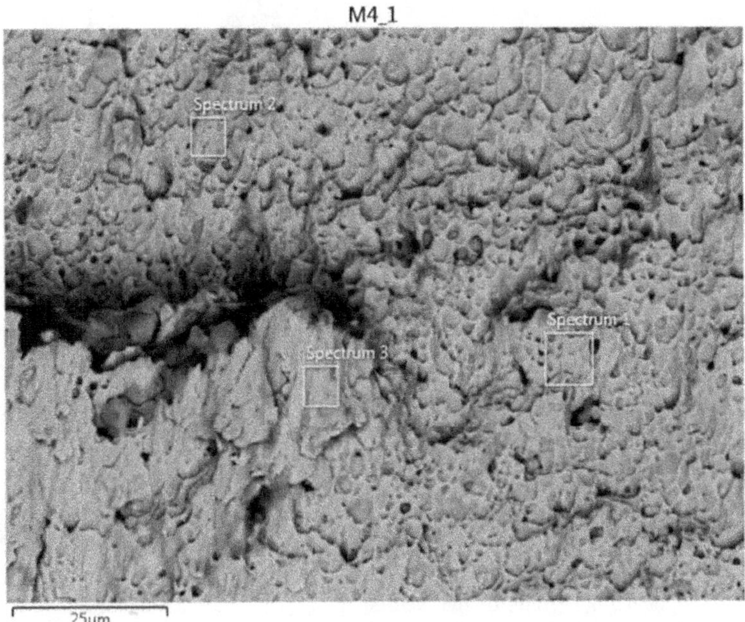

FIGURE 8.23 Elemental maps were obtained of the areas as marked on back-scattered electron (BSE) images of the mating fracture surface of M4 by means of an EDS analyser.

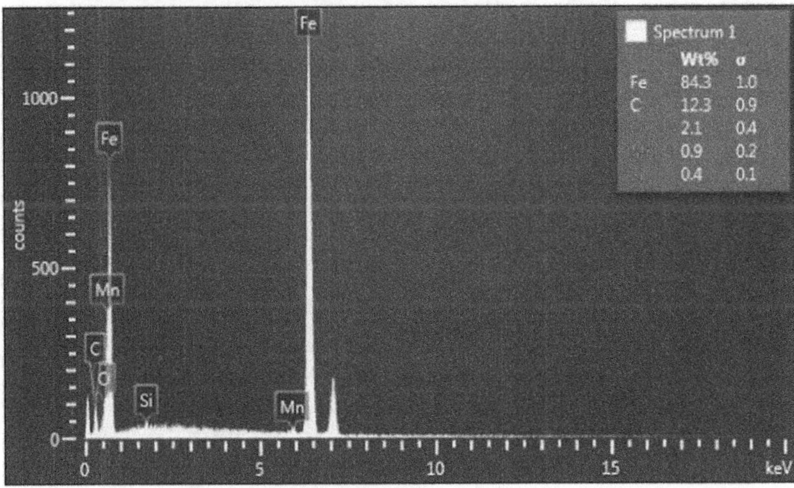

FIGURE 8.24 An elemental map obtained from an EDS analysis of an area no. 1 as marked in Figure 8.23.

8.9 OPTICAL MICROSCOPY

Two representative specimens across the fracture surface for metallographic examinations were obtained by cold-cutting across the length with an abrasive wheel without raising the temperature which included the weld, HAZ and parent material. These specimens were cold-mounted with an epoxy resin using moulds. They were then prepared on cross sections in accordance with ASTM E3 (ASTM E3 2011). Final polishing was carried out to 1-μm diamond suspension, and lastly the

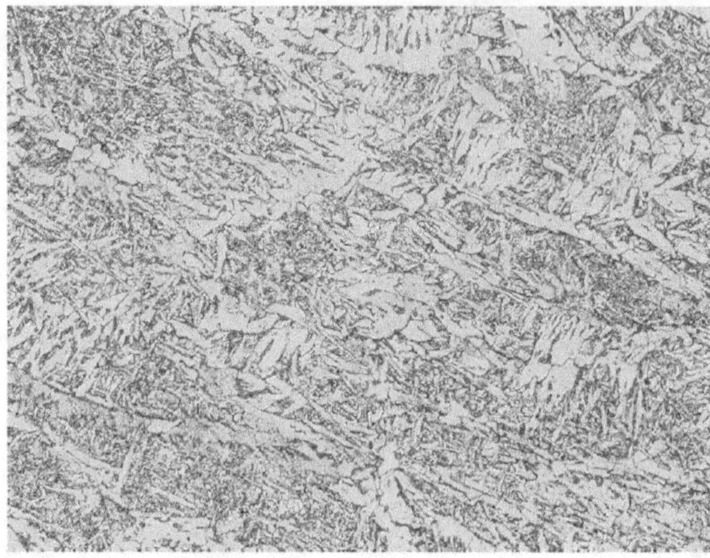

FIGURE 8.25 An optical micrograph of the weld showing a typical cast dendritic weld microstructure located adjacent to the fracture (Mag.: 200×, Etchant: 2% Nital).

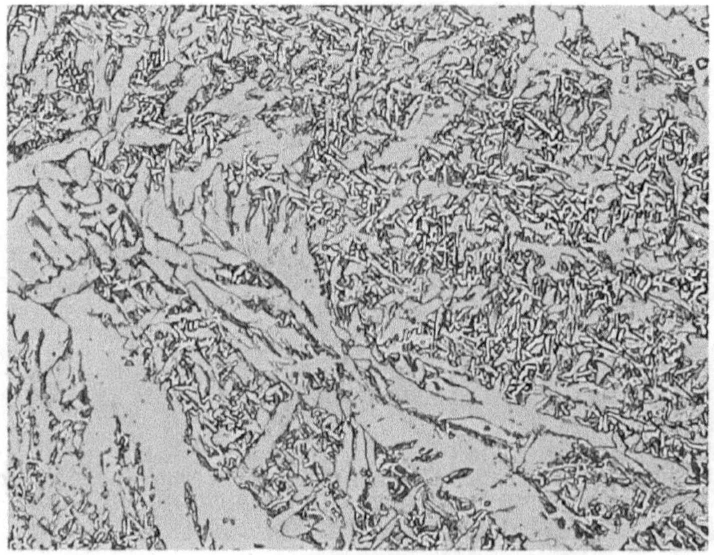

FIGURE 8.26 An optical micrograph of the weld showing a typical cast dendritic weld microstructure located adjacent to the fracture at a higher magnification (Mag.: 500×, Etchant: 2% Nital).

specimens were etched with 2% Nital (i.e., 2% HNO_3 in ethyl alcohol) for 10 seconds to reveal microstructural details. These specimens were examined under a Nikon metallurgical microscope at various magnifications to determine the method of manufacture, microstructural details, discontinuities present and the heat treatment employed.

Optical micrographs of various regions are shown in Figures 8.25–8.30. Figures 8.25 and 8.26 revealed the typical microstructure of the weld consisting of cast dendrites. The HAZ microstructures of the two specimens are given in Figures 8.27 and 8.28 revealing both the coarse and fine

FIGURE 8.27 An optical micrograph of an area adjacent to the weld indicating a coarse-grained (CG) microstructure of the HAZ region showing coarse bainitic microstructure of the boom plate (Mag.: 200×, Etchant: 2% Nital).

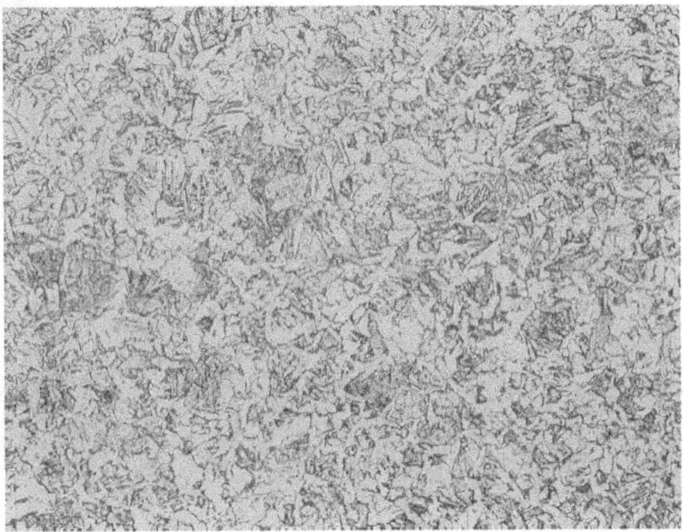

FIGURE 8.28 An optical micrograph of an area next to the weld indicating a fine-grained (FG) microstructure of the HAZ region showing relatively fine bainitic microstructure of the boom plate (Mag.: 200×, Etchant: 2% Nital).

bainitic microstructures (which denote the coarse-grained and fine-grained HAZs), respectively, indicating their locations with respect to the weld.

Figures 8.29 and 8.30 revealed microstructures of the two parent materials: the boom plate and the cylindrical drum across its length (i.e., the transverse section). These microstructures consisted of grains of ferrite and pearlite aligned across the direction of rolling/deformation that were generally used for manufacturing such carbon/manganese steel grades. No microstructural discontinuities were observed in the above microstructures.

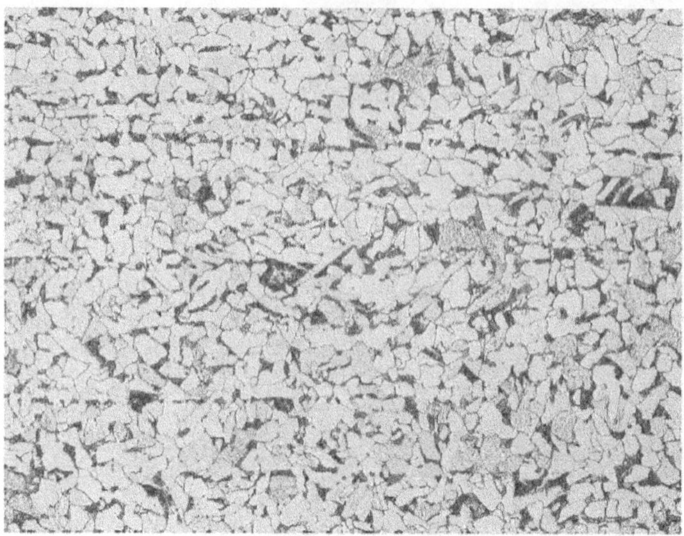

FIGURE 8.29 An optical micrograph of boom plate material consisting of pearlite and ferrite grains across the direction of rolling (Mag.: 200×, Etchant: 2% Nital).

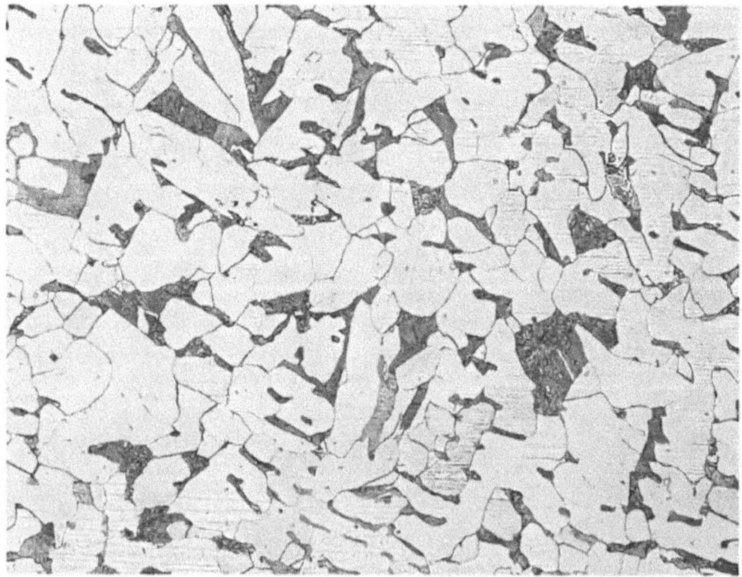

FIGURE 8.30 An optical micrograph of cylindrical drum material consisting of pearlite and ferrite grains across the thickness of the drum which would have been hot forged (Mag.: 200×, Etchant: 2% Nital).

8.10 CONCLUSIONS

1. Fractographic analysis carried out on fracture surfaces by both visual/stereomicroscopy and SEM has revealed that the failure of the boom was generally attributed to fatigue cracking. These fatigue cracks were most likely initiated at the root of the fillet weld between the cylindrical drum and the top plate due to lack of fusion on both sides of the weld, as evidenced adjacent to the fracture face and a weld macro across the weld.

2. Fractographic study has also revealed a number of ratchet marks on the fracture surface indicating where fatigue cracks were initiated and at a later stage progressed toward the other side of the weld across its length leading to the final catastrophic failure with a little amount of plastic deformation as a result of a gross overload. The catastrophic failure or fast fracture had occurred when 80%–85% of the area of the weld had fractured and the remaining 15%–20% of the cross sectional area could no longer bear on the structure causing a gross overload fracture.
3. The weld assessments carried out on two representative samples (which included fillet welds between the side plate and both top and bottom plates) obtained from areas adjacent to the failure/fracture path following the Australian standard, AS2205.5.1-2019 specification, did not reveal any discontinuities.

   Another weld macro/micro assessment carried out on the fillet weld between the cylindrical drum and the bottom plate revealed lack of fusion on both sides of the weld at its root including the presence of slag inclusions within the weld. It is important to note that lack of fusion at the weld root generally initiates fatigue cracks in a cyclic stress field during a normal operation of an excavator.
4. Chemical analyses carried out on two representative specimens from the top and side boom plates indicated both these plates were manufactured from hot-rolled steel Grade SS400 as per JIS G3106 Japanese Industrial Standard specification. Also, this particular steel grade would generally be suitable for the intended structural application such as requiring a tensile strength between 400 and 500 MPa.
5. Results of Vickers hardness measurements indicated that the hardness of both the parent plate materials (i.e., the top plate and the side plate) were about 145 and 154 VHN which are equivalent to tensile strengths of 465 and 485 MPa, as converted following the Australian standard AS5016-2004 specification (a conversion table provided for low-carbon steel). Both these tensile strength levels would comply with the tensile strength requirement for a SS400 grade carbon steel.

   Also, the hardness levels of both the HAZ (170 VHN) and the weld (195 VHN) were within the acceptable levels and are common for welding these structural materials and have complied with the requirements of AS/NZS 1554.1-2004 standard specification. However, in absence of any information on given standard(s), it was not possible to verify the compliance.
6. Results of tensile test carried out on standard test piece obtained from the top boom plate have indicated that the 0.2% proof stress and the ultimate tensile strength (UTS) were about 445 and 337 MPa, respectively, while the percentage elongation at fracture was 38%.

   All these mechanical properties have complied with the requirements for Grade SS400 structural steel as per JIS G3101 Japanese Industrial Standard specification.
7. SEM work revealed micrographs (SE images) indicating, in general, features related to fatigue fracture within the weld area while the final fracture had taken place with small amount of plastic deformation with formation of shallow dimples at this region.
8. Elemental maps/spectra obtained by means of EDS on selected areas of the weld and suspected slag inclusions within the weld, in general, indicated the presence of major elements, for example, iron (Fe), carbon (C) and oxygen (O). Elemental peaks (small) in the spectra of other minor elements, for example, manganese (Mn), silicon (Si), etc., appeared from the constituents of the weld metal.

   Spectra obtained from suspected slag inclusions had large elemental peaks of both iron (Fe) and oxygen (O) which would have originated from iron oxide (FeO), a major constituent of slag. These slag inclusions present in the weld are generally formed due to incorrect manipulation of the electrode at difficult-to-access weld areas and/or improper cleaning between interrun passes and uncleaned surfaces before welding. Such slag inclusions in the weld would be detrimental to weld properties and decrease overall joint strength.

9. Optical microscopy carried out on three representative samples revealed typical cast dendritic weld microstructure on the fracture face. The HAZs next to the welds had coarse- and fine-grained bainitic microstructures depending upon their locations with respect to the weld. The parent plate materials had layers of ferrite and pearlite microstructures across rolling direction while the cylindrical drum, in general, had equiaxed ferrite and pearlite grains. No microstructural discontinuities were observed in the above microstructures.

8.11 RECOMMENDATIONS

1. It is recommended to better understand the weldability group numbers from chemical compositions of the two materials (i.e., the plate and the cylindrical drum) that are to be welded together with a clear knowledge of total heat input and the preheat temperature as calculated from their individual thicknesses. This would improve the weld quality.
2. Slow cooling after welding and/or sticking to a preheat temperature which is appropriate for the workpiece thicknesses may generally be recommended. Sometimes, an appropriate postweld heat treatment for relieving internal stresses may also be recommended, particularly where extreme service conditions due to cyclic loading exist.
3. It may also be recommended to carry out welding with fully qualified weld procedure for a rigid weld design (i.e., a place which is difficult to access from one side). Selecting a suitable welding current, welding technique (i.e., an electrode angle) and welding speed (i.e., weld travel speed) with a judicious electrode diameter-to-amperage ratio may also be required in order to remove discontinuities such as lack of fusion at the weld root and entrapment of slag inclusions in the weld.
4. Slag inclusions appeared in weld can also be removed with proper manipulation of the welding consumable and/or cleaning between interrun passes and weld surfaces. A WPS (containing information such as steel grade, filler metal classification, amperage range, preheat and interpass temperature, etc.) and an approved procedure qualification record (PQR) may be documented to produce an acceptable sound weld.

REFERENCES/SPECIFICATIONS

AS 1391, 2020. *"Metallic materials – Tensile testing – Method of test at room temperature."*
AS 1554.1, 2014. *"Welding of steel structures."*
AS 1817.1, 2003. *"Metallic materials – Vickers hardness test – test method."*
AS 2205.5.1, 2019. *"Methods for destructive testing of welds in metal."*
AS 5016, 2004. *"Metallic materials – conversion of hardness values."*
ASTM E3, 2011. *"Methods for preparation of metallographic specimens."*

APPENDIX 1: MECHANICAL TESTING OF WELDED SPECIMEN

TABLE A1
Information Received from the Client

	Information Received			
Identification	PQR/WPS	Not supplied	Weld #	NA
	Welder	Not supplied	Welder ID	Not supplied
	Joint type	Fillet weld	Process	Not supplied
	Thickness (mm)	12	Position	Not supplied
Material specification	Carbon steel			
Welding consumable	NA			
Test specification	Report findings			

TABLE A2
Tests Performed as Planned

	Tests Conducted				
Abbreviations	Visual	MT	PT	UT	RT
	NR	NR	NR	NR	NR
C – complied	Macro	Tensile	Bend	Impact	Nick Break
DNC – did not comply	CR	CR	NR	NR	NR
E – evaluate property	Fillet break	Hardness	Corrosion	Ferrite test	Chemical
P – process requirements	NR	NR	NR	NR	NR
CR – client requirement	Shear	H/T	Microscopy		
NR – not required	NR	NR	NR		
NP – not performed					

Test details: Refer to the following pages for individual test details.
Signatory: JPC.

APPENDIX 2: MACROSCOPIC EXAMINATION

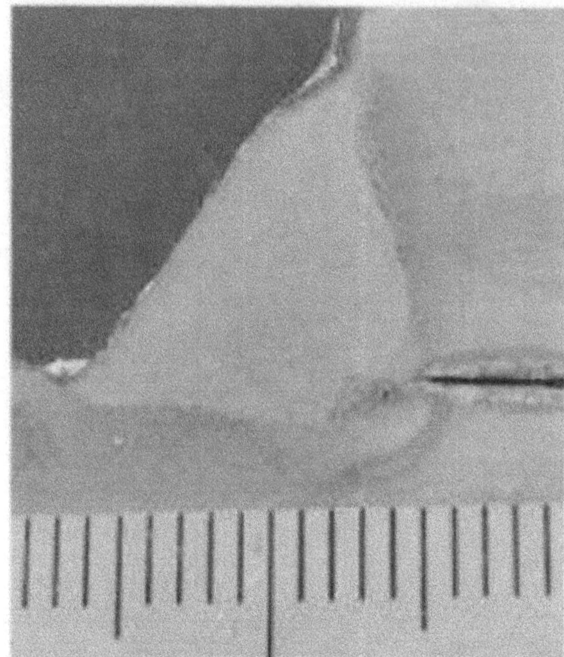

Figure 1 Macro Specimen 1

FIGURE A1.1 Macro specimen 1.

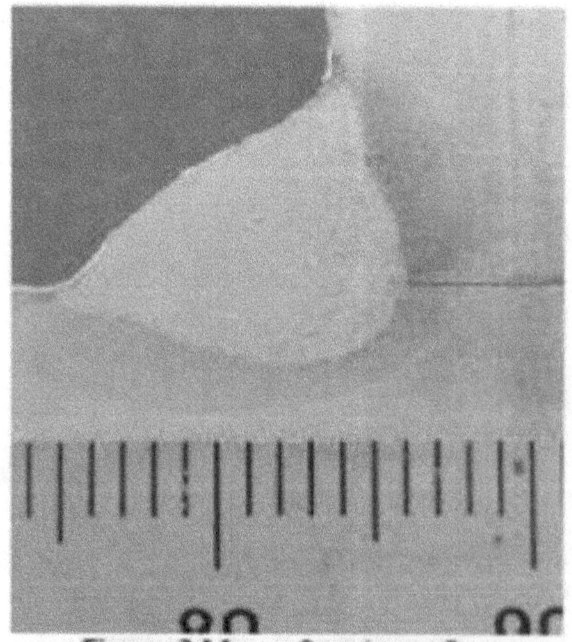

Figure 2 Macro Specimen 2

FIGURE A1.2 Macro specimen 2.

Test Method: AS2205.5.1:2019 (ISO 17639:2003) (AS 2205.5.1 2019).

Specimen Location: Orientation – Cross section.

Surface Preparation: Cold-cutting followed stage grinding and polishing to P#1000 grit SiC-coated abrasive paper.

Etchant Type: 10% Nital (nitric acid in ethyl alcohol).

Etching Method: Swabbing.

Weld Passes: 1.

Leg Length: M1 (8.5 × 10) mm, M2 (9 × 9) mm.

Throat Thickness: M1 6 mm, M2 6 mm.

Magnification: Evaluation – X5.

Discontinuities: Discontinuities were not evident in the test specimens.

Comments: Nil.

Personnel: JPC.

Test Results: No defects detected.

Hardness Testing:

Test Details:

Test Method	AS2205.6.1:2003(R2018)	
Hardness Type	Indentation rows	
Hardness Spacing	0.5 mm spacing	
Indenter	Diamond pyramid	
Test Load	10 kgf	
Test Personnel	JPC	

TABLE A2.1
Results of Hardness Measurements

Sample ID	Zone	S. No.	Harness Measurements (VHN) HV10 (Load 10 kgf)	Average Hardness (Load 10 kgf)
Top plate traverse (T2)	Weld	1	195	192
		2	192	
		3	191	
	HAZ	1	169	166
		2	168	
		3	162	
	Parent	1	144	145
		2	145	
		3	146	
Side plate traverse (T1)	Weld	1	199	202
		2	203	
		3	204	
	HAZ	1	176	172
		2	171	
		3	170	
	Parent	1	152	154
		2	153	
		3	158	

VHN, Vickers hardness number.
Test results: As reported above.
Remarks: Nil.

9 Tribological Behaviour of Polymer Matrix and Metal Matrix Composites

Ch. Sri Chaitanya
VR Siddhartha Engineering College

Babar Pasha Mahammod, G. Chitti Babu, and R. Narasimha Rao
National Institute of Technology Warangal

CONTENTS

9.1 INTRODUCTION

High-performance composite materials have applications in transportation, aviation, military defence, etc., due to their excellent properties like high strength-to-weight ratio and low cost. The components that are used in applications like transportation and aviation industry have generally relative motion and contact with the other components. The relative motion between the contacting bodies causes surface failure in the component due to the wear of the component leading to lower life.

The toughened composites show better mechanical properties. Hence, initial studies are focused on the mechanical properties (Van Velthem et al. 2016; Khalili et al. 2017) of the toughened composites like the impact strength and compression strength (Rao et al. 2019). The impact response of the sandwich structures with toughened epoxy matrix was studied by Irven et al. (2019). The epoxy matrix was toughened by addition of two types of nanoparticles, core shell rubber and silica. The addition of core shell rubber decreased the modulus while the modulus was increased by addition of silica particles. Both the toughening processes improved the fracture toughness of the structure. Addition of brominated epoxy (Sheinbaum et al. 2019) also improved the mechanical properties like strength and toughness and also thermal properties like glass transition temperature. However, it was observed that the shear properties of the composites with blends of brominated epoxy and conventional epoxy resins are negatively impacted. The toughened composites are manufactured by mixing the necessary ingredients and allowed to cure. The curing temperature and the curing rate also affect the properties of the toughened composite materials. The study by Chen et al. (2020) makes an effort to understand the effect of the curing parameters on the properties of the toughened composites. It was reported that the heating rate during curing influences the particle aspect ratio, volume fraction, inter- and intra-layer thicknesses. The shear modulus was observed to be reduced with the increase in the curing temperature. Similarly, the performance of toughened composites with respect to the cryogenic mechanical properties was reported by Qu et al. (2021).

DOI: 10.1201/9780429330575-9

The recent studies about the tribological properties of the composite materials that are toughened by addition of the nano- and microparticles are summarised in this work. Two types of wear mechanisms are considered for the report, and they are sliding wear in dry contact conditions and the erosion wear by air jet mixed with abrasive particles.

9.2 SLIDING WEAR BEHAVIOUR

The sliding wear occurs when two interacting surfaces have relative motion between them. Sliding wear is quantified as wear rate, that is, material loss per unit sliding distance. The effect of the tribological parameters for sliding wear like the sliding distance, sliding velocity and applied pressure on the wear behaviour is reported. Pin on disc tribometer shown in Figure 9.1 is generally used to test the sliding wear behaviour of a material. The pin and the disc materials are selected based on the potential application of the tested material. The pin used in the experimentation generally has lower hardness as compared to that of the disc material. This reduces the damage to the disc, and the disc can be reused after polishing its surface. The wear rate of the sample is obtained by measuring its weight before and after the experimentation. The loss of weight of the sample during the experimentation is used to calculate the volume loss, and the volume loss per unit sliding distance was reported as per Eq. (9.1):

$$\text{Wear rate} = \frac{W_i - W_f}{\rho L} \tag{9.1}$$

W_i and W_f are weights of the sample before and after the experimentation, ρ is the density of the sample and L is the sliding distance. The sliding wear of a material can be reduced by either

FIGURE 9.1 Top view of pin on disc tribometer showing counter surface and pin holder.

improving the lubrication between the interacting surfaces or increasing the wear resistance of the material. Current focus is to review the work carried out by the researchers to increase the wear resistance of the materials.

The tribological performance of polymer (PTFE, PPS, PEEK)-based composites manufactured using different types of fillers like titanium dioxide, zirconia, silica, graphite, bronze and alumina was reviewed by Fredrich et al. (2005). These materials were used in the matrix at different concentrations, and their specific wear rate, that is, volume of material loss per unit sliding distance per unit applied load, was reported. It was observed that the composites with lower-size particles showed better properties compared to the large-sized particles. The composites with lower-sized particles, that is, particle sizes in range of tens of nanometres, achieve significant improvement of wear resistance with low concentrations of 1%–3%. This effect was observed in the composites with inorganic fillers. The addition of organic fillers like carbon fibres also improved the tribological performance of composite materials. The combination effect was observed when both the inorganic particle fillers and short carbon fibres were used together in manufacturing the composite materials.

The effect of the size of the reinforcement particles on the sliding wear properties of metal composites can be obtained by comparing the studies which used the same reinforcements of different sizes. The specific wear rate of the aluminium 7075 matrix-based composites (Al-Salihi et al. 2019) with the alumina nanoparticles as reinforcements is shown in Figure 9.2. Aluminium composites manufactured using the SiC and Gr as reinforcements were studied for their tribological properties.

Reinforcements with microparticles were used by Kaushik and Rao (2016a,b,c, 2017), and with nanosized particles by Prasad Reddy et al. (2017, 2019a,b). Both the studies showed improvement of the tribological behaviour when compared to the base alloy. The improvements were attributed to the increment of hardness of the composite material due to the addition of SiC particles and lubrication effect from the graphite particles. The comparison of the dimension of the particles used for the above studies is shown in Table 9.1.

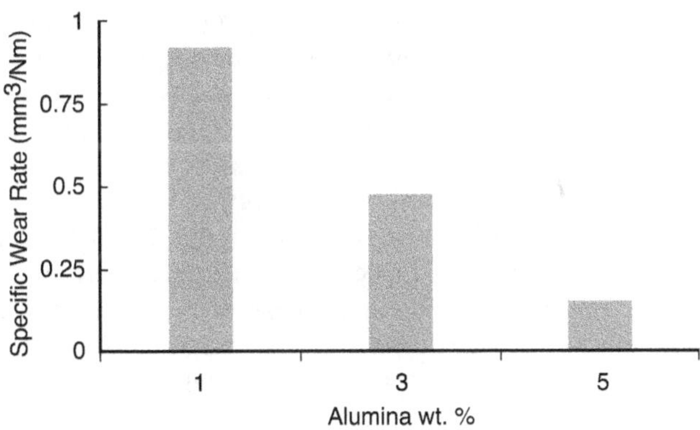

FIGURE 9.2 Specific wear rate of aluminium 7075 matrix composites with nano-alumina as reinforcements.

TABLE 9.1
Average Particle Size of SiC and Gr Reinforcements used by Kaushik et al. and Prasad Reddy et al.

Author	SiC Particle Size	Gr Particle Size
Kaushik et al. (Kaushik and Rao 2017)	32 μm	40 μm
Prasad Reddy et al. (Prasad Reddy, Vamsi Krishna, Rao 2017)	50 nm	500 nm

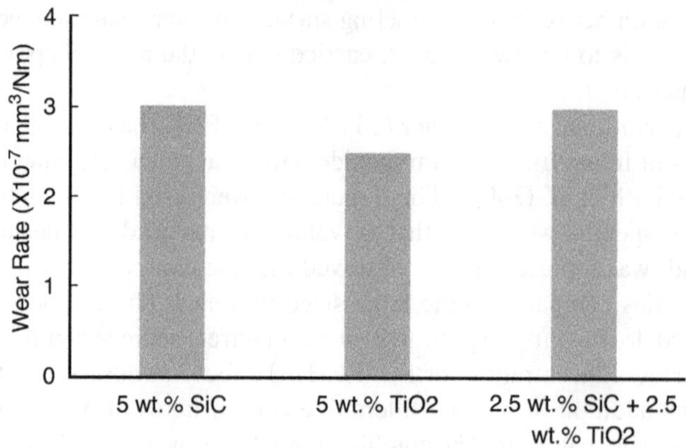

FIGURE 9.3 The wear rate of Al6061 matrix composite material with the addition of the SiC and TiO_2 as reinforcements.

The average wear of the composite materials was estimated to be reduced by 20%–40% by the addition of the reinforcements to the alloy. Although the reduction of the wear seems to be in the same range, the effect of this reduction was achieved by the composition of 2 wt.% when nanoparticles were used while 10 wt.% of reinforcements were required for microparticles. This observation was in line with the polymer composite materials reviewed by Fredrich et al. (2005). The similar responses were seen when TiO_2 is used as the reinforcement in the aluminium matrix alloys (Tamilarasan et al. 2016; Srivastava and Tiwari 2012; Kumar and Rajadurai 2016; Wu et al. 2018; Chang et al. 2005; Baghery et al. 2010; Çomaklı et al. 2014).

The difference in the hardness between the contacting surfaces dictates the wear in those materials. During the sliding wear, the material with the lower hardness exhibits more material loss. The lower the hardness of the target material, the higher will be the hardness. Hence, the harder particles are used as reinforcements in the composite materials to reduce its wear rate. The addition of harder ceramic materials like silica and TiO_2 (Padmavathi et al. 2019) to the aluminium alloy is shown in Figure 9.3. The TiO_2 particles have higher hardness compared to SiC particles. The same trend is visible in the wear rate reported by Padmavathi et al. (2019).

From the studies of Sri Chaitanya and Narasimha Rao (2019, 2020), the aluminium-based and epoxy-based composites with alumina- and silica-based cenospheres as reinforcements are compared. The addition of the reinforcements improved the tribological performance in terms of sliding wear rate, but the improvement is around 30% for the metallic composite with 10 vol.% addition of the cenospheres while the polymer composites improved by only 18%. This shows the effect of the reinforcements is higher in metal matrix composites over polymer-based composites.

9.3 EROSION WEAR BEHAVIOUR

When a component is moving in atmosphere, the transportation vehicles have to endure the impact of the dust and other particles in atmospheric air. These particles when impacted on the surface of the component fracture the material surface and remove certain amount of material from the surface. This loss of material reduces the aesthetics of the surface, interferes with the aerodynamic design of the surface and may lead to the failure of the component.

The erosion resistance of a material is studied using the air-jet erosion test rig. The parameters like the angle of impact of the abrasive particles, the pressure of the air jet and the quantity of the abrasive particles impinge on the sample surface per unit time are used to quantify the wear rate of the sample. The weight of the sample was measured before and after the experimentation. The wear

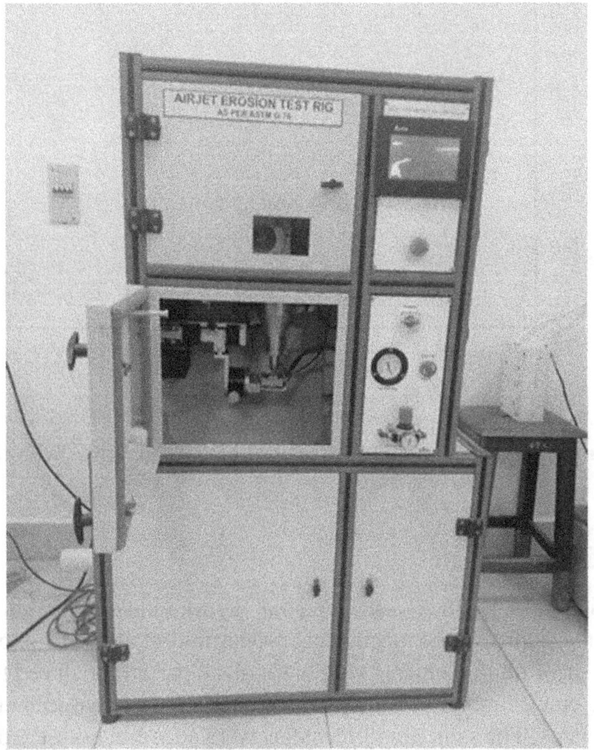

FIGURE 9.4 Air-jet erosion test rig.

rate is reported as the weight loss of the sample unit weight of the abrasive particles. The test rig generally used for the air-jet erosion test is shown in Figure 9.4.

The air-jet erosion properties of the epoxy-based composite materials are with different reinforcements like graphitic carbon nitride (Raghavendra et al. 2019) and carbon black (Ojha et al. 2014). The addition of the reinforcements improved the wear resistance of the composites. The higher volume fraction of the reinforcement in composite improved the resistance of the composite materials. The other parameters like the abrasive particles, speed of the air jet and the angle of the impingement have major influence over the erosion wear properties. The response of the composite to these parameters can be used to define the behaviour of them. For example, the impingement angle at which the wear is maximum defines the elasticity of the material. Although the effect of the additives to the polymer matrix is lower compared to the other parameters, the influence of them is not negligible.

Boggarapu et al. (2020) reviewed the erosion wear properties of various polymer-based composite materials. The velocity of the erodent particles that are impinging the composite material showed to have highest impact over the erosion wear with the relation between them following power law. The addition of the harder particles as reinforcements improved the erosion wear resistance. Egg-shell-based composites were manufactured by Manoj et al. (2018), and their erosion resistance behaviour was studied. These composites have optimal composition of the reinforcement near 4 wt.%. Also, the treatments to the egg shells like boiling them reduced the wear resistance of the final composite. The epoxy composites with SiO_2 as reinforcement also showed similar behaviour. The addition of the reinforcements reduced the material loss due to erosion wear, but the effect is visible only for certain concentration, and then, the material loss tends to increase. Figure 9.5 shows the same for epoxy matrix SiO_2 composites (Abenojar et al. 2017).

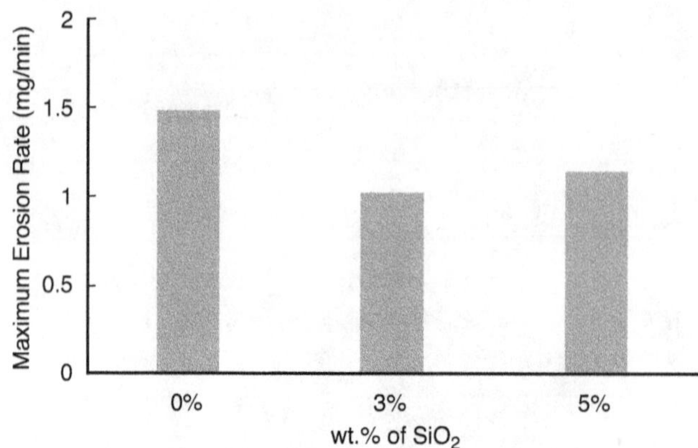

FIGURE 9.5 The maximum erosion rate (mg/minute) of epoxy composites with the addition of SiO_2 as reinforcement.

9.4 CONCLUSIONS

In the review of the literature, it can be concluded that the addition of the reinforcements to toughen the composites not only improves the mechanical properties but also will improve the tribological properties. The addition of reinforcements with lower-sized particles reduced the wear rate at lower concentrations as the concentration of reinforcements required increased with the increase in the size of the reinforcements. The same reinforcements were added to both the polymer matrix composites and the metal matrix composites, and the effect is larger on the metal matrix composites when compared to the polymer matrix composites. Similarly, the addition of the reinforcements to the composites improved the wear rate, but up to a certain optimum value, and further, the properties deteriorated. The erosion behaviour of the composite materials at different impingement angles can be used to predict the elastic behaviour of the composites.

REFERENCES

Abenojar, J., J. Tutor, Y. Ballesteros, J.C. del Real, and M.A. Martínez, 2017. Erosion-wear, mechanical and thermal properties of silica filled epoxy nanocomposites. *Composites Part B: Engineering* 120: 42–53.

Al-Salihi, H.A., A.A. Mahmood, and H.J. Alalkawi, 2019. Mechanical and wear behavior of AA7075 aluminum matrix composites reinforced by Al2O3 nanoparticles. *Nanocomposites* 5: 67–73.

Baghery, P., M. Farzam, A.B. Mousavi, and M. Hosseini, 2010. Ni–TiO2 nanocomposite coating with high resistance to corrosion and wear. *Surface and Coatings Technology* 204: 3804–3810.

Boggarapu, V., R. Gujjala, and S. Ojha, 2020. A critical review on erosion wear characteristics of polymer matrix composites. *Materials Research Express* 7: 022002.

Chaitanya, C.S. and R.N. Rao, 2019. Surface failure of syntactic foams in sliding contact. *Materials Today: Proceedings* 15: 63–67.

Chang, L., Z. Zhang, C. Breidt, and K. Friedrich, 2005. Tribological properties of epoxy nanocomposites I. Enhancement of the wear resistance by nano-TiO2 particles. *Wear* 258: 141–148.

Chen, C., A. Poursartip, and G. Fernlund, 2020. Cure-dependent microstructures and their effect on elastic properties of interlayer toughened thermoset composites. *Composites Science and Technology* 197: 108241.

Çomakli, O., T. Yetim, and A. Çelik, 2014. The effect of calcination temperatures on wear properties of TiO2 coated CP-Ti. *Surface and Coatings Technology* 246: 34–39.

Friedrich, K., Z. Zhang, and A.K. Schlarb, 2005. Effects of various fillers on the sliding wear of polymer composites. *Composites Science and Technology* 65: 2329–2343.

Irven, G., A. Duncan, E. Rolfe, D. Carolan, A. Fergusson, J.P. Dear, 2019. Impact response of composite sandwich structures with toughened epoxy matrices. In *ICCM22 2019, Melbourne*. Engineers Australia: 1895–1906.

Kaushik, N.C. and R.N. Rao, 2016a. Effect of applied load and grit size on wear coefficients of Al 6082–SiC–Gr hybrid composites under two body abrasion. *Tribology International* 103: 298–308.

Kaushik, N.C. and R.N. Rao, 2016b. Effect of grit size on two body abrasive wear of Al 6082 hybrid composites produced by stir casting method. *Tribology International* 102: 52–60.

Kaushik, N.C. and R.N. Rao, 2016c. The effect of wear parameters and heat treatment on two body abrasive wear of Al-SiC-Gr hybrid composites. *Tribology International* 96: 184–190.

Kaushik, N.C. and R.N. Rao, 2017. Influence of applied load on abrasive wear depth of hybrid Gr/SiC/Al-Mg-Si composites in a two-body condition. *Journal of Tribology* 139: 1–9.

Khalili, P., K.Y. Tshai, D. Hui, and I. Kong, 2017. Synergistic of ammonium polyphosphate and alumina trihydrate as fire retardants for natural fiber reinforced epoxy composite. *Composites Part B: Engineering* 114: 101–110.

Kumar, C.A.V. and J.S. Rajadurai, 2016. Influence of rutile (TiO2) content on wear and microhardness characteristics of aluminium-based hybrid composites synthesized by powder metallurgy. *Transactions of Nonferrous Metals Society of China* 26: 63–73.

Ojha, S., S.K. Acharya, R. Gujjala, 2014. Characterization and wear behavior of carbon black filled polymer composites. *Procedia Materials Science* 6: 468–475.

Padmavathi, K.R., R. Ramakrishnan, K. Palanikumar, 2019. Wear properties of SICP and TIO2P reinforced aluminium metal matrix composites. *Indian Journal of Engineering and Materials Sciences* 26: 51–58.

Panchal, M., G. Raghavendra, M.O. Prakash, and S. Ojha, 2018. Effects of environmental conditions on erosion wear of eggshell particulate epoxy composites. *Silicon* 10: 627–634.

Prasad Reddy, P. Vamsi Krishna, and R.N. Rao, 2017. Al/SiC NP and Al/SiC NP /X nanocomposites fabrication and properties: A review. *Proceedings of the Institution of Mechanical Engineers, Part N: Journal of Nanomaterials, Nanoengineering and Nanosystems* 231: 155–172.

Prasad Reddy, P. Vamsi Krishna, and R.N. Rao, 2019a. Two-body abrasive wear behaviour of AA6061–2SiC-2Gr hybrid nanocomposite fabricated through ultrasonically assisted stir casting. *Journal of Composite Materials* 53: 2165–2180.

Prasad Reddy, P. Vamsi Krishna, R.N. Rao, 2019b. Tribological behaviour of Al6061–2SiC-xGr hybrid metal matrix nanocomposites fabricated through ultrasonically assisted stir casting technique. *Silicon* 11: 2853–2871.

Qu, C.-B., T. Wu, et al., 2021. Improving cryogenic mechanical properties of carbon fiber reinforced composites based on epoxy resin toughened by hydroxyl-terminated polyurethane. *Composites Part B: Engineering* 210: 108569.

Raghavendra, G., P. Pratap Naidu, and G. Raghavendra, 2019. Erosion behaviour of graphitic carbon nitride (g-C3N4) reinforced epoxy composites. *IOP Conference Series: Materials Science and Engineering*: 577.

Rao, Y.A., K. Ramji, P.S. Rao, and I. Srikanth, 2019. Effect of A-MWCNTs and ETBN toughener on impact, compression and damping properties of carbon fiber reinforced epoxy composites. *Journal of Materials Research and Technology* 8: 896–903.

Sheinbaum, M., L. Sheinbaum, O. Weizman, H. Dodiuk, and S. Kenig, 2019. Toughening and enhancing mechanical and thermal properties of adhesives and glass-fiber reinforced epoxy composites by brominated epoxy. *Composites Part B: Engineering* 165: 604–612.

Sri Chaitanya, R.N. Rao, 2020. Tribological behavior of cenosphere-filled epoxy syntactic foams in dry sliding conditions. *Journal of Tribology* 142: 1–7.

Srivastava, S. and R.K. Tiwari, 2012. Synthesis of epoxy-TiO2 nanocomposites: A study on sliding wear behavior, thermal and mechanical properties. *International Journal of Polymeric Materials and Polymeric Biomaterials* 61: 999–1010.

Tamilarasan, T.R., R. Rajendran, U. Sanjith, G. Rajagopal, and J. Sudagar, 2016. Wear and scratch behaviour of electroless Ni-P-nano-TiO2: Effect of surfactants. *Wear* 346: 148–157.

Van Velthem, P., W. Ballout, J. Horion, Y-A. Janssens, et al., 2016. Morphology and fracture properties of toughened highly crosslinked epoxy composites: A comparative study between high and low Tg tougheners. *Composites Part B: Engineering* 101: 14–20.

Wu, H., J. Zhao, X. Cheng et al., 2018. Friction and wear characteristics of TiO2 nano-additive water-based lubricant on ferritic stainless steel. *Tribology International* 117: 24–38.

10 Carbon Nanotube Hybrid Fabric – Manufacturing and Applications Including Structural and Wearables

V. Ng, D. Chauhan, R. Noga, M. Chitranshi, A. Pujari, A. Kubley, A. Bhattacharya, and S. Grinshpun
University of Cincinnati

S. Fialkova
North Carolina Agricultural & Technology

W. J. Williams and S. Kilinc-Balci
National Institute for Occupational Safety & Health

J. Campbell
Pike Township Fire Department

B. J. Jetter
Glendale Ohio Fire Department

V. N. Shanov and M. J. Schulz
University of Cincinnati

CONTENTS

DOI: 10.1201/9780429330575-10

10.1 INTRODUCTION: BACKGROUND AND DRIVING FORCES

Wearable technology is an emergent field where previously unsolvable problems related to clothing and the body are now being solved through embedded textile technology. Carbon, the sixth most abundant element in the universe, has been used since prehistoric times as charcoal, coal, and diamond for tools, fuel, and filtration. Therefore, the integration of nanomaterials, such as carbon nanotubes (CNTs), into textiles for wearable applications including sensors, supercapacitors, electronic devices, and filtration are just a few new ways to investigate the use of carbon. Various publications describe the emerging use of nanomaterials in textiles [1–5]. A review of two example areas is discussed in the next section.

10.1.1 Review of Filtering Applications of CNT Fabric

A novel hybridization method to develop a filter with nanofiber, for example, was used to synthesize a CNT/polymer hybrid nonwoven fabric by electrospinning polyethylene oxide nanofibers onto aligned CNT sheets. The CNT sheets (also known as CNT fabric) are formed by wrapping multiple layers of CNT films onto a rotating mandrel [6]. To understand the hybrid properties of the fabric, two processing steps were taken. The first was calendering, in which fabric was passed between two or more rollers (calenders) under pressure or heat. The second was consolidation, where the hybrid fabric was placed under two megapascals of pressure. During the consolidation process, no heat was applied for 5 minutes, and the fabric was calendared at 70°C. This method was used for assembling smaller objects to achieve desired structures and properties. This process was dependent upon the mechanical, thermal, and other energy properties to consolidate and create bonding between the materials. All the consolidated hybrid samples showed improved strength as the applied pressure created more direct contact between the CNTs and the polymer. Calendering also helped increase the tensile strength dramatically. Due to the high-strength and large surface area of CNT sheets, the specific strength increased with increasing volume fraction of the CNTs in the fabric. The hybrid fabric exhibited good particle filtration and barrier properties.

In another approach, continuous aligned CNT sheets, which are suitable as nonwoven fabrics due to the alignment and high surface area, were produced and sandwiched between polyimide (PI) membranes. These composited materials were thermally bonded by melting electrospun polyetherimide (PEI) nanofibers to make a hybrid filter. The filters were fabricated in two different types of structure (shown in Figure 10.1) [7]. The first structure (T, together) consisted of CNT layers stacked together on top of each other. The second structure (S, separated) consisted of CNT layers separated by a layer of PEI. For the T structure, one layer of PEI and three layers of PI were used on the top and bottom to hold layers of CNT sheets together. Due to strong van der Waals force between the CNTs, PEI wasn't needed in between the CNT layers. Similarly, for the S structure filters, one layer of PEI and three layers of PI were used to hold the CNT sheets together. However,

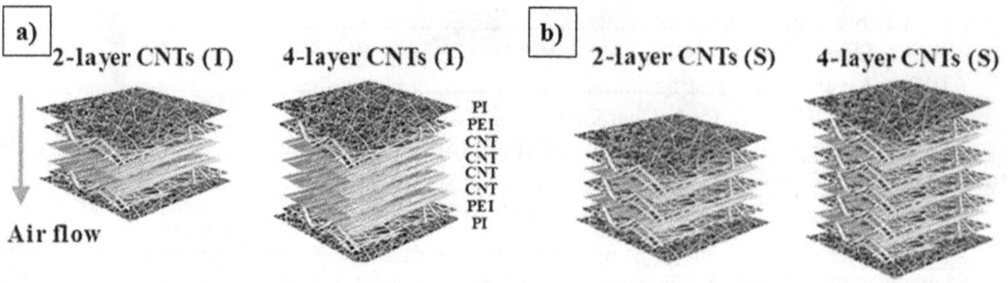

FIGURE 10.1 Diagram of PI (yellow), PEI (green), and CNT sheet (white) hybrid filters. Two and four layer: (a) together (T) structures and (b) CNT sheets separated (S) structures [7].

in the S structure, the CNT layers were separated by PEI nanofiber. Filters with two-layers and four-layers of CNTs were tested. It was found that filters with two-layers showed higher penetration of particles in both structures (T and S) as compared to the four-layered CNT structure. Filters with four-layers of CNT have a high quality factor (Q-factor) [8,9], that is based on filtration efficiency and pressure drop. A high Q factor is indicative of good filtration performance with high filtration efficiency, and reasonable pressure drop. Filters with four-layers in the T and S structures reached high-efficiency particulate air filter level performance. Filtration performance of the hybrid filters was also tested at elevated temperatures, and the result showed that the hybrid filters could be used up to 200°C.

Another example are filters made with a hydrophilic ultrafiltration membrane, which was fabricated using atomic layer deposition of ZnO (zinc oxide) nanoparticles on the CNT sheet. The wettability of the membrane was transformed from hydrophobic to hydrophilic with an increase in the number of atomic layer deposition cycles. The deposition of ZnO nanoparticles increased the wettability and mechanical stability of the CNT membranes [10].

10.1.2 Review of Structural Applications of CNT Fabric

Integrating CNTs with carbon fibers (CFs) is an approach to developing improved composite fabric for structural applications. Carbon microfiber laminated composites are widely used in various structural applications due to their lightweight and high-strength properties. The versatile properties of laminated composites have attracted research attention for applications in various fields [11]. In CF composites, CF is used as a woven or unidirectional fabric, tow, chopped fibers, or mesh, depending upon the required properties for the application [12]. CF serves as a reinforcement, and the epoxy polymer acts as a matrix in laminated composites. The epoxy polymer enables lamination of the CF fabrics by acting as a binder to achieve the required thickness and stiffness for structural applications. The polymer matrix's primary role is to transfer stress between fibers, provide support under compression loading by preventing fiber buckling, provide a barrier for the adverse environment, and protect the fiber's surface from mechanical abrasion. However, the low fracture toughness of the brittle matrix makes laminated composites prone to crack initiation and propagation. The most common failure mechanisms in laminate composites originate in the polymer matrix, such as matrix cracking, fiber matrix interface debonding, and interlaminar delamination [13].

The use of CNTs in laminated composite enhances their resistance to crack initiation and propagation [14]. CNT powders are commonly used as nanofillers to assist laminated composites' fracture toughness and electrical conductivity [15,16]. The integration of CNT sheets in the laminated composite manufacturing process is tedious. The lack of proper resin infiltration of the CNT sheet in laminated composites results in defect sites [11,17]. The integration of CNT powders and sheets into laminated composites is an area of continuing research with prospects for industrialization [18,19].

Overall, the properties of CNT have opened new doors to flexible and wearable applications in heat protection, filtration, and sensing. A discussion of CNT membranes used for water, air, and bioaerosol filtering is given in our previous works [5,20–22]. The properties of CNT materials were customized by synthesizing CNTs at high-temperatures and integrating other materials (such as continuous CFs or CF tissue) into the CNT film during the CNT production process. CF materials, for example, can be layered with the CNT fabric to increase mechanical strength or filtering performance of the CNT fabric. This chapter examines the structure of CNTs manufactured from the substrate-grown nanotube synthesis method and the floating catalyst chemical vapor deposition (FCCVD) method. The structure of CNT-CF hybrid fabric was made in a single-step FCCVD process.

10.2 MANUFACTURING CNT FABRIC

In this section, the following materials were studied: (i) chopped CF (CF, with no CNT), (ii) CNT array (CNT-Arr), (iii) pristine CNT (CNT-Pris), (iv) CNT with CF tissue (CNT-CFT), and (v) CNT integrated with chopped CF (CNT-CF). Acronyms in parenthesis are used herein. Sample (i) was used as purchased. Sample (ii) was synthesized via catalytic CVD (CCVD). Samples (iii–v) are CNT fabrics synthesized from FCCVD. Samples (iv–v) are CNT hybrid fabrics. Pristine means an unmodified sample produced without any additives and postprocessing via the FCCVD method.

Both the tissue and the chopped CF were purchased from commercial manufacturers. The tissue is a CF tissue/veil (CST, The Composite Store, Item#: CF607), and weighs approximately 0.5 ounces per square yard (oz/yd^2) and 127 μm thickness) [23]. The fibers in the tissue were randomly oriented. The chopped (and milled) CF (Zoltek, Item#: PX35) is a powdered material [24]. The CF continuous tows are made up of 50,000 filaments processed from a polyacrylonitrile precursor. The density of CF is about 2 g/mL.

There are a variety of methods in which CNTs are manufactured, such as arc discharge, fluidized bed, CVD, and laser ablation [5,25–40]. Among these, the most common way of synthesizing CNTs is CVD [5,41]. Two types of CVD methods were employed. The first was the CCVD method (shown in Figure 10.2a). The second was the FCVVD method (shown in Figure 10.2b).

The CCVD method is traditionally used to grow vertically aligned free-standing, spinnable CNT arrays via a substrate used in a batch process [5,22,42–44]. The substrate is a silicon wafer coated with a metal catalyst film. When the film is annealed, it forms nanoparticles (such as Fe, Ni, Cu). The substrate is placed inside a high-temperature furnace. The CNT arrays are grown on the substrate using a carbon gas source (such as ethylene, acetylene, methane, xylene), a carrier gas (Ar,

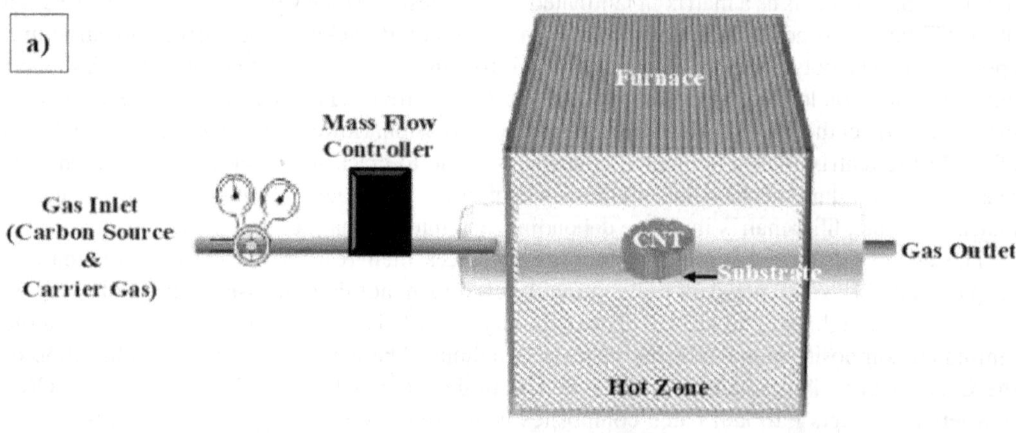

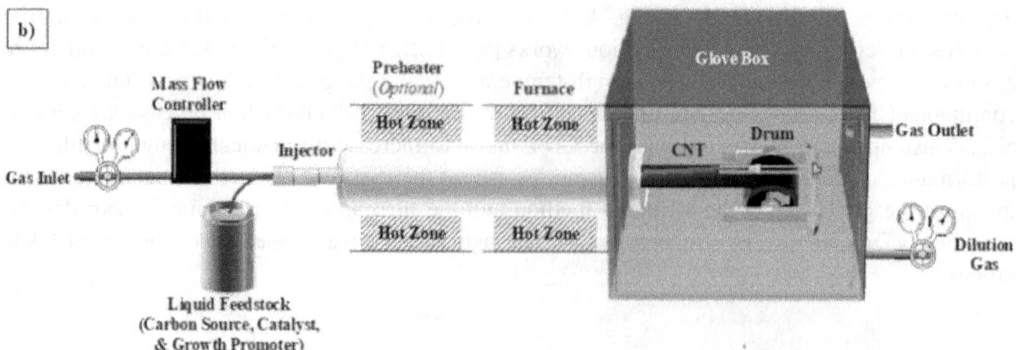

FIGURE 10.2 Schematic of CNT manufacturing processes: (a) CCVD and (b) FCCVD.

H_2, N_2), and assisted with water vapor. The array fabric is drawn from the spinnable CNT array in a postprocessing step and densified every 20 layers with acetone (typically the fabric in totality is 100 layers that is10 μm thick).

The second method is the FCVVD method, which typically produces long CNTs with length >1 mm intertangled in a web. The web is wrapped onto a rotating and translating drum to fabricate CNT fabric. The macroscale CNT sheets have a variety of versatile properties, which makes them a good candidate for hybridization [5,11,21].

In general, there are three different ways of combining CNT and chopped CF.

i. *Formation of CNT – CF tissue sheet.* Using the FCCVD method, the CFT is combined with the CNT web during the synthesis process. The CFT is placed on a Teflon sheet on a drum that rotates and translates. The CNT web is wrapped onto the CFT in the glove box. This produced CNT-CFT two-layered sheet or fabric.
ii. *Formation of CNT – chopped CF sheet.* The chopped CF was dispersed in a solvent, such as acetone. The CNT sock was wrapped onto a Teflon sheet on the drum. The solvent with CF was sprayed onto the CNT sock that was wrapped layer by layer to form a CNT-CF hybrid fabric. The solvent also densifies the CNT fabric.
iii. *Formation of CNT – continuous CF sheet.* Continuous CF in the tows or spread tow tape can also be wrapped onto the CNT layer simultaneously as the CNT fabric is formed. These different variations of CNT hybrid fabric can be used to form laminated composites with improved properties.

Sheet can be a thin film, 3–10 μm thick. Fabric also describes a sheet, but typically fabric is a thicker material.

10.3 CNT FABRIC CHARACTERIZATION

The CNT fabric materials were characterized using thermogravimetric analysis (TGA) and scanning electron microscopy (SEM) with electron-dispersive spectroscopy (EDS). Flame testing was also performed in which the time to burn-through a fabric sample was measured. The temperature of the flame was measured using a FLIR T640 Thermal Imaging IR Camera. A FEI Apreo SEM was used for characterizing air filtration and CF composite materials, which provides the material's microscopic surface morphological information. In addition, the thermal degradation of composites, for both CF and CNT-CF, was analyzed with Netzsch STA409 TGA instrument.

10.3.1 Flame Testing

A butane hand torch was used to test the resiliency (resistance to burn-through) of carbon nanomaterials. Ring stands and clamps were used to hold the samples and torch in place, with the torch nozzle 7 cm from the surface of the sample, with the flame just impinging on the sample surface. The flame test was carried out for 60 seconds or less. The experiment was monitored using a digital video camera to capture the physical damage and temperature data. The data were recorded using a FLIR T640 Thermal Imaging IR Camera, as shown in Figure 10.3.

Flame testing was conducted on two samples: CNT-Arr and CNT-Pris (Figure 10.4). The tested CNT-Pris fabrics had a unique two-phase composition. In the case of CNT-Pris, the two phases were intertwined CNT webs and about 20% Fe residual catalyst. The latter oxidized under the flame in the ambient air environment, thus producing ceramic-like iron oxide. This transition was observed on the sample after burning.

The iron oxide was active and easily formed on the fabric surface in the presence of oxygen (appearance of orangish spots). This promoted localizing the damage on the CNT sheet and reducing the diameter of the hole created by the flame.

FIGURE 10.3 Flame test setup: (a) schematic and (b) experimental. The testing was performed inside of a well-ventilated fume hood. The sample was suspended and held by clamps. The flame source was horizontally placed 7 cm from the sample. A video camera and an IR camera were used to record data results.

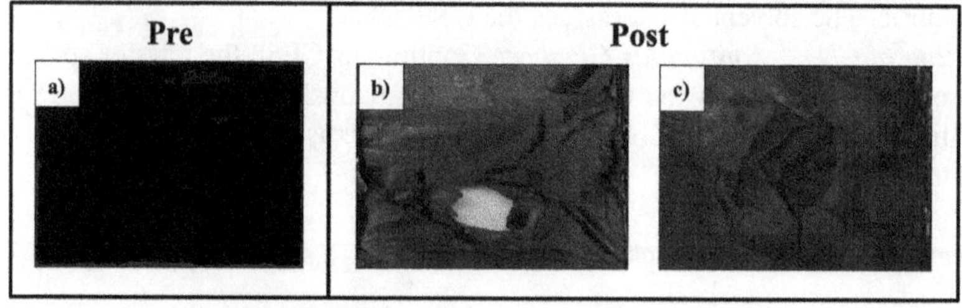

FIGURE 10.4 Nanomaterial samples: (a) pre- and (b & c) postflame test. (a) Image of unmodified CNT fabrics. (b) Sample is CNT-Arr, CNT sheet drawn from spinnable [substrate] array, 100 layers. A hole formed after 25 seconds of exposure time to flame. (c) Sample is CNT-Pris, pristine CNT sheet from floating catalyst CVD, which did not burn-through after being exposed for 60 seconds to the flame.

The characterization confirmed formation of iron oxide during the test. The testing also showed high flame resistance of CNT-Pris fabric. Finally, burn-through tests were conducted with CNT-Arr fabric made by drawing CNT forests that withstood 25 seconds of heating until a hole became visible. This sample also maintained a lower burn-through temperature, at around 500°C, than CNT-Pris fabric. Overall, the CNT-Pris fabric performed better than the CNT-Arr fabric, which may be due to the residual iron catalyst. We believe that the intertwined CNT-Pris webs with iron catalyst helped to dissipate the heat from the surface, thus reducing damage from the flame. In general, the thermo-oxidation of CNT-Pris fabric may play a key role in improving flame resistance properties of composites that can act as a fire shield.

10.3.2 Air Filtration Testing

Filtering applications require a porous surface, high surface area, small pore size, and good mechanical strength. CNT-Pris fabric is thicker than 10 mm and is mostly impermeable to air. Thin CNT fabric can be layered onto CFT to form a more breathable material. Particles such as granulated activated carbon (GAC) can be added to the CNT fabric to make the fabric more breathable. CNT fabric can filter by capturing toxic chemicals or smoke particles on the surface of the material or in the interior of the fabric. As such, CNTs are a suitable candidate for filtering applications. Cigarette smoke contains many toxic chemicals and is used to test the filter. Generally, cigarette smoke generates "aerosol particles in the gas phase" which yields over 7,000 chemical compounds [45]. The

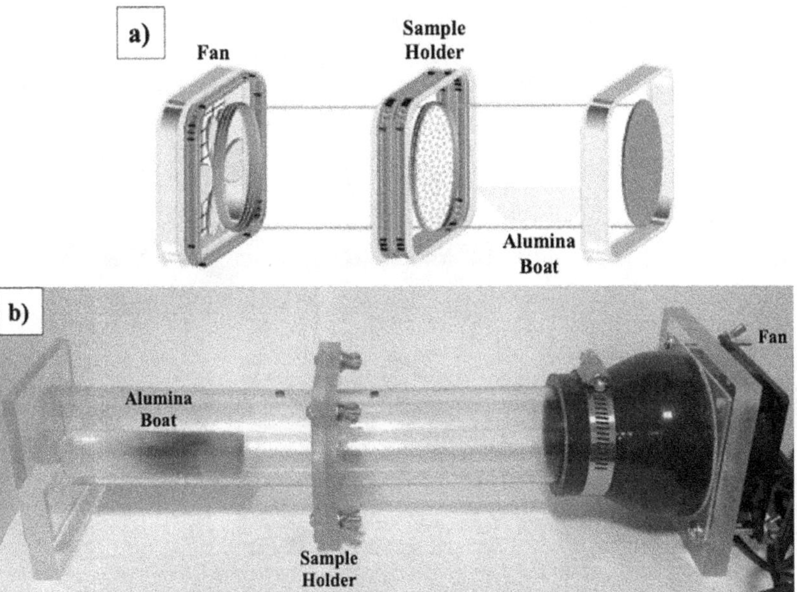

FIGURE 10.5 Air filtration test setup: (a) schematic and (b) experimental. Soot is generated from burning a cigarette placed inside an alumina boat (inlet). The testing filter material is the sample, which is placed in the middle of the sample holder. A fan is used to mimic body movement (outlet). Gas and particulates generated from the soot generator travel from the inlet to the outlet.

main chemical constituents are C, N, O, and H. Tobacco ash is released in two forms: (i) particulate and (ii) gas phases. In the particulate phase typically, the following are generated: "carboxylic acids ($-CO_2H$), phenols (C_6H_6O), water (H_2O), nicotine, cyanide (CN), formaldehyde (CH_2O), polycyclic aromatic hydrocarbons (PAHs)", etc., While, in the gas phase, the following are observed: "nitrogen (N_2), oxygen (O_2), carbon monoxide (CO), carbon dioxide (CO_2), acetaldehyde (CH_3CHO), methane (CH_4), HCN, nitric acid (HNO_3), ammonia (NH_3), hydrogen sulfide (H_2S), formaldehyde (CH_2O), PAHs, volatile organic compounds", etc. According to A. W. Dumas's Tobacco Ash [46], he found "about 20%–25% potassium (as K_2O), 3%–5% phosphates (P), 25%–30% calcium (Ca), 5%–7% magnesium (Mg), 1% sodium (Na), 2% chlorine, and 6% sulfates".

An experimental air filter test setup was developed at the University of Cincinnati (UC) as shown in Figure 10.5. Two tubes (15¼ cm (length) by 5 cm (outer diameter)) were connected between two sample holders. On one end, was the inlet, where the tube was capped. At the other end, was the outlet, where a controlled motorized fan was connected. The fan was used to mimic natural air flow. The inlet tube contained an alumina boat, where material was burned to generate soot.

Two 5-cm-by-5-cm nanomaterial samples were tested, which were CNT-Pris and CNT-CFT. The thicknesses and filter results were analyzed by SEM (Figure 10.6). The samples in Figure 10.6a–c are the CNT-Pris (thickness avg = 70 mm). The samples in Figure 10.6d–f are the CNT-CFT (CNT thickness avg = 40 mm and the CFT thickness avg = 150 mm).

Figure 10.6a, imaged at 350 magnification, shows debris along the cross-section area. Figure 10.6b, imaged at 15,000 (=15k) magnification, displays CNT bundles and catalyst (and possibly some debris/ash). Also, it was difficult to determine if the circled area was either CNT webbing or an agglomeration of ash particulates. Figure 10.6c, imaged at 25k magnification, reveals CNT bundles (and possibly debris and/or catalyst). The image shown is a higher magnification of the circled area from Figure 10.6b. From the image, it can be observed that the debris was engulfed in between CNT bundles

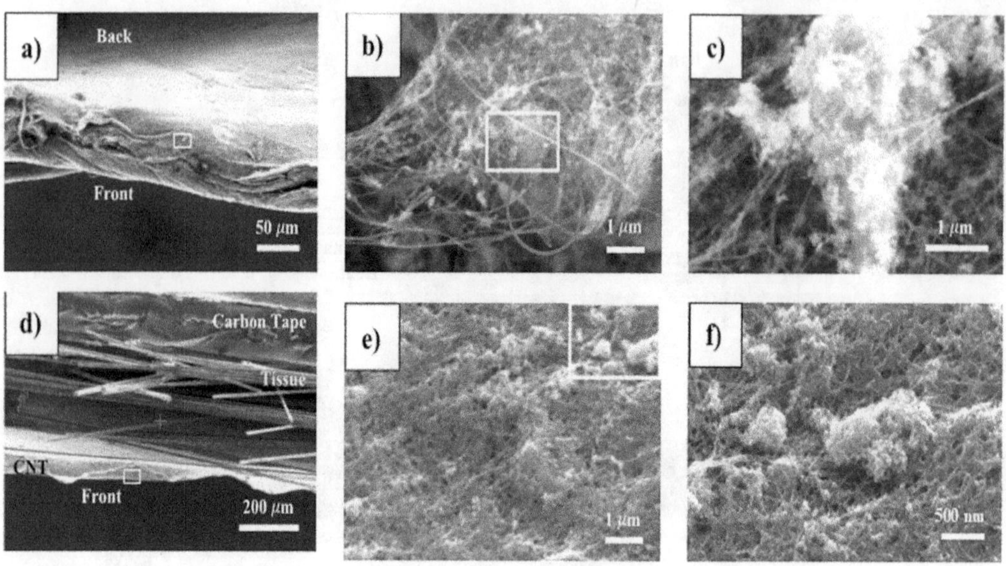

FIGURE 10.6 Cross-section view of filtration results of nanomaterials synthesized from FCCVD. (a–c) CNT-Pris fabric at 350, 15k, and 25k magnification, respectively. (d–f) CNT-CFT hybrid fabric using SEM at 120, 15k, and 35k magnification, respectively. Bright white clumps are soot particulates. The white square is the area of interest imaged at a higher magnification.

with catalyst. The size of the ash agglomeration may be inaccurate due to charge effects during testing. In this case, the charge effect is caused by the nonconductive soot, which may dictate the filter's efficiency.

Figure 10.6d, imaged at 120 magnification, shows three distinct layers. The bottom layer is the CNT (bright solid). The middle layer is the CFT (looks like uncooked spaghetti), which seems unaffected (i.e., no debris). The top layer is the carbon tape (used to mount the sample). The image shows the tissue layer has delaminated from the CNT. Figure 10.6e, imaged at 15k magnification, shows a greater number of debris that settled on the material (bright clumps). CNT bundles are observed, as well. Figure 10.6f, imaged at 35k magnification, confirmed the bright clumps to be either debris (and/or particulate) generated from soot in the filter experiment. Thus, SEM with EDS was performed (Figure 10.7) to qualitatively measure the mass fractions of elements present in a sample [47].

In Figure 10.7a, EDS shows "adsorption" of the following chemical elements: C, O, N, Fe, S, Na, Si, Mg, Al, P, Cl, and various unidentifiable peaks. While in Figure 10.7b, EDS shows C, O, N, Fe, S, Na, and Mg. These peaks are indicative of an accumulation of particulate from tobacco ash. Since EDS is a qualitative measurement, the chemical concentration is unable to be determined. We believe the observed spectra were caused by air passing around or through the fabric samples. The air flow was generated by a fan during testing.

Overall, both the CNT-Pris and the CNT-CFT performed well. Debris was observed throughout the cross-section area of the samples, which was supported by SEM/EDS. The difference between these two samples was the addition of the CFT. It is believed that the tissue layer (which delaminated from the CNT layer) increased the breathability of the filter material.

10.3.3 CNT-CF Composite Characterization

The CNT-CF, which is CNT with chopped CF composite, was successfully synthesized by FCCVD, where the CF was added to the CNT fabric in the densification step, by spraying solvent dispersed

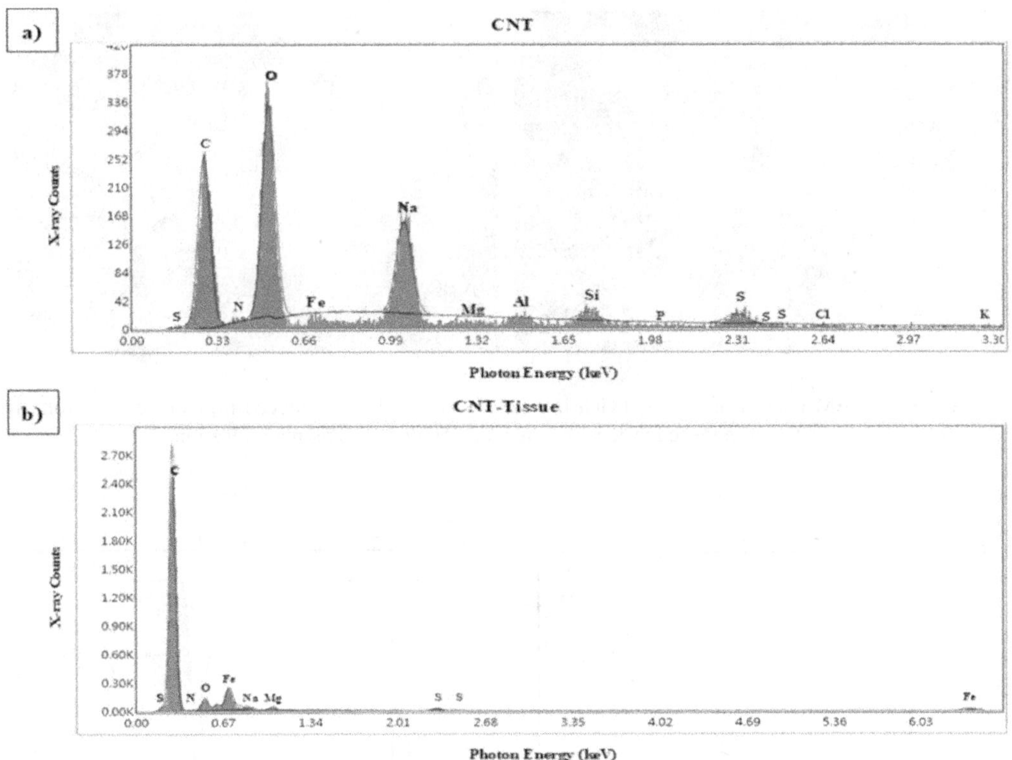

FIGURE 10.7 SEM/EDS filtration results of (a) CNT-Pris fabric and (b) CNT-CFT hybrid fabric. Significant characteristic elemental peaks are identified in each spectrum.

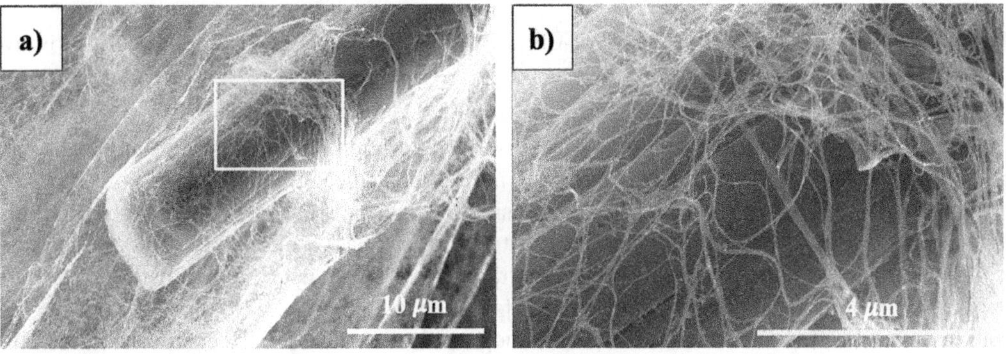

FIGURE 10.8 SEM images of chopped CF encompassed with the multiple CNT bundles, synthesized from FCCVD. The sample was imaged at (a) 3.5k magnification. The white square is the area of interest; imaged at a higher magnification (b) 15k.

with CF onto the CNT as it wound on the drum. Figure 10.8 shows the microscopic image of the CF integrated into the CNT bundles.

Similarly, CNT-CF was produced in a single step during the FCCVD process by interleaving CNT fabric, where CF tows were inserted between the CNT fabric layers (Figure 10.9). Figure 10.9 shows the SEM image of the CNT-CF fabric by hand layup and resin infusion methods.

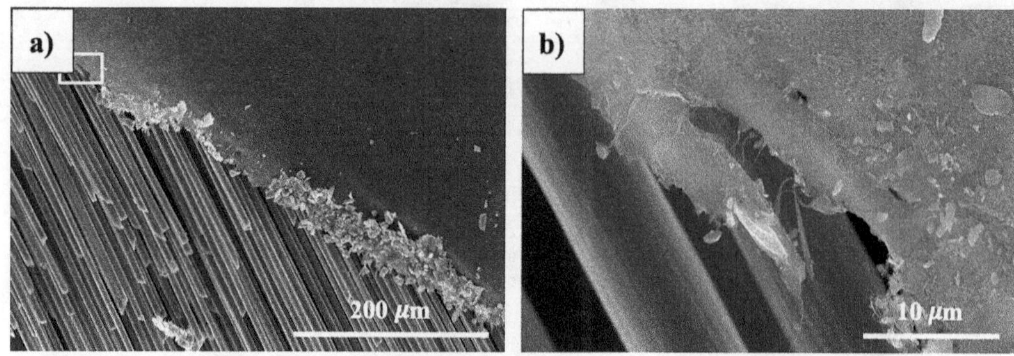

FIGURE 10.9 SEM images of CF sandwich in between CNT fabric, produced in a single step from the FCCVD method. CNT-CF composite fabric is shown at (a) 250 and (b) 3.5k magnification.

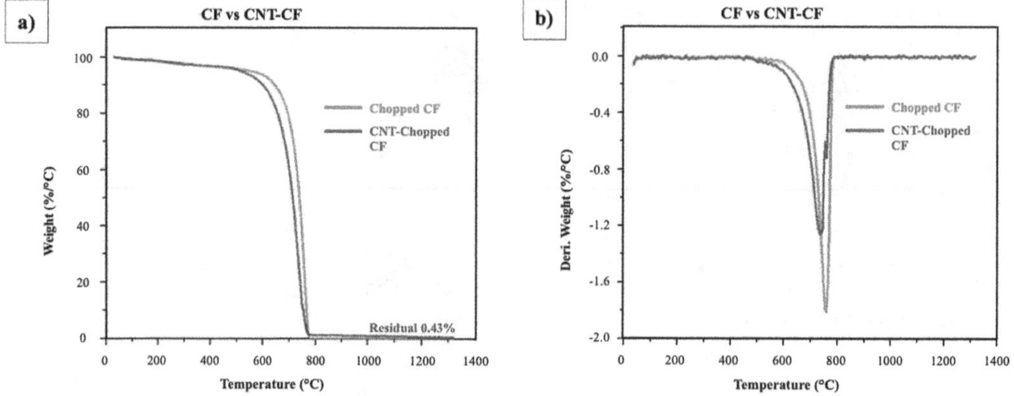

FIGURE 10.10 TGA spectra of CF and CNT-CF. (a) Thermogram of CF and CNT-CF. (b) First derivative of the thermogram.

The CNT bundles stick to the CF tows to resist the separation of CNT fabric from the CF. The porosity in the CNT fabric will allow good resin infiltration and can evade delamination.

The low thermal stability of laminated composites' resin matrix makes them unsuitable for high-temperature structural applications, such as reentry space vehicles which experience temperatures as high as 2,500°C. The resin matrix begins degrading at ~250°C, which makes high-temperature applications of laminated composite challenging [48]. The integration of CNTs enriches the thermal stability of CFRC. Figure 10.10a shows the thermogram for CF and CNT-CF composite fabric. The first derivative curve provides the information on the change in weight percentage with respect to the change in temperature, in °C (Figure 10.10b). A sample mass of 3–4.5 mg was heated from room temperature to 1,300°C at a heating rate of 5°C/minute in air at a flow rate of 100 mL/minute.

Under oxidizing conditions, thermal degradation of CF and CNT-CF commenced at ~600°C. At ~780°C, the CF completely decomposed without leaving any residue. Whereas, CNT-CF, a minimal amount of residue, remained, which may have originated from the iron catalyst used in synthesizing CNTs. From the TGA, it is believed that the CNT-CF composite fabric was as thermally stable as CF. Therefore, reducing the matrix volume fraction and increasing the filler volume fraction may increase the laminated composites' operating temperature. From Figure 10.10b, the first derivative curve, the weight reduction was higher for CF compared to CNT-CF, because the change in mass per temperature is higher for CF than for CNT-CF.

10.3.4 Investigating High-Strength and High-Stiffness Composites

High-strength and high-stiffness composites are important to reduce structural weight in applications that involve motion or dynamics. Applications can benefit from lightweight structural materials, such as spacecraft, aircraft, sporting equipment, and automobiles. An approach being investigated is presented here to increase the strength and stiffness of polymeric fiber-reinforced composites. The approach was to deposit the CNT sock as a thin film on microfiber (MF) spread tow tape. The CNT-MF spread tow tape can then be used to lay up composite structures. Spread tow tape consists of aligned MFs held together by a polymer binder material. The tape is extremely strong and stiff in the fiber direction; but it has very low strength and stiffness in the transverse in-plane direction. Depositing CNT film onto the MF tape may add strength and stiffness perpendicular to the fiber direction. The theory behind the increased properties is briefly explained. Individual CNTs in the CNT film are ~1 mm long, and MFs are typically 7 mm diameter. Therefore, individual CNTs, which are randomly oriented, can bridge tens of MFs. The surface MF in the spread tows will compress the CNT film and tend to grip the undulating CNT. The CNT film will then have to be pulled out from in between the MF tows to fail the composite. The resistance to pulling out the CNT from in-between MF layers will increase the strength of the spread tow tape perpendicular to the fiber direction. The CNT-MF spread tow taped can be layered (e.g., 0° and 90°) to form a composite. Because of the additional strength of the CNT film, which can displace some of the polymer matrix, the composite will have modestly increased properties. Fabrication and testing of CNT-MF spread tow composites is in an early stage of investigation. A tape winder was used to deposit CNT film on CF spread tow tape (Figure 10.11).

The thinner the tow, the greater the increase in properties, in theory. CF tape is shown in the example, but other fibers such as fiberglass, Kevlar, and other fibers in tape form can also be used. Spread tow tape can be layered with minimal resin pockets that occur in woven fabric used in composites. Spread tow tape has a thin epoxy binder holding the CF together. Layering CNT onto the tape and curing at room temperature produces a lightweight composite strip as shown in Figure 10.12. Nanoparticles (NPs) can also be integrated into the CNT being deposited onto the CF tape. Summarizing, this section proposes that CNT film customized by the addition of NPs can be layered onto CF tape. The tape can be used to produce a strong composite structural material with enhanced electrical and thermal properties. The toughness of the composite may also be improved by integration of the CNT.

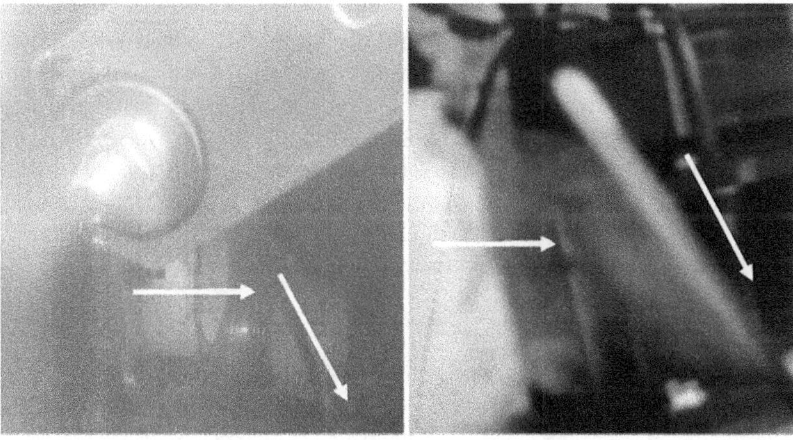

FIGURE 10.11 CF spread tow (20 mm wide, 40 gsm) tape being coated with CNT film in a continuous manufacturing process. The CNT exits the process tube horizontally and coats the CF tape which travels downward from a spool of tape.

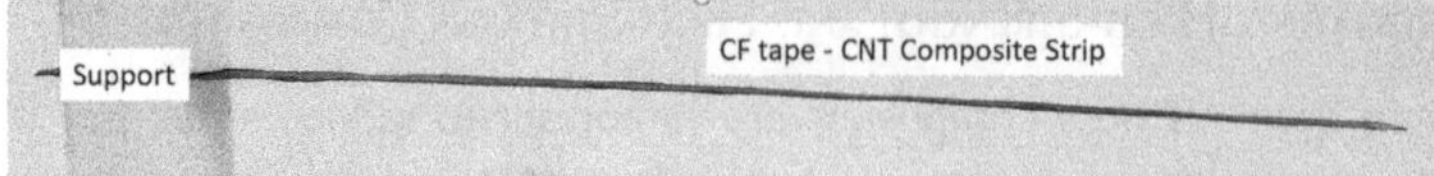

FIGURE 10.12 CNT-CF cantilever strip. The strip was produced by layering CNT onto CF spread tow tape.

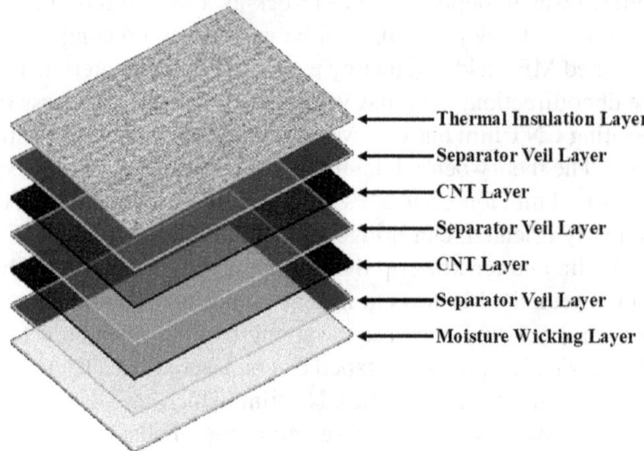

FIGURE 10.13 The CNT composite fabric: from the top, is a thermal insulation layer, separator veil layer, CNT layer, veil, CNT layer, veil, and moisture wicking layer (for heat resistance, moisture wicking, and capillary action).

10.4 APPLICATION EXAMPLES

10.4.1 IsoCool Fabric: A CNT Composite Textile

Firefighters (FFs) are facing greater risks (such as thermal stress, carcinogens from smoke or gas inhalations, and fire ground injuries) possibly due to present-day building and home-furnishing materials [49]. Such materials, for example plastics, tend to burn hotter and release toxic fumes. FFs also experience heat stress at a high rate, which puts their health at risk and decreases their ability to perform in service. To address the most important priorities FFs are facing, UC is developing a CNT heat dissipating and filtering fabric called IsoCool fabric. The fabric spreads heat, wicks moisture, resists flame, and filters toxic chemicals and particulates from the air. It can also provide electromagnetic shielding. An example of the IsoCool construction is shown in Figure 10.13.

GAC particles can also be integrated into the IsoCool fabric to improve the filtering of toxic gases and particles from the air, and to keep the toxic gases and smoke particles from penetrating the garment and reaching the skin. The addition of the CNT may provide biological and human health relevance (specially to protect FFs from serious hyperthermic collapse) due to the CNT's heat dissipation feature. Optimization of the composite fabric design, simulation, and testing the cooling performance of the fabric is being performed. Composite (layered) multifunctional fabrics with a CNT layer inside may have many more commercial applications.

10.4.2 Heat-Resistant Gloves

First responders and other professionals, such as mechanics, chefs, and researchers, use heat-resistant gloves to protect themselves from heat hazards, and contact from hot objects. Due to CNTs high thermal conductivity in the plane of the fabric, incorporating CNTs into a glove (shown in Figure 10.14) may spread heat and reduce hot spots. The CNT fabric is thermally insulating out-of-plane [50] to reduce heat entering the glove. Figure 10.14a shows a Lincoln Electric leather welding gloves (Model#: KH962, Weight = 14 oz) [51], which provide forearm protection, absorb perspiration, and provide heat resistance to the wearer. The composite CNT glove provides heat resistance.

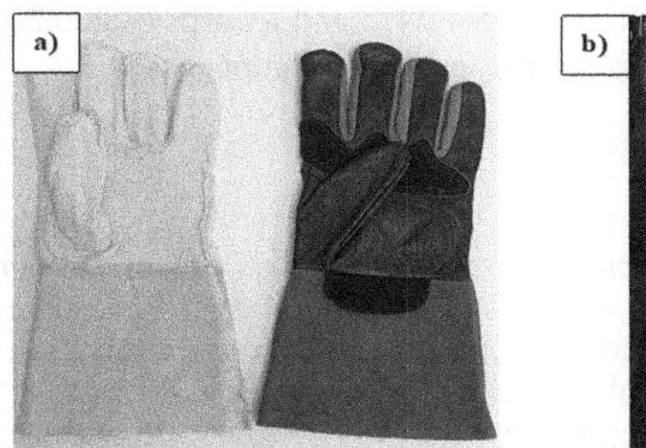

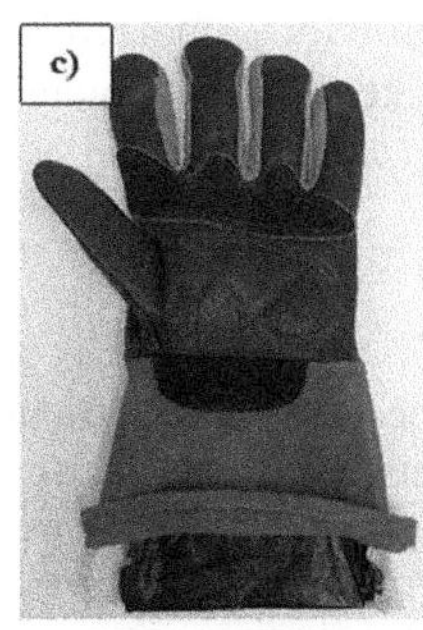

FIGURE 10.14 Heat-resistant CNT glove: (a) deconstructed commercial leather welding glove, (b) CNT-Pris fabric, and (c) CNT glove composite [51].

10.5 CONCLUSIONS & FUTURE WORK

This chapter describes synthesis and testing the flame resistance and air filtration performance of various CNT-based fabrics. CNT-Arr was synthesized from CCVD, while CNT-Pris, CNT-CFT, and CNT-CF were made in a one-step process through FCCVD method. We also presented two possible FF applications, IsoCool fabric and a heat-resistant glove, envisioned with CNT to improve wearer health and safety.

The flame test evaluated the flame resistance of the CNT-Arr fabric and the CNT-Pris fabric, which have a unique two-phase composition. The CNT-Arr does not have significant catalyst. The catalyst promoted localizing the flame damage on the CNT sheet and reduced the diameter of the hole created by the flame. The iron catalyst contained within or bonded on the outside of the nanotube underwent phase change from iron to iron oxide (a refractory material). CNTs having high thermal conductivity showed good flame resistance, which spread heat and cooled the flame area. Addition of NPs in the fabric may increase its flame resistance. Recommended future testing would include testing thicker samples that vary across the range of 10–100 μm for both CNT-Arr and CNT-Pris. Moreover, graphene sheet should also be considered for use in flame-resistant fabric development, including varying the composite compositions, due to their high thermal conductivity.

The filter test was used to assess the filtration capabilities of fabric materials. From the SEM images, both the CNT-Pris fabric and CNT-CFT composite fabric collected particulates on their surface. Thus, if incorporated within a filter application, the CNT fabric may act as a filter or a shield to prevent toxic gases and smoke particles from reaching the skin. From the testing, the CNT fabric was semi-permeable depending on the thickness and therefore may or may not allow contaminants to penetrate through the layer. The next steps would be making the CNT fabric more breathable, which may generate additional filtration efficiency in any application. Also, if the CNT fabric contained NPs, it may inactivate certain viruses and prevent microbial growth.

The integration of CF into CNT sheet improved the overall thermal stability. The reduced mass change with respect to temperature of 38% showed that the CNT-CF had a greater resistance to burning than CF.

Finally, for structural applications, depositing CNT sock directly onto CF spread tow tape produced a tape material that is expected to increase the strength and stiffness of polymeric composites. Future work is aimed to reduce the thickness of spread tow tapes, coat the thin tapes with CNT sock, and to form polymeric composites by layering CNT-CF tape on a mandrel. The CNT sock reduced the volume fraction of the polymer matrix material.

ACKNOWLEDGMENTS

This chapter was supported by the National Institute for Occupational Safety and Health (NIOSH) through the University of Cincinnati Education Research Center (ERC) Grant #T42OH008432. It was, also, broadly supported by ONR Award #N00014-15-1-2473 and I-Corps@Ohio 2017 Cohort. We would like to thank Ronald Hudepohl for machining parts for filtration test.

REFERENCES

[1] *Carbon Nanotube Fibers and Yarns: Production, Properties and Applications in Smart Textiles*, Woodhead Publishing, Cambridge, MA, 2020.

[2] B.S. Shim, W. Chen, C. Doty, C. Xu, N.A. Kotov, Smart electronic yarns and wearable fabrics for human biomonitoring made by carbon nanotube coating with polyelectrolytes, *Nano Letters* 8(12) (2008) 4151–4157.

[3] A.M. Abdelkader, N. Karim, C. Vallés, S. Afroj, K.S. Novoselov, S.G. Yeates, Ultraflexible and robust graphene supercapacitors printed on textiles for wearable electronics applications, *2D Materials* 4(3) (2017) 035016.

[4] S. Pan, H. Lin, J. Deng, P. Chen, X. Chen, Z. Yang, H. Peng, Novel wearable energy devices based on aligned carbon nanotube fiber textiles, *Advanced Energy Materials* 5(4) (2014) 1401438.

[5] *Nanotube Superfiber Materials (Second Edition)*, 2nd ed., Elsevier, Cambridge, MA, 2019.

[6] O. Yildiz, K. Stano, S. Faraji, C. Stone, C. Willis, X. Zhang, J.S. Jur, P.D. Brandford, High performance carbon nanotube-polymer nanofiber hybrid fabrics, *Nanoscale* 7 (2015) 16744–16754.

[7] Q. Wang, O. Yildiz, A. Li, K. Aly, Y. Qiu, Q. Jiang, D.Y.H. Pui, S.-C. Chen, P.D. Bradford, High temperature carbon nanotube – Nanofiber hybrid filters, *Separation and Purification Technology* 236 (2020) 116255.

[8] M. Yasuda, K. Takei, T. Arie, S. Akita, Oscillation control of carbon nanotube mechanical resonator by electrostatic interaction induced retardation, *Scientific Reports* 6 (2016) 22600.

[9] D.Y. Choi, S.-H. Jung, D.K. Song, E.J. An, D. Park, T.-O. Kim, J.H. Jung, H.M. Lee, Al-coated conductive fibrous filter with low pressure drop for efficient electrostatic capture of ultrafine particulate pollutants, *ACS Applied Materials & Interfaces* 9(19) (2017) 16495–16504.

[10] J. Feng, S. Xiong, Z. Wang, Z. Cui, S.-T. Sun, Y. Wang, Atomic layer deposition of metal oxides on carbon nanotube fabrics for robust, hydrophilic ultrafiltration membranes, *Science* 550 (2018) 246–253.

[11] Y. Song, D. Chauhan, G. Hou, X. Wen, M. Kattoura, C. Ryan, V. Shanov, Carbon nanotube sheet reinforced laminated composites, *Thirty-First Technical Conference, Proceedings of the American Society for Composites*, 2016.

[12] H. Tang, Z. Chen, O. Avinesh, H. Guo, Z. Meng, C. Engler-Pinto, H. Kang, X. Su, Notch insensitivity in fatigue failure of chopped carbon fiber chip-reinforced composites using experimental and computational analysis, *Composite Structures* 244(15) (2020) 112280.

[13] J.J. Andrew, S.M. Srinivasan, A. Arockiarajan, H.N. Dhakal, Parameters influencing the impact response of fiber-reinforced polymer matrix composite materials: A critical review, *Composite Structures* 224 (2019) 111007.

[14] R. Guzman de Villoria, P. Hallander, L. Ydrefors, P. Nordin, B.L. Wardle, In-plane strength enhancement of laminated composites via aligned carbon nanotube interlaminar reinforcement, *Composites Science and Technology* 133 (2016) 33–39.

[15] A. Bisht, K. Dasgupta, D. Lahiri, Investigating the role of 3D network of carbon nanofillers in improving the mechanical properties of carbon fiber epoxy laminated composite, *Composites Part A: Applied Science and Manufacturing* 126 (2019) 105601.

[16] R.P. Behera, P. Rawat, S.K. Tiwari, K.K. Singh, A brief review on the mechanical properties of carbon nanotube reinforced polymer composites, *Materials Today: Proceedings* 22 (2020) 2109–2117.

[17] S. Ma, Y. He, L. Hui, L. Xu, Effects of hygrothermal and thermal aging on the low-velocity impact properties of carbon fiber composites, *Advanced Composite Materials* 29(1) (2020) 55–72.

[18] Huntsman, Miralon, 2019. https://www.miralon.com. (Accessed 10 March 2021).

[19] Boronite, 2021. https://www.boronite.com. (Accessed 02 February 2021).

[20] M.J. Schulz, G. Hou, V. Ng, M. Rabiee, M. Cahay, S. Chaudhary, D. Lindley, D. Chauhan, M. Paine, D. Vijayakumar, C. Xu, Z. Yin, K. Haworth, Y. Liu, M. Sundaram, W. Li, D. Mast, V.N. Shanov, Science to commercialization of carbon nanotube sheet and yarn, *WSEAS Transactions on Applied and Theoretical Mechanics* 12 (2017) 41–50.

[21] M.J. Schulz, S. Kanakaraj, D. Mast, V.N. Shanov, D. Chauhan, G. Hou, V. Ng, C. Xu, R.D. Chen, A. Kubley, X. Hou, R. Kleismit, *Carbon Nanotube Hybrid Material Fabric, Composite Fabric, and Personal Protective Apparel and Equipment*, 2019.

[22] Y. Yun, V.N. Shanov, Y. Tu, S. Subramaniam, M.J. Schulz, Growth mechanism of long aligned multiwall carbon nanotube arrays by water-assisted chemical vapor deposition, *The Journal of Physical Chemistry B* 110(47) (2006) 23920–23925.

[23] *CST_TheCompositesStore, Carbon Fiber Tissue/Veil*, 2018. https://www.cstsales.com/a-carbon-fiber-tissue-veil.html. (Accessed 10 March 2021).
[24] Zoltek, *Zoltek PX35*, 2021. https://zoltek.com/production/px35/. (Accessed 05 March 2021).
[25] M.S. Dresselhaus, G.F. Dresselhaus, R. Saito, Physics of carbon nanotubes, *Carbon* 33(7) (1995) 883–891.
[26] H.-M. Cheng, F. Li, G. Su, H.Y. Pan, L.L. He, X. Sun, M.S. Dresselhaus, Large-scale and low-cost synthesis of single-walled carbon nanotubes by the catalytic pyrolysis of hydrocarbons, *Applied Physics Letters* 72(25) (1998) 3282–3284.
[27] O. Zhou, H. Shimoda, B. Gao, S. Oh, L. Fleming, G. Yue, Materials science of carbon nanotubes: Fabrication, integration, and properties of macroscopic structures of CNTs, *Accounts of Chemical Research* 35(12) (2002) 1045–1053.
[28] Y.-L. Li, I.A. Kinloch, A.H. Windle, Direct spinning of carbon nanotube fibers from chemical vapor deposition synthesis, *Science* 304(5668) (2004) 276–278.
[29] A. Barreiro, C. Kramberger, M.H. Rümmeli, A. Grüneis, D. Grimm, S. Hampel, T. Gemming, B. Büchnera, A. Bachtoldb, T. Pichlera, Control of the single-walled carbon nanotube mean diameter in sulphur promoted aerosol-assisted chemical vapour deposition, *Carbon* 45(1) (2007) 55–61.
[30] I. Khatri, T. Soga, T. Jimbo, S. Adhikari, H.R. Aryal, M. Umeno, Synthesis of single walled carbon nanotubes by ultrasonic spray pyrolysis method, *Diamond and Related Materials* 18(2–3) (2009) 319–323.
[31] G. Hou, R. Su, A. Wang, V. Ng, W. Li, Y. Song, L. Zhang, M. Sundaram, V.N. Shanov, D. Mast, D.S. Lashmore, M.J. Schulz, Y. Liu, The effect of a convection vortex on sock formation in the floating catalyst method for carbon nanotube synthesis, *Carbon* 102 (2016) 513–519.
[32] A. Szabó, C. Perri, A. Csató, G. Giordano, D. Vuono, J.B. Nagy, Synthesis methods of carbon nanotubes and related materials (review), *Materials* 3(5) (2010) 3092–3140.
[33] J. Prasek, J. Drbohlavova, J. Chomoucka, J. Hubalek, O. Jasek, V. Adamc, R. Kizek, Methods for carbon nanotubes synthesis – Review, *Journal of Materials Chemistry* 21 (2011) 15872–15884.
[34] S. Shahidi, B. Moazzenchi, Carbon nanotube and its applications in textile industry – A review, *The Journal of the Textile Institute* (2018).
[35] H. Zhu, C. Xu, D. Wu, B. Wei, R. Vajtai, P.M. Ajayan, Direct synthesis of long single-walled carbon nanotube strands, *Science* 296(5569) (2002) 884–886.
[36] B.I. Kharisov, O.V. Kharissova, *Carbon Allotropes: Metal-Complex Chemistry, Properties and Applications*, Springer Nature Switzerland AG, Cham, 2019.
[37] G. Lee, M. Sung, J.H. Youk, J. Lee, W.-R. Yu, Improved tensile strength of carbon nanotube-grafted carbon fiber reinforced composites, *Composite Structures* 220 (2019) 580–591.
[38] H. Yu, Z. Li, G. Luo, F. Wei, Growth of branch carbon nanotubes on carbon nanotubes as support, *Diamond and Related Materials* 15(9) (2006) 1447–1451.
[39] E. López-Honorato, P.J. Meadows, P. Xiao, Fluidized bed chemical vapor deposition of pyrolytic carbon – I. Effect of deposition conditions on microstructure, *Carbon* 47(2) (2009) 396–410.
[40] K. Fujisawa, H. Kim, S. Go, H. Muramatsu, T. Hayashi, M. Endo, T. Hirschmann, M.S. Dresselhaus, Y. Kim, P. Araujo, A review of double-walled and triple-walled carbon nanotube synthesis and applications, *Applied Sciences* 6(4) (2016) 109.
[41] G. Hou, D. Chauhan, V. Ng, C. Xu, Z. Yin, M. Paine, R. Su, V.N. Shanov, D. Mast, M.J. Schulz, Y. Liu, Gas phase pyrolysis synthesis of carbon nanotubes at high temperature, *Materials & Design* 132 (2017) 112–118.
[42] S. Charanjeet, M.S.P. Shaffer, I.A. Kinloch, A.H. Windle, Production of aligned carbon nanotubes by the CVD injection method, *Physica B, Condensed Matter* 323(1–4) (2002) 339–340.
[43] K. Kuwana, H. Endo, K. Saito, D. Qian, R. Andrews, E.A. Grulke, Catalyst deactivation in CVD synthesis of carbon nanotubes, *Carbon* 43(2) (2005) 253–260.
[44] K.M. Samant, S.K. Haram, S. Kapoor, Synthesis of carbon nanotubes by catalytic vapor deposition (CVD) method: Optimization of various parameters for the maximum yield, *Pramana – Journal of Physics* 68(1) (2007) 51–60.
[45] NIOSH, CDC, *How Tobacco Smoke Causes Disease: The Biology and Behavioral Basis for Smoking-Attributable Disease: A Report of the Surgeon General*, National Institute for Occupational Safety and Health (NIOSH) Education and Information Division; Centers for Disease Control (CDC) and Prevention, Atlanta, GA, 2010.
[46] A.W. Dumas, Tobacco ash, *Journal of the National Medical Association* 29(3) (1937) 103–104.
[47] D.E. Newbury, N.W.M. Ritchie, Is scanning electron microscopy/energy dispersive X-ray spectroscopy (SEM/EDS) quantitative?, *Scanning* 35(3) (2012) 141–168.
[48] D. Grund, M. Orlishausen, I. Taha, Determination of fiber volume fraction of carbon fiber-reinforced polymer using thermogravimetric methods, *Polymer Testing* 75 (2019) 358–366.

[49] NFPA, R. Campbell, B. Evarts, *United States Firefighter Injuries in 2019*, 2020. https://www.nfpa.org/-/media/Files/News-and-Research/Fire-statistics-and-reports/Emergency-responders/osffinjuries.pdf. (Accessed 29 April 2021).

[50] B. Kumanek, D. Janas, Thermal conductivity of carbon nanotube networks: A review, *Journal of Materials Science* 54 (2019) 7397–7427.

[51] *Lowe's, Lincoln Electric Red and Black Premium Leather Welding Gloves*, 2021. https://www.lowes.com/pd/Lincoln-Electric-Red-and-Black-Premium-Leather-Welding-Gloves/1001573608?cm_mmc=shp-_-c-_-prd-_-tol-_-google-_-lia-_-217-_-welding-_-1001573608-_-0&placeholder=null&ds_rl=1286981&gclid=EAIaIQobChMIjuSF3Yau8AIVhM3ICh2MGwXCEAkYBiABEgIIkfD_BwE&gclsrc=aw.ds. (Accessed 18 April 2021).

11 Toughening Mechanisms in Mullite-Based Composites

Pramod Koshy, Bernadette Pudadera, Tina Majidi, Vicki Zhong, Vienna C. Wong, Xiaoran Zheng, Naomi Ho, Theodora K.-H. Cheng, and Charles C. Sorrell
UNSW Sydney

CONTENTS

11.1 MULLITE

Mullite is an aluminosilicate and the only stable binary phase in the alumina–silica (Al_2O_3-SiO_2) phase diagram. Its orthorhombic crystal structure consists of edge-sharing AlO_6 octahedra chains cross-linked by oxygen vacancy-filled (Si, Al)O_4 tetrahedra (Schneider and Komarneni 2005; Schneider, Fischer, and Schreuer 2015). Mullite is part of a general series of aluminosilicates with the formula $Al_2(Al_{2+2X}Si_{2-2X})O_{10-X}$. The most common forms of mullite are $3Al_2O_3{\cdot}2SiO_2$ (3/2 mullite), where $X = 0.25$, and $2Al_2O_3{\cdot}SiO_2$ (2/1 mullite), where $X = 0.40$ (Schneider and Komarneni 2005; Schneider, Fischer, and Schreuer 2015; Schneider, Schreuer, and Hildmann 2008). Two other forms, namely, 4/1 mullite ($X = 0.67$) and 9/1 mullite ($X = 0.825$) have been produced using the sol–gel method. The 3/2 form of mullite (72 wt% Al_2O_3) is considered stoichiometric mullite, and its representative structure is shown in Figure 11.1.

Owing to its crystal structure, mullite possesses low thermal expansion, low thermal conductivity, and good creep resistance, which have proven favourable for numerous applications. Mullite has a refractoriness of 1,850°C–1,865°C (equivalent to a PCE of 38–39), which makes it ideal for high-temperature applications (Aramaki and Roy 1962); (https://www.ortonceramic.com/files/2676/File/Orton-Cone-Chart-F-022-14.pdf). Furthermore, excellent chemical and mechanical stabilities at high temperatures have allowed for the widespread use of mullite refractories in the steel and glass industries (Schneider, Schreuer, and Hildmann 2008). However, the actual maximal temperature of use of mullite is lower than the refractoriness temperature owing to the presence of glass between the mullite grains (Morell 1985), which results in undesirable failure owing to creep deformation at temperatures ≥1,200°C. A percolated (scaffold) form of 3/2 mullite was fabricated using Si-rich aluminosilicate precursors, which showed no deformation on heating to 1,600°C (Koshy et al. 2018, 2021). This unique form of mullite thus shows remarkable high-temperature stability, which would enable their use as refractory aggregates and shapes. In addition to the most common application as refractories, with advances in processing and characterisation, mullite and mullite-based composites have emerged as promising material candidates in advanced structural, electronic, and optical

DOI: 10.1201/9780429330575-11

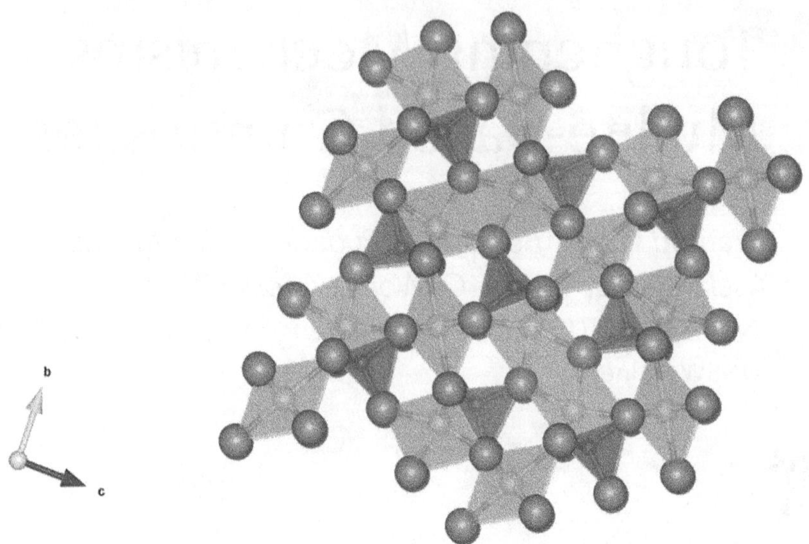

FIGURE 11.1 Representative structure of mullite with Al(VI) atoms (light blue), Al(VI)/Si(IV) atoms (purple), and O atoms (red).

applications (Aksay, Dabbs, and Sarikaya 1991) such as engine components, coatings, substrates for packaging, and composite materials.

Mullite is found naturally, but the reserves are limited (Bowen and Greig 1924). Therefore, alternative synthesis techniques are required to produce mullite. Common techniques include (i) solid-state diffusion-controlled reactions, to achieve stoichiometric compositions; (ii) crystallisation from aluminosilicate melts, which tend to produce needle-shaped crystals of alumina-rich compositions; and (iii) through the use of sol–gel precursors to produce compositions with extremely high alumina contents (Davis and Pask 1971; Fischer et al. 2012). The most commonly produced mullite is sinter mullite (3/2 mullite), which is produced by sintering Al- and Si-containing precursors. 2/1 mullites are produced by melting the precursors at very high temperatures (>2,000°C), followed by cooling to recrystallise the mullite; these fused mullites show higher Al_2O_3 contents (78 wt%). Mullite produced using the sol–gel method is known as chemical mullite and has alumina/silica stoichiometries of 4/1 and 9/1. This method involves the use of Al- and Si-containing sol–gel precursors and requires comparatively lower synthesis temperatures (700°C–1,000°C) (Schneider and Komarneni 2005).

11.2 MULLITE-BASED COMPOSITES

Mullite is generally used in composites with other oxide and non-oxide ceramics to enhance its properties for different applications. Zirconia (ZrO_2), alumina (Al_2O_3), and silicon carbide (SiC) are materials commonly used in composites with mullite. Table 11.1 compares the properties of mullite with ZrO_2, Al_2O_3, and SiC. Table 11.2 further lists the potential/current applications of the resulting mullite-based composites. Mullite possesses a lower maximal use temperature than other materials, and thus, the addition of alumina and zirconia can enhance the thermal stability of the composite, and the addition of SiC can improve the thermal stability but only for applications in non-oxidising conditions.

Zirconia is commonly recognised as possessing high fracture toughness owing to transformation toughening mechanisms. Therefore, the formation of a mullite-based composite with zirconia would result in improved fracture toughness compared to mullite alone. Slight improvements to fracture toughness may be obtained through the addition of Al_2O_3, particularly if debonding at the

TABLE 11.1
Properties of Mullite in Comparison with Common Ceramics Present in Mullite-Based Composites

Properties	Mullite $3Al_2O_3\ 2SiO_2$	Alumina Al_2O_3	Zirconia ZrO_2	Silicon Carbide SiC
Crystal structure	Orthorhombic	Hexagonal (Corundum)	Monoclinic (m) Tetragonal (t) Cubic (c)	Hexagonal (alpha) Cubic (beta)
Melting temperature (°C)	1,850	2,070	2,600	2,730
Bulk density (kg/m^3)	3,200	3,900	5,600	3,160–3,250
Young's modulus (GPa)	190–210	390	250	390–700
Tensile strength (MPa)	200	500	200	–
Fracture toughness (MPa·m$^{1/2}$)	2.5	2.5–4.5	6–12	2.6–2.8
Specific heat capacity (J/g·K)	0.7–0.9	0.7–0.8	0.4–0.5	0.7–0.75
Coefficient of thermal expansion (× 10^{-6}/°C)	4.5–6.0	12.0	9.0–11.0	2.4
Thermal conductivity (W/mK)	5–7	40	1.7–2.7	3.2–4.9

Source: Data from Schneider and Komarneni 2005; Schneider, Fischer, and Schreuer 2015; Schneider, Schreuer, and Hildmann 2008; Osendi and Baudin 1996; Hildmann and Schneider 2004; Shackelford and Doremus 2008; Kingery, Bowen, and Uhlmann 1976; Kern et al. 1969; Harris 1995; Degueldre et al. 2003; Nevitt et al. 1990; Philippe and Niepce 2007.

TABLE 11.2
Current and Potential Applications of Mullite-Based Composites

Composite	Application
Mullite–zirconia	Hypersonic vehicles, high-temperature structural applications, dental and biomedical applications, packaging, and optical applications
Mullite–alumina	Thermal insulation, structural ceramics, and refractories
Mullite–SiC	Solar thermal storage and refractories
Mullite–cordierite	Refractories

Source: Schneider and Komarneni 2005; Schneider, Fischer, and Schreuer 2015; Schneider, Schreuer, and Hildmann 2008; Bodhak, Bose, and Bandyopadhyay 2010; Rocha-Rangel et al. 2005; Rodrigo and Boch 1985; Denry and Kelly 2014; Bella, Hamidouche, and Gremillard 2021; Lin and Tsang 2003.

interfaces occurs. As shown in Table 11.1, an important factor to consider in selecting the secondary phase is the coefficient of thermal expansion since large differences in expansion coefficients can result in interfacial stresses and crack formation during heating and cooling processes either during sintering or during use. The addition of either ZrO_2 or Al_2O_3 can result in high interfacial stresses when added with mullite owing to the significant differences in thermal expansion coefficients, which can result in cracking and loss of contact between grains, and this is commonly seen in applications involving high temperatures and frequent thermal cycling.

11.3 FABRICATION OF MULLITE-BASED COMPOSITES

The methods of ceramic composite fabrication usually involve the same starting steps. Raw materials are mixed with water and subjected to different milling processes, such as ball milling, attrition milling, or planetary milling. Then, the resultant powders are mixed with water, organic binders, and/or sintering additives. The resultant mixtures are compacted at high pressures or extruded to the desired shape and then sintered at high temperatures (Zhang et al. 2016; Bodhak, Bose, and Bandyopadhyay 2010). For mullite–zirconia (MZ) composites, mullite powder is mixed with stabilised tetragonal/cubic zirconia polycrystalline powders and then sintered at 1,400°C–1,600°C (Bodhak, Bose, and Bandyopadhyay 2010; Rodrigo and Boch 1985). Alternate routes involve alumina (Al_2O_3) and zircon ($ZrSiO_4$) powders as precursors (Rodrigo and Boch 1985; Khor and Li 1998). Sillimanite, andalusite, and fly ash have also been used as precursors in combination with zircon/zirconia (Ying, Shao-gang, and Zhai 2010; Prusty et al. 2012; Kumar et al. 2015; Bouchetou et al. 2019). A thermal plasma melting process is used to fabricate zirconia-toughened mullite using zircon sand and high-purity alumina (Bhattacharjee, Singh, and Galgali 2000). In the preceding work, varying the alumina contents resulted in differences in the sizes of ZrO_2 grains, with larger grains (5–10mm) seen in Si-rich conditions, while smaller grains (<1 mm) are seen in Al-rich compositions. MZ composites are fabricated using spark plasma sintering using mixtures of zircon, aluminium, and α-alumina, which are then heated initially to oxidise aluminium and then sintered at 1,350°C–1,460°C at 40MPa pressure (Rocha-Rangel et al. 2005). Fibrous composites of mullite and zirconia are fabricated using starting fibres of these materials with organic binders and sintering aids to form a slurry, which is then extruded, dried, and sintered at 1,400°C (Zhang et al. 2016). Investigations have been conducted to determine the impact of different precursors; mullite–zirconia composites are prepared using three different combinations of precursors, namely, mullite+zirconia powders, alumina+amorphous silica+zirconia, and alumina+zircon, followed by sintering at 1,520°C. The use of mullite+zirconia powders as starting materials resulted in elongated mullite grains owing to glass phase formation; however, the samples were not dense. The use of alumina+amorphous silica+zirconia precursors enhanced liquid-phase sintering, leading to the formation of zirconia grains dispersed between equiaxed mullite grains. Contrastingly, the use of alumina+zircon precursors resulted in irregular mullite grains with zirconia grains dispersed in between them (Koyuma et al. 1994).

For the preparation of mullite–alumina (MA) composites, mixtures of silica and alumina are usually milled, compacted, and sintered at 1,600°C (Sadik, Amrani, and Albizane 2014; Aksel 2003). In addition to pure silica and alumina, other aluminosilicates such as kaolin, sillimanite, andalusite, and bauxite can be used; for example, kaolin and aluminium hydroxide mixtures are sintered at 1,600°C to form these composites (Bella, Hamidouche, and Gremillard 2021). The sol–gel method has been used to fabricate MA composites, with aluminium nitrate and tetraethyl orthosilicate used as the alumina and silica sources. These are heated initially at low temperatures to remove the solvents followed by fibre drawing from the sol and calcination at 1,000°C (Wei, Duan, and Wu 2011). The sol–gel method has been used for producing just the alumina sol, and then mullite fibres were added to the mixture, followed by sintering at 1,550°C (Yang et al. 2019). Aluminium sec-butylate is commonly used as the Al-precursor, and silicon tetrachloride is used as the Si-precursor for the sol–gel process; these can be used individually to form the respective powders or in combination to form mullite (Yang et al. 2019; Fischer, Schneider, and Schmucker 1994). Mullite–SiC composites are generally fabricated by mixing powder mixtures of mullite or aluminosilicate precursors (bauxite, kaolin, andalusite) with SiC, followed by dry pressing, heating to remove excess carbon at 600°C, and final sintering at 1,500°C–1,600°C (Bhattacharjee, Singh, and Galgali 2000). Hot pressing at 1,600°C/40MPa has been used to produce dense composites (Gustaffon, Falk, and Pitchford 2009), while spark plasma sintering at 1,300°C–1,400°C has been conducted on mullite powders containing 10–20vol% SiC whiskers to produce these composites (Huang et al. 2004). In addition to binary mixtures, ternary-phase composites have also been produced. Additions of zirconia/zircon, alumina/refractory bauxite, and silica have been shown to assist in the formation of zirconia–alumina–mullite composites; one study used plasma melting (Bhattacharjee, Singh, and Galgali 2000), while another study used direct sintering at 1,450°C–1,650°C for 5hours (Vazquez Carbajal et al. 2012) for the fabrication process. Quaternary compositions were prepared by further addition of SiC particles to these compositions in amounts of 10–30 vol% with sintering carried out at 1,600°C (Majidian, Ebadzadeh, and Salahi 2011). Fibrous and porous mullite–zirconia fibre composites have been fabricated with a quasi-layered structure with 5–15 vol% ZrO_2 fibres, followed by vacuum squeezing and sintering at 1,400°C (Zhang et al. 2016). Similar composites with 20–60 vol% ZrO_2 were further impregnated with an Al_2O_3-SiO_2 aerogel to improve densification and strength (Zhang et al. 2017). In another study, the laser floating zone method was used to melt a mullite (70 wt%)–zirconia (30 wt%) mixture, followed by extrusion (Carvalho et al. 2014). Coating of mullite fibres is another approach that can be used to modify the interfacial bonding with the matrix; $NdPO_4$ (Boccaccini et al. 2005; Kaya et al. 2002), carbon, and boron nitride (BN) (Liu et al. 2018) have been used as coatings on mullite fibres. SiC–mullite porous composites (Jinga et al. 2014) have been fabricated by sintering at 1,250°C–1,450°C using mixtures of kaolin, aluminium hydroxide, SiC, and graphite; the graphite burnout during heat treatment was used for pore formation.

There have been limited studies on other mullite-based composites. TaB_2/mullite (Yeh and Kao 2014) composites were fabricated by combustion synthesis at 1,250°C–1,600°C, while mullite composites with cordierite for refractory compositions were fabricated by conventional sintering mixtures of alumina, magnesia, and silica through a commercial process (temperature not provided) (Boccaccini et al. 2005). Zirconium diboride (ZrB_2) and carbon nanotubes (CNTs) have also been added to form composites with mullite with the processing carried out using hot pressing techniques (Orooji et al. 2019). It should be noted that the works listed previously employed composites composed of 3/2 mullite. There are no reports of composites of mullite prepared from composites of 2/1, 4/1, or 9/1 mullite. This could be related to the complexity of processing conditions for the fabrication of these mullite types. Mullite 2/1 as previously mentioned is fabricated from melt processing, which limits the chances of forming composites owing to the additional processing step for mullite fabrication. Al-rich mullite compositions such as 4/1 and 9/1 are produced by sol–gel processing. This involves the use of aluminium sec-butylate and silicon tetrachloride in an Al/Si ratio of 9:1 to form a viscous gel, which is then annealed at temperatures of 700°C–1,000°C. Heat treatment formed a mixture of high–Al-containing mullite of varying compositions (83–97 wt%), along with γ-Al_2O_3. Structural refinement of the XRD patterns revealed that the most stable compositions are potentially 4/1 (formula of $Al_{3.33}Si_{0.67}O_{1.33}$) and 9/1 (formula of $Al_{3.65}Si_{0.35}O_{1.175}$). These specific formulae are devised based on the number of T_2O, T_3O, and T_4O units present in the chain to form a stable structure. T_2O represents sillimanite-like dimers of tetrahedra, while T_3O and T_4O groups represent the AlO_4 tetrahedra, which are linked to the central octahedral chain (Fischer, Schneider, and Schmucker 1994; Schneider and Komarneni 2005). In these Al-rich mullites, the excess Al atoms are incorporated into the interstitial spaces within the crystal structure. Regarding thermodynamic stability, increasing the heat treatment temperature above 1,000°C starts to cause a lowering of the Al content in the mullite, which is accompanied by the formation of increasing amounts of Al_2O_3, leading to a decomposition to form stoichiometric 3/2 mullite at 1,650°C (Fischer, Schneider, and Voll 1996). This indicates a lower stability of these forms of mullite for high-temperature applications since several composite fabrication routes typically involve temperatures exceeding 1,000°C; the instability of these phases limits their use for ceramic composite applications.

11.4 GENERAL TOUGHENING MECHANISMS IN COMPOSITES

Ceramics show rapid failure and brittle fracture owing to the presence of cracks, pores, and other defects in the microstructure. Therefore, the strength of the material is closely related to the crack morphology and fracture toughness using Eq. (11.1) (Steinbrech 1992):

$$\sigma_C = \frac{1}{Y}\frac{K_{IC}}{\sqrt{a_c}} \tag{11.1}$$

Here,

σ_C = Critical strength (MPa)

Y = Geometry and crack parameter (dimensionless)

K_{IC} = Fracture toughness $\left(\text{MPa} \cdot \text{m}^{0.5}\right)$

a_c = Critical crack size (m)

An increase in crack length during stress loading results in a further increase in fracture toughness owing to the growing crack acting as an impediment to crack propagation. Thus, Eq. (11.1) transforms to Eq. (11.2):

$$\sigma_C = \frac{1}{Y}\frac{K_{R(a)}}{\sqrt{a_c + \Delta a}} \tag{11.2}$$

Here,

$K_{R(a)}$ = Change in toughness with crack growth(R-curve behaviour)

Δa = Change in length of crack during loading

Several general mechanisms can enhance the fracture toughness in composites (as shown in Figure 11.2), some of which are given as follows:

a. **Crack Deflection and Bowing:** The addition of secondary phases in the form of particles or fibres can allow for crack deflection and/or bowing behaviour, which results in improved fracture toughness. When the propagating crack approaches the interface/surface of a strong secondary particle, in order to continue propagation, it needs to either cause the fracture of the ceramic particle or deviate around the particle. This delays the onset of failure of the material and increases its resistance to failure. Hence, the addition of alumina (higher strength than mullite) and addition of zirconia (higher fracture toughness) secondary phases in mullite composites are expected to enhance the toughness. This improvement in toughness can be further enhanced by ensuring a homogenous dispersion of the secondary phase, which is achieved by using sub-micron and nanosized particles (Steinbrech 1992).
b. **Microcracking:** This mechanism involves the formation of microcracks in the region near the crack tip. The formation of these microcracks will increase the stress concentration and lead to potentially crack branching; this reduces the energy at the crack tip and hinders crack propagation. Microcracking is generally seen with the addition of ZrO_2 to composites, with both tetragonal and monoclinic forms seen to contribute to this effect (Steinbrech 1992; Ohji et al. 1998)
c. **Whisker and Fibre Bridging:** The addition of ceramic reinforcement in the form of whiskers or fibres can increase the fracture toughness of the composite. These materials can serve to hold the crack face together and prevent its rapid growth/expansion during loading. This mechanism has been observed with the addition of mullite or zirconia fibres or SiC whiskers to different ceramic matrices. Crack propagation would occur only when the stress at the crack tip overcomes the strength of the bridging fibre/whisker, and this will slow the crack propagation process (Bengisu and Inal 1994).

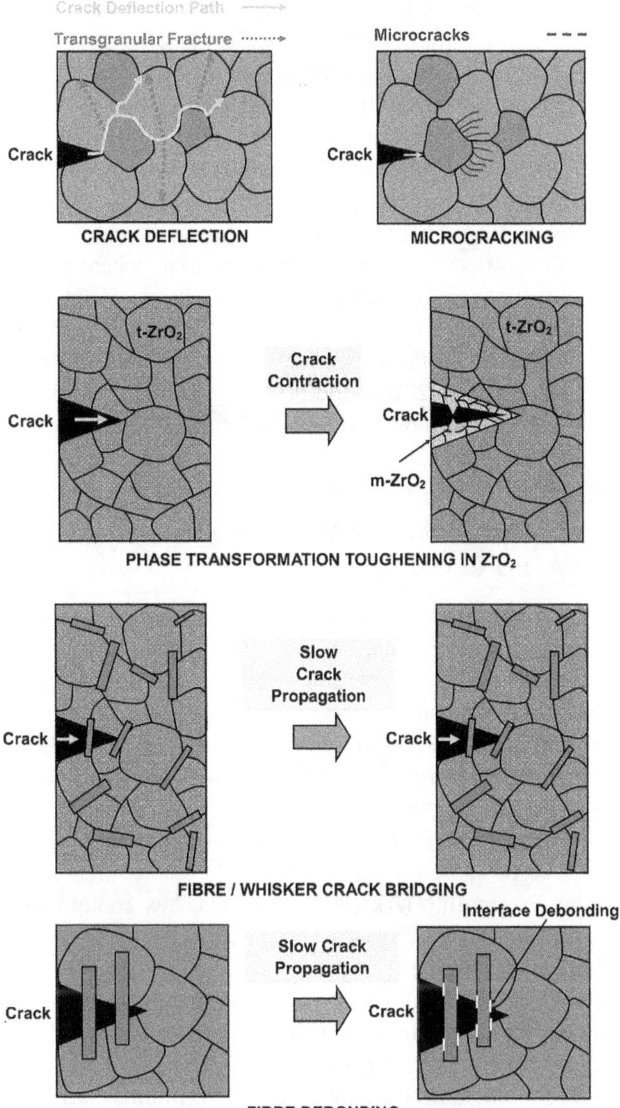

FIGURE 11.2 Typical mechanisms related to fracture toughening of ceramics; top row: crack deflection and microcracking, second row: phase transformation toughening in zirconia, third row: crack bridging by fibres and/or whiskers, and last row: fibre debonding at the interface.

d. **Zirconia Phase Transformation:** Zirconia undergoes a phase transformation from its room-temperature monoclinic form to its high-temperature tetragonal and cubic forms on heat treatment. These transformations are associated with a sudden change in volume during both the heating and cooling stages, leading to crack formation, propagation, and material failure. To minimise this issue, zirconia is typically stabilised through the incorporation of additives such as yttria, magnesia, or ceria; the resultant material is either partially stabilised zirconia (PSZ), where tetragonal precipitates are dispersed in a cubic zirconia matrix, or tetragonal zirconia polycrystals, where the matrix is composed predominantly of fine-grained tetragonal grains (Karihaloo and Andresen 1996).

An important feature of partially or fully stabilised zirconia is that when this stabilised material is subjected to high stresses at the crack tip, the tetragonal grains transform to the monoclinic phase. This transformation is accompanied by lattice expansion, leading to the closure of the crack face and restriction of crack propagation (Karihaloo and Andresen 1996). An unfortunate consequence is that with repeated thermal cycling, all the stabilised grains transform to monoclinic grains at the crack face, and thus, the positive transformation toughening effect is diminished.

e. **Fibre Pull-Out/Debonding:** This mechanism is seen in the case of fibre or whisker reinforcement of ceramic matrices, especially after fabrication, where a gap forms at the interface owing to either differences in the thermal expansion coefficients between the fibre and matrix or due to insufficient reactions between the matrix and fibre. When the propagating crack comes into contact with the debonded interface, it will present a zone-hindering crack propagation (Kabel et al. 2018). Additionally, coating of the fibres with a thin layer of a ceramic phase can present a barrier to crack formation.

11.5 TOUGHENING AND FAILURE MECHANISMS IN MULLITE-BASED COMPOSITES

The mechanisms of toughening in mullite-based composites are complex and vary significantly with the processing conditions and fabrication conditions, forms and amounts of additives, and interactions between the matrix constituents in the composites under application. The presence of mullite fibres would inherently result in an improvement of fracture toughness when added to composites, although equiaxed mullite grains are expected to not have the same positive impact on fracture toughness owing to the difference in morphology. Most mullite-based composites use 3/2 mullite, which is equiaxed, and thus, there is a need for further improvement in the fracture toughness of these composites through the addition of alumina, zirconia, and/or SiC. The effects of these additives are discussed subsequently.

When comparing alumina–mullite (Aksel 2003) composites containing varying amounts of Al_2O_3 (86.1–91.4 wt%) and SiO_2 (8.6–13.9 wt%), followed by sintering at 1,600°C, the strengths were higher for the samples with higher Al_2O_3 contents. The preceding effect was attributed to the formation of fine needle-shaped (~5 mm) mullite particles, which improved the mechanical strength and elastic modulus. In the sample with lower Al_2O_3 contents, coarse and medium-sized alumina and mullite particles were observed, which lowered the strength. Mullite needles improved the resistance to crack propagation; shorter length, needle-like mullite particles led to pull-out and crack deflection due to a weak mullite–alumina interface, and this can improve fracture toughness. The stoichiometric mullite matrix was prepared using the sol–gel method, and to this, Al_2O_3 platelets were added in varying ratios, followed by hot pressing. The addition of 44 vol% of platelets increased the toughness from 2.1 to 5.1 MPa·m$^{1/2}$ for cracks propagating parallel to the hot-pressing axis; the increase was minimal (1.5–2.8 MPa·m$^{1/2}$) in the direction perpendicular to the hot-pressing axis (Zhou et al. 1994). Due to the use of hot pressing, anisotropy is introduced in terms of the microstructural features of the composites, particularly platelet alignment. Therefore, the mismatch in thermal expansion coefficients between Al_2O_3 and mullite would become prevalent in this scenario, causing debonding at the interface owing to stress build-up during heat treatment and cooling during fabrication. In this work, the interfacial bonding was weak between Al_2O_3 and mullite, and this was responsible for high fracture toughness. In another study, the addition of 11.5 vol% sapphire (Al_2O_3) fibres to a mullite matrix (Pearce et al. 1996) increased Young's modulus and the flexural strength. The failure mechanism was identified as delamination, with the cracks deviating around fibre reinforcement. Increasing the fibre volume content to 20 vol% was expected to improve the fracture toughness by increasing the load-carrying capability and through the provision of additional weaker interfaces, resulting in crack deflection and fibre pull-out becoming the dominant mechanisms compared to delamination, although no direct measurements of the fracture toughness were made. This pull-out could result from the expansion mismatch between the matrix and fibre, as well as low reactivity between the mullite and alumina components.

SiC addition has been shown to have a positive impact in improving the fracture toughness of mullite-based composites. SiC whisker-reinforced mullite (Huang et al. 2004) was fabricated by spark plasma sintering of a mixture of SiC whiskers (10–30 vol%) and mullite powder at 1,300°C–1,400°C and 50 MPa. The addition of SiC whiskers had a grain pinning effect, suppressing mullite grain growth and leading to nanometre-sized grains. The increasing addition of SiC was shown to progressively increase the bending strength and fracture toughness, with the latter reaching a constant and highest value between 20 and 30 vol% (4.5 MPa·m$^{1/2}$). The strength increase was attributed to higher densification, while the increase in fracture toughness was likely due to the whisker reinforcement effect and associated crack bridging effects (Jinga et al. 2014). In another study, SiC fibres containing β-SiC and free carbon were used to form three-dimensional fabrics, which were then mixed with mullite sols that were infiltrated into the preform, followed by sintering at 1,000°C to form the composite. CVD coating of the fibres with carbon was performed for a different set of samples, where the volume fraction of the preform was 40%–42% (Han et al. 2017). The results showed that the SiC fibres were chemically bonded to the mullite matrix, resulting in a high interfacial shear strength. The presence of such a strong interface resulted in low toughness values. Deposition of pyrocarbon by CVD at the interface resulted in the weakening of the interface and subsequent fibre debonding and pull-out, which enhanced fracture toughness such that the values were 8.3 MPa·m$^{1/2}$ for the SiC–carbon–mullite composite compared to 0.8 MPa·m$^{1/2}$ for the SiC–mullite composite.

In addition to composites containing two phases, further improvements in toughness can be obtained by having three or four phases in the composite. Mullite–zirconia(–alumina) composites were fabricated using different amounts of alumina and zircon powders (Garrido et al. 2006) to generate (i) a stoichiometric mullite composite, (ii) one with 6 wt% excess alumina, and (iii) with 10 wt% zircon excess; all these samples were subjected to multiple thermal cycling steps, and the fracture toughness values of these composites ranged from 2.4 to 3.2 MPa·m$^{1/2}$. In the sample with excess zircon subjected to the highest number of thermal heating/cooling cycles, a high value of K_{IC} (3.9 MPa·m$^{1/2}$) was observed; this was attributed to the increased zirconia content from zircon decomposition at the high sintering temperature and the high extent of microcracks formed in the microstructure with increasing numbers of thermal cycles. In another study, alumina–mullite–zirconia–silica composites (Majidian, Ebadzadeh, and Salahi 2011) were fabricated using mixtures of alumina and zircon with/without SiC (10–30 vol%) and sintering at 1,600°C. These composites showed high fracture toughness with the values increasing from 7.29 to 8.13 MPa·m$^{1/2}$, with increasing SiC addition from 10 to 20 vol%. The increase in fracture toughness with SiC addition is attributed to the presence of internal residual stresses owing to the thermal expansion mismatch between Al_2O_3 and SiC. Two other contributing factors are the phase transformation toughening effect of tetragonal ZrO_2 and the plastic deformation of monoclinic ZrO_2, which decreased the driving force for transgranular crack propagation. However, further addition of SiC to 30 vol% led to lowering of the hardness and stiffness of the composites, and this could be related to microstructural alteration or lowering of the relative mullite/zirconia contents below the optimal level.

Alumina–zirconia–mullite composites (Vazquez Carbajal et al. 2012) were also fabricated with varying amounts of Al_2O_3, SiO_2, and ZrO_2. These were labelled C1 (51.79 Al_2O_3 + 15.83 SiO_2 + 32.37 ZrO_2), C2 (45.3 Al_2O_3 + 14.4 SiO_2 + 40.3 ZrO_2), and C3 (38.13 Al_2O_3 + 11.5 SiO_2 + 50.36 ZrO_2), and sintering was performed at 1,450°C–1,650°C. For all three sample sets, the samples sintered at 1,650°C showed the best mechanical properties owing to greater densification than those sintered at lower temperatures. Higher temperatures resulted in an increase in the size of the ZrO_2 grains, which increased the fracture toughness (K_{IC}) while decreasing the Vickers microhardness. However, increasing the ZrO_2 content did not increase the fracture toughness with the values ranging from 0.8 to 1.7 MPa·m$^{1/2}$ for C3 samples, while those with higher alumina contents (C1 and C2) showed higher values in the range of 1.4–2.1 and 1.2–1.9 MPa·m$^{1/2}$, respectively. Thus, in the present case, poor interfacial bonding between the different phases could be a factor contributing to higher fracture toughness values observed for the samples with higher alumina contents.

In addition to the previously described common additions to mullite composites, additives such as zirconium diboride (ZrB_2) and CNTs have been used. The mullite powder (prepared using the sol–gel) method was combined with ZrB_2 (10 wt%) and CNTs (1 wt%) and densified by spark plasma sintering at 1,350°C at a pressure of 30 MPa; the resultant sample showed high hardness, flexural strength, and high fracture toughness (~4.2 MPa·m$^{1/2}$). The pull-out and crack bridging mechanisms of the CNTs were revealed to be the major toughening mechanism from microstructural investigations (Orooji et al. 2019).

In addition to dense composites, porous composites have been prepared using mullite fibres, which have been coated using different ceramic compositions. Porous mullite matrix composites with $NdPO_4$-coated mullite woven fibre mats (Nextel 720) were fabricated with 5 wt% zirconia (Boccaccini et al. 2005; Kaya et al. 2002) by sintering at 1,200°C. The samples showed fracture toughness values in the range of 1.8–3.3 MPa·m$^{1/2}$, and the variations were attributed to heterogeneity of the microstructure. $NdPO_4$ coating on the fibres resulted in extensive fibre pull-out during failure, owing to a favourable matrix/fibre bond. Furthermore, the increase in fracture toughness could be attributed to microcracking, fibre matrix interface debonding, and localised fibre fracture. Sintering of these composites at a higher temperature of 1,300°C was observed to lower the fracture toughness, possibly owing to increased densification of the matrix and/or degradation of the coating on the fibre surface.

Using mullite fibres in a mullite matrix to form the composite is another possible route of fabrication (Liu et al. 2018). Fibre mats of mullite coated with boron nitride (BN) were placed in a mullite powder matrix, followed by hot pressing at different temperatures. Increasing the sintering temperatures resulted in grain growth, and fibre damage resulted in the lowering of the flexural strength. The increase in the fibre volume increased fracture toughness but at the expense of the flexural strength. The highest flexural strength was seen at 20 vol% fibre content and sintering at 1,300°C, while the samples sintered at 1,300°C with 30 vol% fibres showed the highest fracture toughness (4.74 MPa·m$^{1/2}$). In these composites, toughening mechanisms included crack deflection and crack branching, along with fibre pull-out (owing to a weak interface between the fibre and matrix). Fracture toughness is an important consideration for refractory materials subjected to thermal cycling applications. Cordierite–mullite refractory samples were tested for thermal cycling performance; these contained unreacted quartz, and the presence of this phase was believed to enable increased fracture toughness owing to its contribution to microcracking toughening. At higher temperatures, the viscous nature of the glass is expected to cause the blunting of the crack tip and limit crack propagation. However, fracture toughness was not measured (Boccaccini et al. 2005).

Mullite–mullite composites are another possible composite material. Mullite–whisker composites produced *in situ* in a mullite grain matrix by chemical treatment showed an increase in fracture toughness from 1.9 to 3.0 MPa·m$^{1/2}$ compared to the base mullite matrix with no whiskers (Meng et al. 1998). This increase is attributed to the whiskers assisting in crack deflection and crack bridging. A summary of the variations in fracture toughness in mullite-based composites is shown in Figure 11.3.

11.6 SUMMARY AND CONCLUSIONS

Investigation of the toughening mechanisms in different mullite-based composites shows that the extent of toughening is highly variable owing to the critical influence of the nature and amount of the secondary phases, as well as the processing conditions. When looking at the summative figure (Figure 11.3), it is clear that the greatest extent of toughening was seen in the case of high amounts of SiC fibre reinforcement (40–42 vol%) combined with pyrocarbon coating on fibres; this resulted in a seven-fold increase in toughness compared to the non-coated fibres, and this enhancement was attributed to fibre debonding mechanisms. Similar mechanistic effects were also observed in other studies where SiC whiskers were used with a mullite matrix. Mullite fibres themselves could be used with a BN coating on their surface to improve fracture toughness; however, there are other

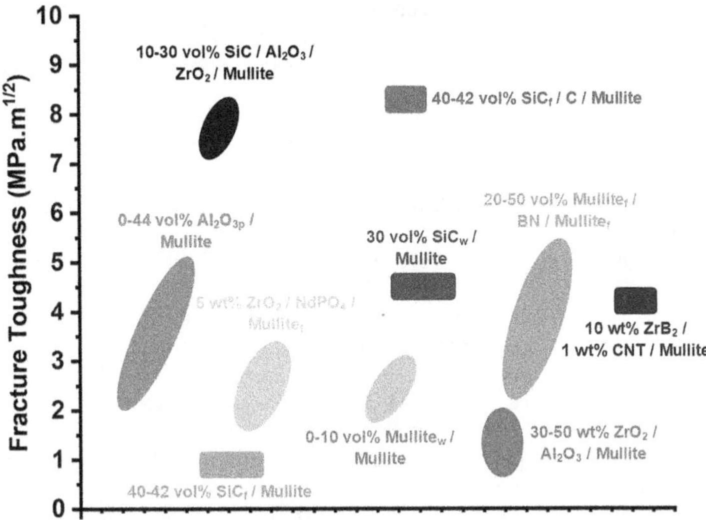

FIGURE 11.3 Summary of fracture toughness data for mullite-based composites.

cases where *in situ* formation of mullite whiskers in a mullite matrix resulted only in a minor toughening effect. In terms of particulate reinforcement, the use of SiC in combination with alumina and zirconia in a mullite matrix resulted in high values for the fracture toughness; this positive effect was attributed to both the transformation toughening impacts of ZrO_2 and weak bonding between SiC and Al_2O_3 particles in the matrix (due to lower reactivity), leading to crack deflection. However, in some other compositions, the addition of ZrO_2 alone did not provide a significant toughening effect, and therefore, the grain size and its distribution are additional factors to consider when designing the composite.

Further consideration in the design of these materials is the total porosity and nature of pores in the composites. Even though the presence of pores is known to lower the mechanical strength and Young's modulus, pores can assist in hindering crack propagation in the ceramic/composite, and thus, in several studies, the impact of the overall porosity on the fracture toughness is another important factor. The total porosity of the composite has a direct correlation to the fabrication technique and processing parameters since these have a critical impact on the extent of densification and bonding between components in the composite. The fabrication route can further impact on the anisotropy of the constituents, which affects matrix properties, particularly in terms of interfacial debonding between the constituents and the resultant fracture toughness. Therefore, direct comparisons between composites of the same composition/components but fabricated using different methods are difficult owing to the complexity of interdependent and independent factors involved.

REFERENCES

Aksay, I.A., D.M. Dabbs, and M. Sarikaya 1991. Mullite for structural, electronic, and optical applications. *Journal of the American Ceramic Society* 74(10): 2343–2358.

Aksel, C. 2003. The effect of mullite on the mechanical properties and thermal shock behaviour of alumina–mullite refractory materials. *Ceramics International* 29: 183–188.

Aramaki, S., and R. Roy 1962. Revised phase diagram for the system Al_2O_3—SiO_2. *Journal of the American Ceramic Society* 45(5): 229–242.

Bella, M.L., M. Hamidouche, and L. Gremillard 2021. Preparation of mullite-alumina composite by reaction sintering between Algerian kaolin and amorphous aluminum hydroxide. *Ceramics International* 47(11): 16208–16220.

Bengisu, M. and O.T. Inal 1994. Whisker toughening of ceramics: Toughening mechanisms, fabrication, and composite properties. *Annual Reviews in Materials Science* 24: 83–124.

Bhattacharjee, S., S.K. Singh, and R.K. Galgali 2000. Preparation of zirconia toughened mullite by thermal plasma. *Materials Letters* 43(1–2): 77–80.

Boccaccini, A.R., S. Atiq, D.N. Boccaccini, I. Dlouhy, and C. Kaya 2005. Fracture behaviour of mullite fibre reinforced–mullite matrix composites under quasi-static and ballistic impact loading. *Composites Science and Technology* 65:325–333.

Boccaccini, D.N., C. Leonelli, M.R. Rivasia, M. Romagnolia, and A.R. Boccaccini 2005. Microstructural investigations in cordierite–mullite refractories. *Ceramics International* 31: 417–432.

Bodhak, S., S. Bose, and A. Bandyopadhyay 2010. Microwave sintering of mullite and mullite zirconia composites. *Innovative Processing and Manufacturing of Advanced Ceramics and Composites* 212: 95–103.

Bouchetou, M.L., J. Poirier, L. Arbelaez Morales, T. Chotard, O. Joubert, and M. Weissenbacher 2019. Synthesis of an innovative zirconia-mullite raw material sintered from andalusite and zircon precursors and an evaluation of its corrosion and thermal shock performance. *Ceramics International* 45(10): 12832–12844.

Bowen, N.L. and J.W. Greig 1924. The system: Al_2O_3. SiO_2. *Journal of the American Ceramic Society* 7(4): 238–254.

Carvalho, R.G., F.J. Oliveira, R.F. Silva, and F.M. Costa 2014. Mechanical behaviour of zirconia–mullite directionally solidified eutectics. *Materials & Design* 61: 211–216.

Davis, R.F. and J.A. Pask 1971. Mullite. In *Refractory Materials* 5: 37–76.

Degueldre, C., P. Tissot, H. Lartigue, and M. Pouchon 2003. Specific heat capacity and Debye temperature of zirconia and its solid solution. *Thermochimica Acta* 403(2): 267–273.

Denry, I., and J.R. Kelly 2014. Emerging ceramic-based materials for dentistry. *Journal of Dental Research* 93(12): 1235–1242.

Fischer, R.X., A. Gaede-Köhler, J. Birkenstock, and H. Schneider 2012. Mullite and mullite-type crystal structures. *International Journal of Materials Research* 103(4): 402–407.

Fischer, R.X., H. Schneider, and D. Voll 1996. Formation of aluminium rich 9:1 mullite and its transformation to low alumina mullite upon heating. *Journal of the European Ceramic Society* 16(2): 109–113.

Fischer, R.X., H. Schneider, and M. Schmucker 1994. Crystal structure of Al-rich mullite. *American Mineralogist* 79(9–10): 983–990.

Garrido, L.B., E.F. Aglietti, L. Martorello, M.A. Camerucci, and A.L. Cavalieri 2006. Hardness and fracture toughness of mullite–zirconia composites obtained by slip casting. *Materials Science and Engineering A* 419: 290–296.

Gustaffon, S., Falk, L.K.L. and J.E. Pitchford 2009. Development of microstructure during creep of polycrystalline mullite and a nanocomposite mullite/5 vol.% SiC. *Journal of the European Ceramic Society* 29: 539–550.

Han, S., L.W. Yang, H.T. Liu, X. Sun, R. Jiang, W.G. Mao, and Z.H. Chen 2017. Micro-mechanical properties and interfacial engineering of SiC fiber reinforced sol-gel fabricated mullite matrix composites. *Materials & Design* 131: 265–272.

Harris, G.L. 1995. *Properties of Silicon Carbide*, INSPEC. The Institution of Electrical Engineers, London, 5.

Hildmann, B., and H. Schneider 2004. Heat capacity of mullite-new data and evidence for a high-temperature phase transformation. *Journal of the American Ceramic Society* 87(2): 227–234.

https://www.ortonceramic.com/files/2676/File/Orton-Cone-Chart-F-022-14.pdf.

Huang, Z., Z. Shen, L. Lin, M. Nygren, and D. Jiang 2004. Spark-plasma sintering consolidation of SiC-whisker-reinforced mullite composites. *Journal of the American Ceramic Society* 87(1): 42–46.

Jinga, Y., X. Denga, J. Li, C. Bai, and W. Jiang 2014. Fabrication and properties of SiC/mullite composite porous ceramics. *Ceramics International* 40: 1329–1334.

Kabel, J., P. Hosemann, Y. Zayachuk, D.E.J. Armstrong, T. Koyanagi, Y. Kutoh, and C. Deck 2018. Ceramic composites: A review of toughening mechanisms and demonstration of micropillar compression for interface property extraction. *Journal of Materials Research* 33(4): 424–439.

Karihaloo, B.L. and J.H. Andresen 1996. Transformation toughening materials in mechanics of transformation toughening and related topics. *North-Holland Series in Applied Mathematics and Mechanics* 40: 9–33.

Kaya, C ., E.G. Butler, A. Selcuk, A.R. Boccaccini, and M.H. Lewis 2002. Mullite (Nextel™ 720) fibre-reinforced mullite matrix composites exhibiting favourable thermomechanical properties. *Journal of the European Ceramic Society* 22(13): 2333–2342.

Kern, E.L., D.W. Hamill, H.W. Deem, and H.D. Sheets 1969. Thermal properties of β-silicon carbide from 20 to 2000C. In *Silicon Carbide*, Elsevier, 25–32.

Khor, K.A., and Y. Li 1998. Effects of mechanical alloying on the reaction sintering of $ZrSiO_4$ and Al_2O_3. *Materials Science and Engineering: A* 256(1–2): 271–279.

Kingery, W.D., H.K. Bowen, and D.R. Uhlmann 1976. Chapter 16 In *Introduction to Ceramics*. John Wiley & Sons.

Koshy, P., N. Ho, V. Zhong, L. Schreck, S. Alex Koszo, E.J. Severin, and C. Christopher Sorrell 2021. Fly ash utilisation in mullite fabrication: Development of novel percolated mullite. *Minerals* 11(1): 84.

Koshy, P., S. A. Koszo, E. Severin, and C. Christopher Sorrell 2018. High-performance refractory ceramics of percolated mullite from waste materials. *American Ceramic Society Bulletin* 97(6): 29–33.

Koyuma, T., S. Hayashi, A. Yasumori, and K. Okada 1994. Preparation and characterisation of mullite-zirconia composites from various starting materials. *Journal of the European Ceramic Society* 14: 295–302.

Kumar, P., M. Nath, A. Ghosh, and H. Sekhar Tripathi 2015. Enhancement of thermal shock resistance of reaction sintered mullite–zirconia composites in the presence of lanthanum oxide. *Materials Characterization* 101: 34–39.

Lin, Y.-J., and C.-P. Tsang 2003. Fabrication of mullite/SiC and mullite/zirconia/SiC composites by 'dual' in-situ reaction syntheses. *Materials Science and Engineering: A* 344(1–2): 168–174.

Liu, D., P. Hu, C. Fang, and W. Han 2018. Fabrication of unidirectional continuous fiber-reinforced mullite matrix composite with excellent mechanical property. *Ceramics International* 44: 13487–13494.

Majidian, H., T. Ebadzadeh, and E. Salahi 2011. Effect of SiC additions on microstructure, mechanical properties and thermal shock behaviour of alumina–mullite–zirconia composites, *Materials Science and Engineering A* 530: 585–590.

Meng, J., S. Cai, Z. Yang, Q. Yuan, and Y. Chen 1998. Microstructure and mechanical properties of mullite ceramics containing rodlike particles. *Journal of the European Ceramic Society* 18: 1107–1114.

Morell, R. 1985. *Handbook of Properties of Technical & Engineering Ceramics—Part 1*. Data Reviews, Section 1. HMSO, London, 255.

Nevitt, M.V., Y. Fang, and S.-K. Chan 1990. Heat capacity of monoclinic zirconia between 2.75 and 350 K. *Journal of the American Ceramic Society* 73(8): 2502–2504.

Ohji, T., Y.K. Jeong, Y.-H. Choa, and K. Niihara 1998. Strengthening and toughening mechanisms of ceramic nanoparticles. *Journal of the American Ceramic Society* 81(6): 1453–1460.

Orooji, Y., M.R. Derakhshandeh, E. Ghasalia, M. Alizadeh, M.S. Asl, and T. Ebadzadeh 2019. Effects of ZrB_2 reinforcement on microstructure and mechanical properties of a spark plasma sintered mullite-CNT composite. *Ceramics International* 45: 16015–16021.

Osendi, M.I., and C. Baudin 1996. Mechanical properties of mullite materials. *Journal of the European Ceramic society* 16(2): 217–224.

Pearce, D.H., J. Janczak, A.R. Boccaccini, C.B. Ponton, and L. Rohr 1996. Characterisation of a pressureless sintered sapphire fibre reinforced mullite matrix composite. *Materials Science and Engineering A* 214: 170–173.

Philippe, B., and J.-C. Niepce 2007. *Ceramic Materials Processes, Properties and Applications*. Vol. 98, ISTE Ltd.

Prusty, S., D.K. Mishra, B.K. Mohapatra, and S.K. Singh 2012. Effect of MgO in the microstructure formation of zirconia mullite composites from sillimanite and zircon. *Ceramics International* 38(3): 2363–2368.

Rocha-Rangel, E., Díaz-de-la-Torre, S., Umemoto, M., Miyamoto, H. and Balmori-Ramírez, H., 2005. Zirconia–mullite composites consolidated by spark plasma reaction sintering from zircon and alumina. *Journal of the American Ceramic Society* 88(5): 1150–1157.

Rodrigo, P.D.D. and P. Boch 1985. High purity mullite ceramics by reaction sintering. *International Journal of High Technology Ceramics* 1(1): 3–30.

Sadik, C., I.E.E. Amrani, and A. Albizane 2014. Processing and characterization of alumina–mullite ceramics. *Journal of Asian Ceramic Societies* 2(4): 310–316.

Schneider, H., J. Schreuer, and B. Hildmann 2008. Structure and properties of mullite—a review. *Journal of the European Ceramic Society* 28(2): 329–344.

Schneider, H., R.X. Fischer, and J. Schreuer 2015. Mullite: Crystal structure and related properties. *Journal of the American Ceramic Society* 98(10): 2948–2967.

Schneider, H., and S. Komarneni 2005. Basic properties of mullite. *Mullite*: 141–225.

Shackelford, J.F., and R.H. Doremus 2008. *Ceramic and Glass Materials*. Springer-Verlag, Berlin.

Steinbrech, R.W. 1992. Toughening mechanisms for ceramic materials. *Journal of the European Ceramic Society* 10: 131–142.

Vazquez Carbajal, G.I., J.L. Rodriguez Galicia, J.C. Rendon Angeles, J. Lopez Cuevas, and C.A. Gutierrez Chavarria 2012. Microstructure and mechanical behavior of alumina–zirconia–mullite refractory materials, *Ceramics International* 38: 1617–1625.

Wei, W., W. Duan, and X. D. Wu 2011. Preparation and thermal stability of zirconia-doped mullite fibers via sol-gel method. *Progress in Natural Science: Materials International* 21(2): 117–121.

Yang, M., X. Luo, J. Yi, X. Zhang, and Z. Peng 2019. Fabrication of fibrous mullite-alumina ceramic with high strength and low thermal conductivity. *Journal of Wuhan University of Technology-Material Science Edition* 34(6): 1415–1420.

Yeh, C.L. and W.C. Kao 2014. Preparation of TaB/TaB_2/mullite composites by combustion synthesis involving aluminothermic reduction of oxide precursors. *Journal of Alloys and Compounds* 615: 734–739.

Ying, L.I., Shao-gang, and Y. C. Zhai 2010. Preparation and sintering properties of zirconia-mullite-corundum composites using fly ash and zircon. *Transactions of Nonferrous Metals Society of China* 20(12): 2331–2335.

Zhang, R., X. Hou, C. Ye, and B. Wang 2017. Enhanced mechanical and thermal properties of anisotropic fibrous porous mullite–zirconia composites produced using sol-gel impregnation. *Journal of Alloys and Compounds* (2017) 699: 511–516.

Zhang, R., X. Hou, C. Ye, B. Wang, and D. Fang 2016. Fabrication and properties of fibrous porous mullite–zirconia fiber networks with a quasi-layered structure. *Journal of the European Ceramic Society* 36(14): 3539–3544.

Zhou, Y., J. Vleugels, T. Laoui, and O. Van der Biest 1994. Toughening of sol-gel derived mullite matrix by Al_2O_3 platelets. *Journal of Materials Science Letters* 13(15): 1089–1091.

12 Generalised Model of Trends in Mechanical Properties of Toughened Sphere-reinforced Polymer Matrix Composites

I.I. Kabir, C.C. Sorrell, and G.H. Yeoh
UNSW Sydney

S. Bandyopadhyay
UNSW Sydney
Aus Defence DSTO MRL

CONTENTS

12.1 INTRODUCTION

Polymer matrix composites (PMCs), metal matrix composites, and ceramic matrix composites have emerged as important materials in a range of applications (Ralph, Yuen, and Lee 1997). Of these, PMCs are advantageous owing to their low cost, ease of manufacture, low weight, high elasticity, and recyclability (Tapper et al. 2018). In 2020, the global market for PMCs was estimated to be US$ 4.77 billion, with a market segregation as shown in Figure 12.1 (Aspray et al. 2017); (https://www.researchandmarkets.com/reports/4751797/plastic-market-size-share-and-trends-analysis). The most common polymeric matrix used is the thermoset epoxy, while thermoplastics such as high-density polyethylene (HDPE) and elastomers such as rubber are less commonly used (Kessler 2012). The most common fillers are the oxides alumina, α-quartz, and glass; calcium carbonate; and carbon (Chung 2019). The principal mechanical properties of these materials are summarised in Table 12.1.

Although increased elastic modulus (E), tensile strength (σ_t), and fracture toughness (K_{Ic}) are generally targeted through the addition of fibrous or plate reinforcements (Ashby et al. 2018; Shackelford 2000), spherical reinforcements have been established in the marketplace because they are inexpensive; they are available in small, graded, and monomodal sizes; and their flow characteristics lend them to the ready achievement of blending homogeneity (Wyatt and Dew-Hughes 1974; Tang et al. 2013b).

DOI: 10.1201/9780429330575-12

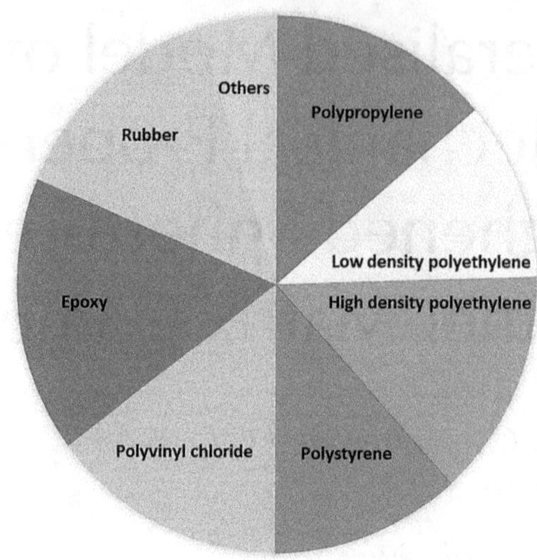

FIGURE 12.1 2020 global market segregation for PMC matrices.

TABLE 12.1
Summary of mechanical properties of common matrix and filler materials used in PMCs

Matrix		**Mechanical properties**			**Refs.**
		Elastic modulus (MPa)	**Tensile strength (MPa)**	**Fracture toughness (MPa·m½)**	**Ashby et al. (2018), Shackelford (2000), Wyatt and Dew-Hughes (1974)**
Epoxy		68–80	7.58–96.5	0.3–0.5	
HDPE		550–1000	20–37	1	
Polypropylene		1200–1700	50–70	3	
Polystyrene		3000–3300	35–68	2	
Polyvinyl chloride		2400–3000	40–60	2.4	
Rubber		2.89	-	0.9	
Reinforcements		**Elastic Modulus (GPa)**	**Tensile Strength (MPa)**	**Fracture Toughness (MPa·m½)**	**Refs. Ashby et al. (2018), Shackelford (2000), Wyatt and Dew-Hughes (1974)**
Alumina (99%)		380	>400	2.5–5.0	
α-Quartz	// *a* axis	78	155	<1.0	
	// *c* axis	105	–	–	
Glass	Soda–lime–silica	73	50–100	<1.0	
	Fly ash	74	155	<1.0	
	Cenospheres	74	155	<1.0	
	Precipitated silica	74	50–100	<1.0	
$CaCO_3$	Limestone	52	10	<1.0	
	Calcite	48	8	<1.0	
Carbon	Graphite	4–28	4.8–76	0.4–2.4	
	Nanotube	270–950	150,000	≥2.7	

12.2 EXPERIMENTAL PARAMETERS

The mechanical properties of sphere-reinforced PMCs are determined principally by a limited range of factors. While the basic properties of the polymer matrix provide a platform, improvements are derived largely through the following experimental parameters:

Particle size
Particle loading
Particle/matrix adhesive strength

While E and σ_t depend mainly on the effectiveness of load transfer from the flexible weak polymer matrix to the stiff strong ceramic, K_{Ic} depends on the adhesive strength, when the failure mode results in cavitation, and the matrix σ_t, when the failure mode is by crazing (i.e., intergranular fracture) (Tang et al. 2013b; Kabir et al. 2016). The principles underpinning these mechanisms have been elucidated as a general model for recycled HDPE reinforced with spherical fly ash particles (~13 μm) (Kabir et al. 2016). In this work, the load–displacement data for a range of studies were synthesised into a series of similar trends as a function of particle loading for the stress–strain regimes indicative of the different failure modes. The starting point for the prediction of the mechanical properties of all composites is the rules of mixtures, which are illustrated in Figure 12.2.

The relevant rules of mixtures models are as follows:

Linear (parallel model) $\phi_C = \phi_A X_A + \phi_B X_B$
Linear (series model) $\phi_C = \phi_A / X_A + \phi_B / X_B$
Arithmetic log $\phi_C = (\log \phi_A) X_A + (\log \phi_B) X_B$
Quadratic $\phi_C = \phi_A X_A + \phi_B X_B + \kappa X_A X_B$

where ϕ is the experimental parameter, X is the volume fraction, A and B refer to each of the two phases, and C refers to the composite. The quadratic model assumes interaction between the matrix and reinforcement and $\kappa > 0$ for (beneficial) synergism and $\kappa < 0$ for (detrimental) antagonism.

However, it is well known that these relations are simplifications that assume that the properties are purely additive and that there is no alteration of the interface (Yosomiya 2020). Consequently, a considerable amount of work has been carried out to model and validate real systems of sphere-reinforced polymers, although these studies are limited to the electrical and thermal properties (Berhan and Sastry 2007; Guo et al. 2014). However, the limited number of

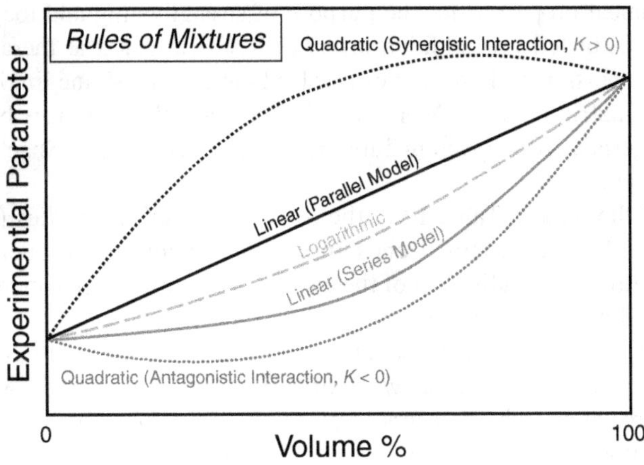

FIGURE 12.2 Rules of mixture approaches for composites.

works reporting experimental data for polymer matrices reinforced with spherical particles or agglomerates reveal general trends only for E and σ_t; there are no studies that examined K_{Ic}. That being said, these three mechanical properties are related using the following equation:

$$K_k = \left(\frac{2E\gamma_i}{1 - \nu^2} \right)^{1/2} \tag{12.1}$$

where γ_i is the fracture energy for rapid crack propagation (i.e., a reflection of σ_t) and ν is Poisson's ratio (typically 0.3 for ceramics) (Morrell 1989). Of E and γ_i, K_{Ic} tracks more closely with γ_i and so, by extension, σ_t, rather than E.

A key aspect of these data is the effect of percolation of the reinforcement particles, which reflects particle loading at the point at which the reinforcing particles come into mutual contact, thereby establishing continuity of both matrix and reinforcement (Li and Kim 2007; Kingery, Bowen, and Uhlmann 1976). Percolation effectively represents the transition between matrix- and reinforcement-dominated behaviour. However, since the mechanical properties of brittle polycrystalline materials are controlled largely by microstructural effects (Kingery, Bowen, and Uhlmann 1976; Han et al. 2018), the trends illustrated reflect this effect, rather than the intrinsic effects of the materials.

12.3 MODELLING AND EXPERIMENTAL STUDIES

Table 12.2 presents representative experimental data for the mechanical properties of sphere-reinforced PMCs.

12.4 TRENDS

The data in Table 12.2 reveal that the most commonly studied sphere-reinforced PMC is epoxy reinforced with glass. Concerning the matrix, this is likely due to the advantages offered by epoxies, which are ease of fabrication, modifiable composition (*viz.*, thermosetting), highly cross-linked structure, high stiffness, high strength, high creep resistance, and elevated temperature stability (Tang et al. 2013a; Adachi et al. 2008; Imanaka et al. 2001; Liu et al. 2011). Concerning the reinforcement, this is likely due to the ready availability of the spherical waste product fly ash, ease of fabrication of spherical silica glass, availability in the nanometre to micrometre size range, low true densities, and low costs. Most of these studies reported the effects of relatively high addition levels on the mechanical properties for the purpose of strengthening and toughening. One study (Sun et al. 2014) investigated low addition levels as it was targeted at therapeutic applications. Also, a significant proportion of these studies involved the use of a silane linking agent in order to improve the particle/matrix bonding. More broadly, the general differences between the relevant properties of the materials are as given in Table 12.3 (Ashby et al. 2018; Shackelford 2000; Wyatt and Dew-Hughes 1974).

Consideration of the data in Table 12.2, the relevant modelling studies (Berhan and Sastry 2007; Guo et al. 2014), and general trends in the fabrication and properties of composites (Kabir et al. 2016) allows generalisation of the effects of the three experimental parameters that affect the mechanical properties, as given in Table 12.4.

Percolation: Table 12.2 shows that most of the studies reveal regular increasing trends as a function of particle loading, although some show maxima. The latter data are consistent with the effects of reinforcement percolation, which occurs in principle when the particle loading is sufficiently high to cause the reinforcement particles to come into mutual contact throughout the volume of the PMC (Berhan and Sastry, 2007). Consequently, most of the studies focussed on particle loadings

TABLE 12.2
Representative experimental data for the mechanical properties of sphere-reinforced PMCs

	Spheres		Parameters examined					
Polymer matrix	Material	Diameter	Volume addition (%)	Young's modulus (GPa)	Tensile strength (MPa)	Fracture toughness (MPa·m$^{1/2}$)	Comments	Ref.
Epoxy	Hollow spheres	-	0	-	-	-	Young's modulus apparently maximised and then decreased, but compressive strength increased with increasing volume fraction of reinforcement	Dzenis and Maksimov (1991)
			10	3.75				
			20	3.99				
			30	3.25				
			40	3.35				
			50	3.30				
			60	3.31				
Poly(aryl ether sulfone)	Soda–lime–phosphate silicate glass spheres	106–125 μm	0	-	-	-	High surface area and surface roughness enhanced load transfer from matrix to reinforcement; failure by shear.	Oréfice et al. (2001)
			20	3.75				
Polypropylene	Hollow glass spheres[a]	20–35 μm	0	-	-	-	Tensile deformation facilitated by the effect of silane coupling agent in increasing interfacial stress transfer	Yağci et al. (2021)
			20	4.74				
Epoxy	Silica glass spheres + CNT[a]	1–12 μm	0	-	-	-	Young's modulus, tensile strength, and fracture toughness increased with increasing volume fraction of mixed reinforcement, which consisted of microscale spheres and gap-filling nanoscale fibers	Fu et al. (2008)
			10	4.51	113			
			18	5.11	119			
			30	7.75	129			
Epoxy	Glass spheres[a]	1–30 mm	0	-	-	-	Tensile strength determined by interfacial failure, not shear of matrix	Leidner and Woodhams (1974)
			35	5.85	70.0			
			40	5.93	58.4			
Rubber	Glass spheres[a]	50 μm	0	-	-	0.90	Static fracture toughness increased, but fatigue resistance was unaffected by increasing volume fraction of reinforcement; shear plasticity of the matrix was unimportant	Al-Ostaz et al. (2004)
			2.5			2.31		
			5.0			3.26		
			7.5			3.59		

(Continued)

TABLE 12.2 (*Continued*)
Representative experimental data for the mechanical properties of sphere-reinforced PMCs

Polymer matrix	Spheres: Material	Spheres: Diameter	Parameters examined: Volume addition (%)	Young's modulus (GPa)	Tensile strength (MPa)	Fracture toughness (MPa·m$^{1/2}$)	Comments	Ref.
Poly(vinyl) alcohol	Paraffin spheres	100–200 μm	0	-	-	-	Paraffin spheres were template to create a pore network in a scaffold; Young's modulus apparently maximised and then decreased with increasing volume fraction of reinforcement	Azimi et al. (1996)
			10	0.0237				
			12	0.0810				
			15	0.0468				
			18	0.0508				
Polystyrene	Grafted spherical silica glass nanoparticles[a]	14 nm	0	-	-	-	Young's modulus increased while increasing ductility through strong interfacial binding between the polystyrene matrix and reinforcement achieved by grafting of polystyrene chains on silica spheres and use of silane coupling agent	Ma and Choi (2001)
			3.0	3.30				
			5.0	3.75				
			7.5	3.95				
			10.0	3.79				
			15.0	4.75				
Epoxy	Polystyrene spheres	16–20 μm	0	-	59.3	1.21	Tensile strength and fracture toughness increased with increasing volume fraction of reinforcement	Maillard et al. (2012)
			2		61.5	1.42		
			5		64.5	1.88		
			8		73.3	2.11		
			15		74.5	2.23		
Polystyrene	Silica glass sphere Janus nanoparticles with hemispherical coating of magnetite-loaded polystyrene[a]	40 nm	0	-	-	-	A slight increase in tensile strength with increasing volume fraction of reinforcement; agglomeration of reinforcement was important	Sun et al. (2014)
			0.25		42.12			
			0.50		42.16			
			1.00		42.56			
			3.00		43.64			
Epoxy	Spherical glass silica particles[a]	300–800 nm	0	-	-	-	Speculation that multiscale particle size of reinforcement increased Young's modulus and tensile strength	Sharifzadeh and Amiri (2020)
			1.0	3.35	83.67			

(*Continued*)

TABLE 12.2 (*Continued*)
Representative experimental data for the mechanical properties of sphere-reinforced PMCs

Polymer matrix	Spheres: Material	Spheres: Diameter	Volume addition (%)	Parameters examined: Young's modulus (GPa)	Tensile strength (MPa)	Fracture toughness (MPa·m$^{1/2}$)	Comments	Ref.
Epoxy	Spherical micro-silica particles	1.56 μm–240 nm	0	3.8	-	1.01	Young's modulus increased, but fracture toughness maximised with an increasing volume fraction of reinforcement and with decreasing particle size. With low volume fraction reinforcement, the dominant effect was particle loading; with high volume fraction reinforcement, and the dominant effect was particle size	Tang et al. (2013a)
			10	4.1		1.75		
			20	5.1		1.95		
			30	6.2		1.35		
			35	7.8		1.46		
Epoxy	Glass spheres	-	0	-	-	-	Fracture toughness and impact strength increased with decreasing particle size owing to enhanced interfacial strength	Adachi et al. (2008)
			5	4.26				
			10	5.08				
			15	5.67				
Epoxy	Spherical silica particles	6–30 μm	0	3.190	-	-	Young's modulus and fracture toughness increased with increasing volume fraction of reinforcement. Fracture toughness increased with increasing particle size owing to decreased interfacial strength, which involved crack pinning and crack blunting	Imanaka et al. (2001)
			6.25	3.201		1.22		
			12.5	3.620		1.45		
			25.0	4.020		1.85		
Epoxy	Spherical glass silica particles	100 nm	0	2.86	42.1	0.95	Young's modulus, tensile strength, and fracture toughness increased with increasing volume fraction of reinforcement. A systematic study on the effects of silica on the fracture toughness behaviour of epoxy was conducted. Toughening mechanisms included reinforcement debonding, cavitation, bridging before reinforcement pull-out, and matrix plastic shearing	Liu et al. (2011)
			2	2.88	45.1	1.01		
			4	2.93	42.0	1.14		
			6	2.98	43.1	1.26		
			8	3.12	42.7	1.39		
			10	3.14	46.5	1.57		
			12	3.20	48.3	1.70		
			20	3.48	54.2	2.11		

(*Continued*)

TABLE 12.2 (*Continued*)
Representative experimental data for the mechanical properties of sphere-reinforced PMCs

Polymer matrix	Spheres: Material	Spheres: Diameter	Parameters examined: Volume addition (%)	Young's modulus (GPa)	Tensile strength (MPa)	Fracture toughness ($MPa \cdot m^{1/2}$)	Comments	Ref.
High-density polyethylene	Fly ash	13 μm	0	0.073	20	-	Young's modulus and tensile strength reached maxima with increasing volume fraction of reinforcement. These trends reflect the progression as follows: 1. Dispersion strengthening and stress concentration of ductile material 2. Debonding and cavitation of plastic material undergoing crazing 3. Fibril failure of plastic material 4.Brittle failure while approaching percolation	Kabir et al. (2016)
			2.5	0.091	22			
			5.0	0.083	21			
			7.5	0.088	22			
			10.0	0.088	19			
High-density polyethylene	Fly ash	14 μm	0	0.490	23	-		Ahmad and Mahanwar (2010)
			5	0.500	23			
			10	0.520	23			
			15	0.575	25			
			20	0.645	28			
			30	0.875	28			
			40	0.610	29			
High-density polyethylene	Fly ash	<106 μm	0	0.560	32	-		Atikler et al. (2006)
			10	0.635	31			
			20	0.640	28			
			30	0.820	25			
			40	1.010	21			

[a] Silane coupling agent used to enhance matrix reinforcement adhesion

TABLE 12.3
General trends in relevant mechanical properties for materials commonly used in sphere-reinforced PMCs

Property	Trends: Reinforcement	Trends: Matrix
E	Glasses > polymers	Thermosets > thermoplastics > elastomers
σ_t	Glasses > polymers	Thermosets ≈ thermoplastics > elastomers
K_{Ic}	Polymers > glasses	Thermoplastics > thermosets > elastomers

TABLE 12.4
Effects of the principal experimental parameters on the fabrication and mechanical properties of sphere-reinforced PMCs

Experimental Parameter	Effects of increasing experimental parameter on: Fabrication	E	σ_t	K_{Ic}
Particle loading	Particles <5 μm increase viscosity slightly; particles >5 μm significantly increase viscosity,	↑	↑	↑
Particle size	thereby decreasing ease of mixing, shape ability, and microstructural homogeneity	↑	↓	↑
Particle/matrix adhesive strength	Increasing adhesive strength significantly increases viscosity, but size and loading effects are converse to those above	↑	↑	↑

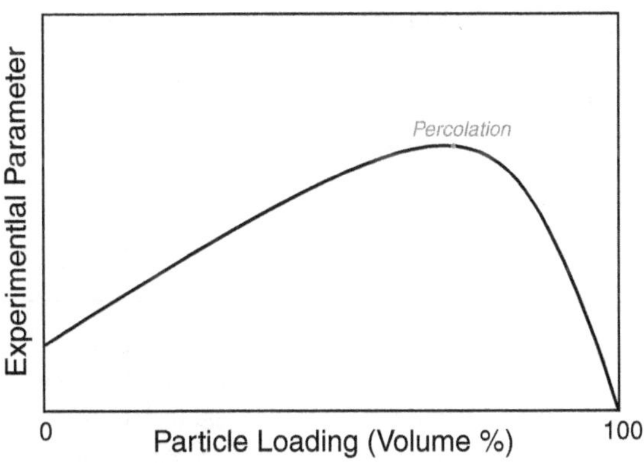

FIGURE 12.3 General trend in mechanical properties of PMCs as a function of particle loading.

insufficient to cause percolation. Figure 12.3 generalises the trends in mechanical properties as a function of particle loading, revealing the role of percolation in establishing a quadratic rule of mixtures and the associated maximum.

Mechanical Failure: Figure 12.4 shows schematics of the typical effects of particle size on crack initiation in and propagation through PMCs, which invariably reveal the transgranular fracture through the weaker polymer matrix, rather than the transgranular fracture through the hard reinforcing phase. The effects of particle loading and particle/matrix adhesive strength can be inferred from the hindrance of crack initiation and extension by high particle loading (i.e., domination of

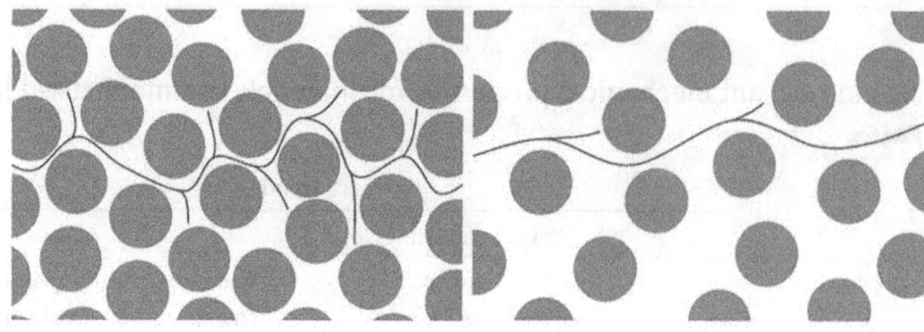

Large Particle Size | High Volume Fraction **Large Particle Size | Low Volume Fraction**

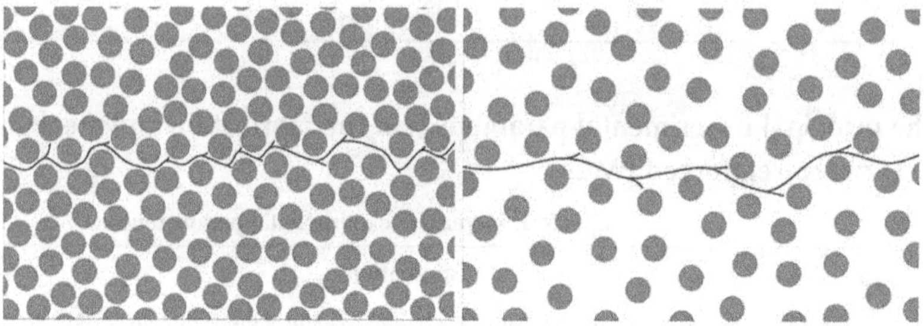

Small Particle Size | High Volume Fraction **Small Particle Size | Low Volume Fraction**

FIGURE 12.4 Schematics of effects of particle size and volume fraction on crack initiation in and propagation through PMCs.

properties by reinforcement) and the hindrance of crack extension by high adhesive strength (i.e., effective stress transfer (Ashby et al. 2018; Shackelford 2000; Wyatt and Dew-Hughes 1974) and constraint on crack opening).

Fracture Toughening: Methods of increasing the fracture toughness of composites are well established (Qin and Ye 2015), which are listed as follows:

Forms of crack deflection	• Crack front pinning • Crack path deflection
Forms of crack branching	• Crack branching • Microcrack formation
Role of adhesive strength	• Particle bridging or crack bridging • Particle pull-out

12.5 MECHANISMS

The mechanisms affecting stiffening (E), strengthening (σ_t), and toughening (K_{Ic}) leading to mechanical failure are summarised briefly in Table 12.5.

TABLE 12.5
Effects of the increasing particle reinforcement fraction on the mechanical properties of sphere-reinforced PMCs

Region	Description	Effect		
		E	σ_t	K_{Ic}
Elastic	Neat polymer and low particle reinforcement fraction exhibit ductile behaviour dominated by dispersion strengthening offset by stress concentration	↑	↑	↑
Plastic	Increasing particle fraction exhibits plastic behaviour dominated progressively by particle reinforcement cavitation, filament formation, and fibril failure, ultimately leading to matrix crazing	↑	↑	↑
Percolation	Transition from dominance of effects from composite to dominance of effects from particle reinforcement	-	-	-
Brittle	Excessive particle fraction exhibits brittle behaviour dominated by particles	↓	↓	↓

12.6 GENERALISED MODEL

The trends in the load–displacement behaviour of sphere-reinforced polymer-based composites discussed previously by the authors (Kabir et al. 2016) reveal the effects of particle loading on the deformation and failure mechanisms of these materials. These can be segregated, as shown in the last three rows of Table 12.2, into the four stages described, which traverse the elastic deformation, plastic deformation, percolation, and brittle failure regions. The present work develops these mechanical concepts from sole consideration of the effects of spherical fly ash particle loading of high-density polyethylene on load–displacement trends (Kabir et al. 2016); that is, the present work extends the examination to spherical particles and polymers more broadly; it extends the parameters considered to particle loading, particle size, and particle/matrix adhesive strength; and it also extends the properties of consideration to Young's modulus, tensile strength, and fracture toughness. These are carried out through the consideration of the data in Table 12.2 and the resultant trends indicated in Tables 12.4 and 12.5. Furthermore, these data and trends are integrated to develop a similar approach to the fracture toughness for all three experimental parameters. Thus, the general effects of particle loading, particle size, and particle/matrix adhesive strength on Young's modulus, tensile strength, and fracture toughness of sphere-reinforced polymer-based composites are shown in Figure 12.5.

The upper portion of the generalised model indicates the response of sphere-reinforced PMCs to changes in the three preceding experimental parameters in terms of the load–displacement behaviour. The increase or decrease in these is indicated by the small black vertical arrows that are followed by the relevant parameter. The large coloured arrows indicate the direction of shift in the data upon the indicated increase or decrease in the parameter. The three coloured curves reflect the general trends for low, medium, and high particle loadings, and the curves are labelled according to the associated failure mechanisms. The corresponding boxes list the typical causes of degradation.

The lower portion of the generalised model indicates the expected response of fracture toughness to the alteration of the three preceding experimental parameters. These data show that fracture toughness increases according to the increase or decrease in the experimental parameter up to the point of percolation, at which point rapid degradation occurs.

The lower portion of the generalised model indicates the expected response of fracture toughness to the alteration of the three preceding experimental parameters. These data show that fracture toughness increases according to the increase or decrease in the experimental parameter up to the point of percolation, at which point rapid degradation occurs.

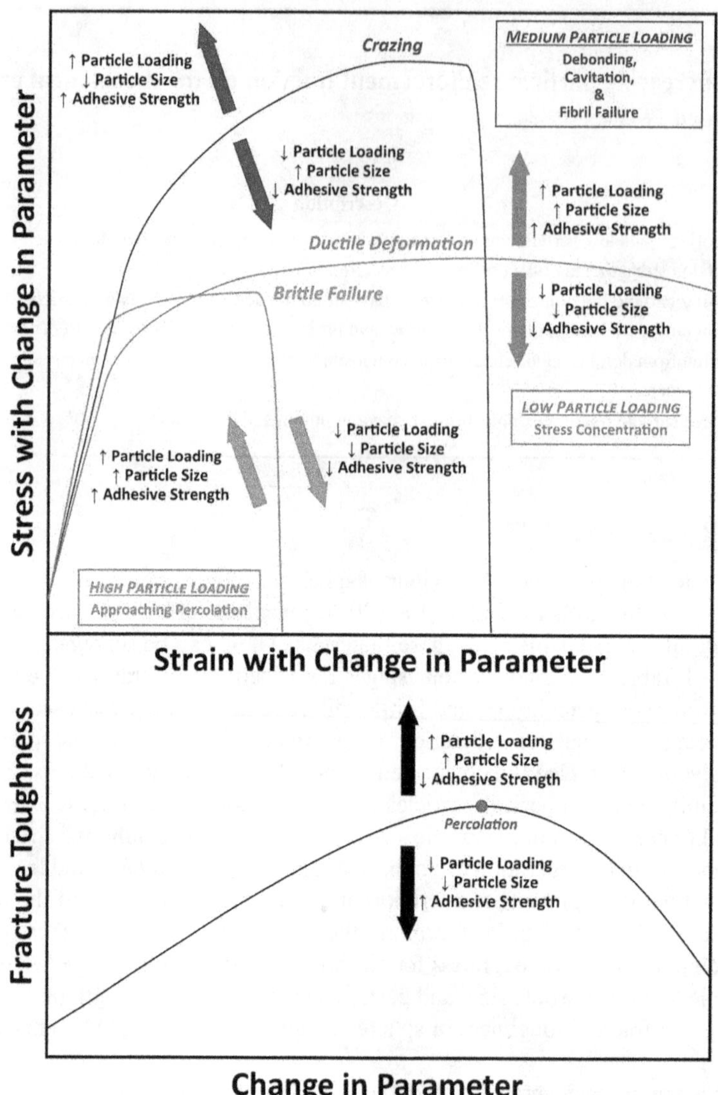

FIGURE 12.5 Effects of experimental parameters on load–displacement curves and fracture toughness.

12.7 CONCLUSIONS

The present work briefly examines the current market for PMCs, surveys the basic mechanical properties of typical ceramic reinforcements and polymeric matrices, and considers the three main experimental parameters that affect the mechanical properties of PMCs, which being particle loading, particle size, and particle/matrix adhesive strength. The different rules of mixture models for two-component composites are given, although the literature reveals that modelling of sphere-reinforced composites is limited to electrical and thermal properties. As the present work considers the effects of the experimental parameters on the mechanical properties, an extensive literature survey was undertaken, and this revealed typical trends in Young's modulus, tensile strength, and fracture toughness. The approach to and achievement percolation of the reinforcing phase are critical because this microstructural condition causes a reversal in trends. The present work concludes with graphical representations of the simultaneous mechanistic effects of the three experimental parameters on the three mechanical properties and fracture toughness.

REFERENCES

Adachi, T., M. Osaki, W. Araki, and S.-C. Kwon. 2008. Fracture toughness of nano-and micro-spherical silica-particle-filled epoxy composites. *Acta Materialia* 56: 2101–2109.

Ahmad, I., and P.A. Mahanwar, 2010. Mechanical properties of fly ash filled high density polyethylene. *Journal of Minerals and Materials Characterization and Engineering* 9(3): 183.

Al-Ostaz, A., H. Al-Moussawi, and L.T. Drzal. 2004. Characterization of the interphase in glass sphere reinforced polymers. *Composites Part B: Engineering* 35: 393–412.

Ashby, M.F., H. Shercliff, and D. Cebon. 2018. *Materials: Engineering, Science, Processing and Design.* Elsevier, Butterworth - Heinemann (Oxford).

Aspray, T.J., M. Dimambro, and J. Steiner. 2017. *Investigation into Plastic in Food Waste Derived Digestate and Soil.* Scottish Environment Protection Agency (Cambridge Eco Ltd).

Atikler, U., D. Basalp, and F. Tihminlioğlu. 2006. Mechanical and morphological properties of recycled high-density polyethylene, filled with calcium carbonate and fly ash. *Journal of Applied Polymer Science* 102(5): 4460–4467.

Azimi, H.R., R.A. Pearson, and R.W. Hertzberg. 1996. Fatigue of hybrid epoxy composites: Epoxies containing rubber and hollow glass spheres. *Polymer Engineering & Science* 36: 2352–2365.

Berhan, L., and A.M. Sastry. 2007. Modeling percolation in high-aspect-ratio fiber systems. I. Soft-core versus hard-core models. *Physical Review E* 75: 041120.

Chung, D.D.L. 2019. A review of multifunctional polymer-matrix structural composites. *Composites Part B: Engineering* 160: 644–660.

Dzenis, Y.A., and R.D. Maksimov. 1991. Prediction of the physical-mechanical properties of hollow-sphere reinforced plastics. *Mechanics of Composite Materials* 27: 263–270.

Fu, S.-Y., X.-Q. Feng, B. Lauke, and Y.-W. Mai. 2008. Effects of particle size, particle/matrix interface adhesion and particle loading on mechanical properties of particulate–polymer composites. *Composites Part B: Engineering* 39: 933–961.

Guo, Z., X. Shi, Y. Chen, H. Chen, X. Peng, and P. Harrison. 2014. Mechanical modeling of incompressible particle-reinforced neo-Hookean composites based on numerical homogenization. *Mechanics of Materials* 70: 1–17.

Han, N.M., Z. Wang, X. Shen, Y. Wu, X. Liu, Q. Zheng, T.-H. Kim, J. Yang, and J.-K. Kim. 2018. Graphene size-dependent multifunctional properties of unidirectional graphene aerogel/epoxy nanocomposites. *ACS Applied Materials & Interfaces* 10: 6580–6592.

Imanaka, M., Y. Takeuchi, Y. Nakamura, A. Nishimura, and T. Iida. 2001. Fracture toughness of spherical silica-filled epoxy adhesives. *International Journal of Adhesion and Adhesives* 21: 389–396.

Kabir, I.I., C.C. Sorrell, M.R. Mada, S.T. Cholake, and S. Bandyopadhyay. 2016. General model for comparative tensile mechanical properties of composites fabricated from fly ash and virgin/recycled high-density polyethylene. *Polymer Engineering & Science* 56: 1096–1108.

Kessler, M.R. 2012. Polymer matrix composites: A perspective for a special issue of polymer reviews. *Polymer Reviews* 52(3): 229–233.

Kingery, W.D., H.K. Bowen, and D.R. Uhlmann. 1976. *Introduction to Ceramics.* Vol. 17, John Wiley & Sons.

Leidner, J., and R.T. Woodhams. 1974. The strength of polymeric composites containing spherical fillers. *Journal of Applied Polymer Science* 18: 1639–1654.

Li, J., and J.-K. Kim. 2007. Percolation threshold of conducting polymer composites containing 3D randomly distributed graphite nanoplatelets. *Composites Science and Technology* 67: 2114–2120.

Liu, H.-Y., G.-T. Wang, Y.-W. Mai, and Y. Zeng. 2011. On fracture toughness of nano-particle modified epoxy. *Composites Part B: Engineering* 42: 2170–2175.

Ma, P.X., and J.-W. Choi. 2001. Biodegradable polymer scaffolds with well-defined interconnected spherical pore network. *Tissue Engineering* 7: 23–33.

Maillard, D., S.K. Kumar, B. Fragneaud, J.W. Kysar, A. Rungta, B.C. Benicewicz, H. Deng, L. Cate Brinson, and J.F. Douglas 2012. Mechanical properties of thin glassy polymer films filled with spherical polymer-grafted nanoparticles. *Nano Letters* 12: 3909–3914.

Morrell, R. 1989. *Handbook of Properties of Technical and Engineering Ceramics. Part 1: An Introduction for the Engineer and Designer.* Her Majesty's Stationery Office, London.

Oréfice, R.L., L.L. Hench, and A.B. Brennan. 2001. Effect of particle morphology on the mechanical and thermo-mechanical behavior of polymer composites. *Journal of the Brazilian Society of Mechanical Sciences* 23: 1–8.

Plastic Market Size, Share & Trends Analysis Report by Product (PE, PP, PU, PVC, PET, Polystyrene, ABS, PBT, PPO, Epoxy Polymers, LCP, PC, Polyamide), by Application, by Region, and Segment Forecast, 2022–2030. 2020. https://www.researchandmarkets.com/reports/4751797/plastic-market-size-share-and-trends-analysis.

Qin, Q., and J. Ye, 2015. *Toughening Mechanisms in Composite Materials.* Elsevier.

Ralph, B., H.C. Yuen, and W.B. Lee. 1997. The processing of metal matrix composites—An overview. *Journal of Materials Processing Technology* 63: 339–353.

Shackelford, J.F. 2000. *Materials Science for Engineers.* Upper Saddle River, NJ.

Sharifzadeh, E., and Y. Amiri. 2020. The effects of the arrangement of Janus nanoparticles on the tensile strength of blend-based polymer nanocomposites. *Polymer Composites* 41: 3585–3593.

Sun, T., Z. Wu, Q. Zhuo, X. Liu, Z. Wang, and H. Fan. 2014. Microstructure and mechanical properties of aminated polystyrene spheres/epoxy polymer blends. *Composites Part A: Applied Science and Manufacturing* 66: 58–64.

Tang, L.-C., Y.-J. Wan, K. Peng, Y.-B. Pei, L.-B. Wu, L.-M. Chen, L.-J. Shu, J.-X. Jiang, and G.-Q. Lai. 2013a. Fracture toughness and electrical conductivity of epoxy composites filled with carbon nanotubes and spherical particles. *Composites Part A: Applied Science and Manufacturing* 45: 95–101.

Tang, Y., L. Ye, Z. Zhang, and K. Friedrich. 2013b. Interlaminar fracture toughness and CAI strength of fibre-reinforced composites with nanoparticles–A review. *Composites Science and Technology* 86: 26–37.

Tapper, R.J., M.L. Longana, H. Yu, I. Hamerton, and K.D. Potter, 2018. Development of a closed-loop recycling process for discontinuous carbon fibre polypropylene composites. *Composites Part B: Engineering* 146: 222–231.

Wyatt, O.H., and D. Dew-Hughes, 1974. *Metals, Ceramics and Polymers: An Introduction to the Structure and Properties of Engineering Materials.* London and New York, Cambridge University Press, 650.

Yağci, Ö., B.E. Gümüş, and M. Taşdemir, 2021. Thermal, structural and dynamical mechanical properties of hollow glass sphere-reinforced polypropylene composites. *Polymer Bulletin* 78: 3089–3101.

Yosomiya, R. 2020. *Adhesion and Bonding in Composites.* CRC Press, Boca Raton, FL.

13 Case Studies on Toughened Composites for Structural and Engineering Applications

Dheeraj K Gara, Raghavendra Gujjala, and Satish Jain
National Institute of Technology Warangal

Shakuntala Ojha and Omprakash
Kakatiya Institute of Technology and Science

CONTENTS

13.1 INTRODUCTION

Composites exhibit low fracture toughness, especially polymer-based composites, but an excellent mechanical strength-to-weight ratio and tribological properties. These properties make them amenable to aerospace, automotive, and biomedical applications. A comprehensive review in the context of the development of composite materials can be cited from the literature (Nicolais et al. 2011; Gay and Hoa 2007). Composite materials are metal–metal, metal–polymer, and metal–ceramics combinations, which give multifaceted properties to the material system. Most widely used composite materials are fiber-reinforced composites (FRCs) composites (Ratna 2008), metal matrix composites (Zhang et al. 2015), ceramic matrix composites (CMCs) (Belmonte et al. 2016), glass FRCs (Sprenger 2020), and polymer matrix composites (Xia et al.2019). They have been significantly used in aerospace, automotive, civil, and structural applications, followed by biomedical applications. However, the heterogeneous characteristics of the composite materials have identified lacunae such as low interfacial and interlaminar strength and non-uniform stress distribution, allowing them to be used in a wide range of applications (Markham and Dawson 1975).

Enhancing the properties such as strength, thermal resistance, and tribological characteristics plays a crucial role and has received a huge attention. Combining a number of materials in stack increases strength but, at the same time, increases the density and volume of the material, thus making them difficult to be used in applications for various industries.

To surmount this complexity, researchers in the scientific community have introduced materials with excellent mechanical and tribological properties as nanoparticles to ensure that the desired property of the material is achieved. But the composition of the constituent is challenging as the actual properties of the desired composite would deviate; hence, an appropriate composition must be configured. These composites which have nanoparticles as dispersants are referred

DOI: 10.1201/9780429330575-13

to as nanocomposites (Mai and Yu 2006; Thostenson, Li and Chou 2005; Paul and Robeson 2008; Komarneni 1992).

As mentioned earlier, composite materials possess low fracture toughness and therefore needs to be combined with metals which have good fracture toughness. But the addition of these metals to the composites would increase the density and may be limited to a proportion. For this, it is suggested that a suitable technique needs to be adopted for the processing of the material so as to ensure that of microstructural configurations of both the materials are compatible. The process of achieving the final composite with desired toughness is referred to as the toughened composite. These toughened composites have greater advantage than the traditional composites, and hence, the current reviews focus on the applications and case studies pertaining to structural engineering, energy storage, and non-engineering applications.

13.2 TOUGHENED COMPOSITES: A PROLOGUE

Toughened composites have been used since the introduction of ceramics and composites in combination. The combination of ceramic with metals have advantages of high thermal resistance and strength-to-weight ratio, as well as excellent corrosion resistance (Rosso 2006; Stefanescu et al. 1988). But the fracture toughness of the composite was completely attributed to the metal interface, rather than the ceramic. As the fracture toughness is too low for ceramics, it was identified that various techniques, such as addition of other materials as coatings or dispersants, could enhance their fracture toughness.

First reported toughened composites were ceramic-toughened composites, which are processed by chemical vapor deposition. The fracture toughness of silica carbide was doubled when methyltrichlorosilane and titanium tetrachloride were coated as vapors on silica carbide. They have also identified addition of organometallic compounds and fluorides is not suitable for toughened composites (Stinton et al. 1984). The chemical treatment of oxides and nitrates of constituents produces better fracture toughness when than the conventional preparation of composites. A methodology pertaining to this context can be inferred from Coblenz and Lewis (1988).

In general, the matrix material such as epoxies, polyesters, and polyimides, and reinforcements such as glass or graphite are brittle in nature. Thus, the damage tolerance is less; that is, fracture toughness is too low. Few works have shown that fracture toughness can be enhanced through the addition of elastomeric modifiers (Raghava 1989). Early works have also revealed that fracture toughness of the epoxy composites can be improved through the addition of elastomer modifiers. Some liquid and solid elastomer modifiers include carboxyl terminated butadiene acrylonitrile, nitrile rubber (Kapgate et al. 2015), acrylic rubber (Balakrishnan et al. 2005), polychloroprene rubber (Subramaniam et al. 2012), chlorinated polyethylene (Mondal et al. 2016), hydrogenated nitrile butadiene (Akulichev et al. 2016), epichlorohydrin polymer (Mahaling, Jana and Das 2005), ethylene propylene diene monomer (Chakraborty et al. 2004), fluoroelastomer (Wei, Jacob and Qiu 2014), natural rubber (Georgopoulou, Kummerlöwe and Clemens 2020), and styrene-butadiene rubber (Yoon et al. 2020).

It was observed that toughened composites exhibit mode I interlaminar fracture toughness, which is the most common delamination failure. This failure occurs due to the degradation of the composite, and one such work is reported in Nasuha, Azmi and Tan (2017). Factors effecting mode I failure is emphasized in detail in the work by Parker (1989). Ceramics can be toughened by using martensite, but it involves tedious and careful examination of the microstructure and simultaneous examination of its property (Kriven 1990; Becher 1991). Various toughening mechanisms can be cited from the literature for CMC systems. Rice (1985) revealed interesting insights on limitations of ceramic composites such as tension toughening and isotropy of toughening through crack impediment techniques to realize the toughening mechanism of the CMCs. Few more interesting concepts may be found in the literature, for example (Walker et al. 2011), toughening in graphene ceramics, cement-based composites (Li and Maalej 1996), ceramic-based nanocomposites (Awaji, Choi and

Yagi 2002), whisker-reinforced ceramics (Becher et al. 1988), carbon nanotube ceramic composites (Xia et al. 2004), and zirconia-based ceramic composites (Chevalier, Olagnon and Fantozzi 1999).

On the other hand, toughening of epoxy-based composites received a significant attention in the scientific community due to versatility of the epoxy resin in almost all the composite systems due to its tailorable properties. In this context, few works can be cited (Carolan et al. 2016; Marouf et al. 2016; Tang et al. 2012; Chandrasekaran et al. 2014; Li et al. 2013; Wong et al. 2017) on toughening of epoxy-based hybrid nanocomposites, epoxy nanocomposites, epoxy-based ternary composites, graphene-based epoxy composites, epoxy with carbon nanotubes, and woven fabric carbon/epoxy composites. Various nanoparticles, such as silica, core shell rubber, and carbon nanotubes, show excellent enhancement in mechanical and tribological properties and fracture toughness of the composite system.

Epoxy- and natural fiber-based composites have now become the focus of the material science community due to their excellent mechanical and biocompatible properties, which play a crucial role and can be tailored as per the property of interest. For instance, biomedical applications do not require high mechanical strength but require low cytotoxicity and biodegradability, whereas for aerospace applications, we do not require biocompatible properties but require stiffness and strength. Hence, in both of these cases, polymers can be tailored by adopting a suitable process by combining various metallic or organic particles in the composite system. Works pertaining to the toughening of epoxy natural fibers include Nair et al. (2019); Kuo et al. (2017); Rangappa et al. (2021); Kinloch et al. (2015); Haq et al. (2008). Epoxy can be toughened in the composite system, and toughening mechanisms, as aforementioned earlier, would be used according to the properties of natural fiber. The rheology of the natural fibers plays a vital role in determining the type of the processing method and toughening mechanism to be adopted.

13.3 STRUCTURAL APPLICATIONS OF THE TOUGHENED COMPOSITES

As discussed earlier, the various elastomer modifiers, which are used to toughen the composite materials, actually decide the applications of the toughened composites to specific requirements. In this context, first, nitrile rubber elastomer toughened composites are used in products that entails oil and fuel resistance such as gaskets, oil seal, diaphragms, kitchen mats, shaft seals, adhesives, and pump liners, as displayed in Figure 13.1.

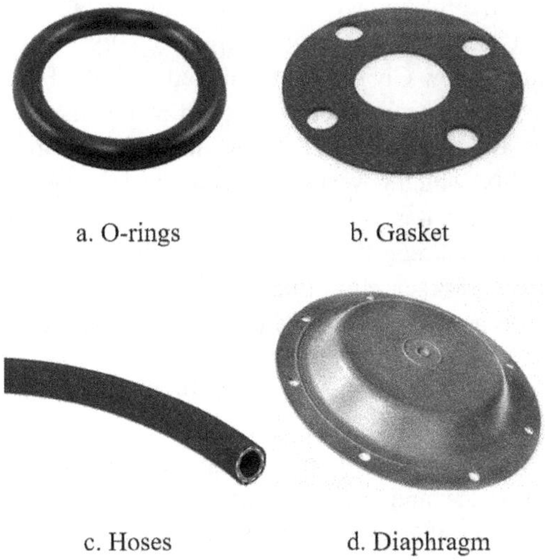

FIGURE 13.1 Nitrile rubber elastomer toughened polymer composite applications.

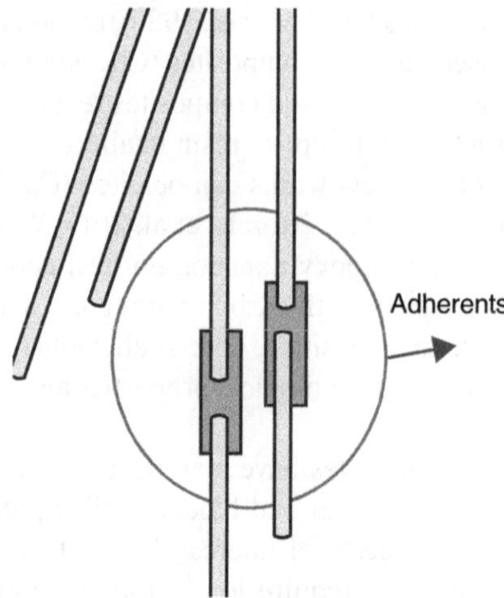

FIGURE 13.2 Creep test device.

The next set of applications of an acrylic rubber pertains to continuous intermittent exposure to heat. It is processed for toughening of composites via curing with carbonyl groups to form various types of monomers. Few of the applications pertaining to acrylic rubber include metal clad shaft seals, spark plug boots, pan seals, transmission seal, and adhesives. Toughened adhesives such as shown in Figure 13.2 below, the creep test device where an auxiliary bondline is applied to one of the two steel adherents to be joined (Dong et al. 2020).

Case study

As a case study, a novel work on multiscale toughening of composites with carbon nanotubes for structural engineering applications is emphasized in the work (Drissi-Habti, El Assami and Raman 2021). The applications include wind blades where offshore wind generation is the actual issue. The type of the configuration increases fracture toughness of the composite, as shown in Figure 13.3. The FRCs possess low fracture toughness, whereas carbon nanotubes possess high strength and fracture toughness. As a result, reinforcing Carbon NanoTubes (CNT) between the laminates of the composites adhering to epoxy will provide excellent fracture toughness.

The authors have discoursed how CNTs have combined with the FRC laminates to improve the fracture toughness and strength. The characteristics and mechanisms were clearly distinguished through atomic simulations and physical process. The atomic structure of CNT has a zigzag axis, which is referred to as chirality, which governs the mechanical strength, as shown in Figure 13.4.

In the next phase of the process, it is crucial to model the CNT and matrix interface, which actually decide the interlaminar shear strength of the derived composite. However, the authors disclosed that there is an ample amount of research in the numerical modeling in the literature, and hence, the information pertaining to atomic-scale finite element method and the statistics were adopted directly. Few works pertaining to this modeling includes atomic-scale finite element modeling of CNTs (Tserpes, Papanikos, Tsirkas 2006; Sinnott et al. 1998; Li and Chou 2006). The interactions of the mechanical properties of the carbon nanotubes at atomic scale is shown in Figure 13.5.

The applications concerning CNT-reinforced composites include various industries, such as acoustics, textile, paper, aerospace, and aircraft industry, as well as automotive parts, sports equipment, energy production, and portable power sources, as shown in Figure 13.6.

One more application is the turbostratic structure with carbon fiber with a diameter range of 5–10 μm and carbon range of 95% to 99%, as disclosed (Hoecker and Karger-Kocsis 1996) in Figure 13.7.

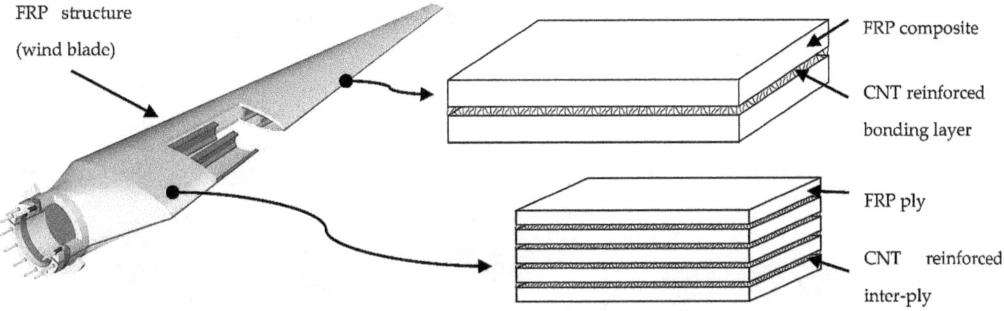

FIGURE 13.3 Processing of toughened FRCs (Drissi-Habti, El Assami and Raman 2021). FRC, fiber-reinforced composite.

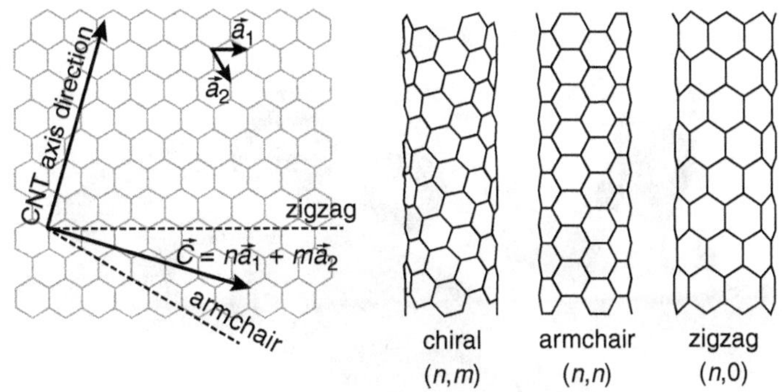

FIGURE 13.4 Atomic structure of the CNT (Drissi-Habti, El Assami and Raman 2021).

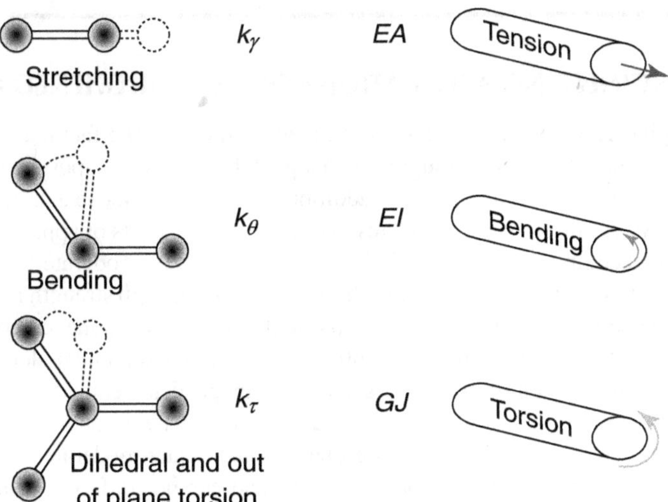

FIGURE 13.5 Atomic interactions between the atoms for the realizing mechanical characteristics (Drissi-Habti, El Assami and Raman 2021).

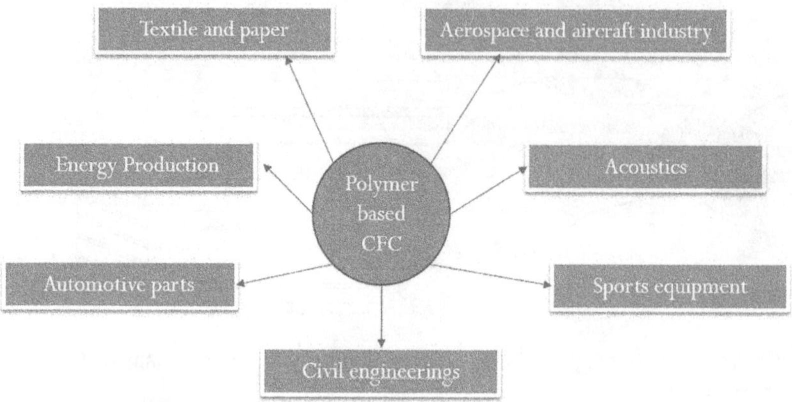

FIGURE 13.6 Applications of toughened composites.

FIGURE 13.7 Turbine components with a turbostratic structure.

13.4 NON-ENGINEERING APPLICATIONS OF THE TOUGHENED COMPOSITES

The term non-engineering applications have been conferred due to the fact that not all the elastomeric modifiers enhance the fracture toughness to a peak level. Few composites may not be apt to enhance their mechanical properties after the addition of elastomers above a certain threshold, and this constrains the versatility of few composites to have broader aspects of applications. In this context, these toughened composites are classified as light to moderate, moderate, and high toughened composites. However, in our earlier section, we have discussed the high strength and fracture toughness composite applications, and here, we focus on moderate to low toughened composite applications. The moderately toughened composite applications are discussed under electronic devices.

Toughened glass matrix composites with graphene nanoparticles embedded are in used aerospace and automotive industries, for example, capacitors, transparent conductors, and lithium ion batteries. Graphene-based composites are categorized mostly under moderately toughened and are amenable to energy storage devices. A work in this context can be conferred from Pumera (2009) who utilized a novel 2D graphene material as energy storage devices, as discussed earlier. In a similar context, graphene-based ultra-capacitors have found their immeasurable applications in energy storage devices (Stoller et al. 2008). The capacitor application, as shown in Figure 13.8, improves the electro- and optic transparency of the toughened composite materials (Wu et al. 2021).

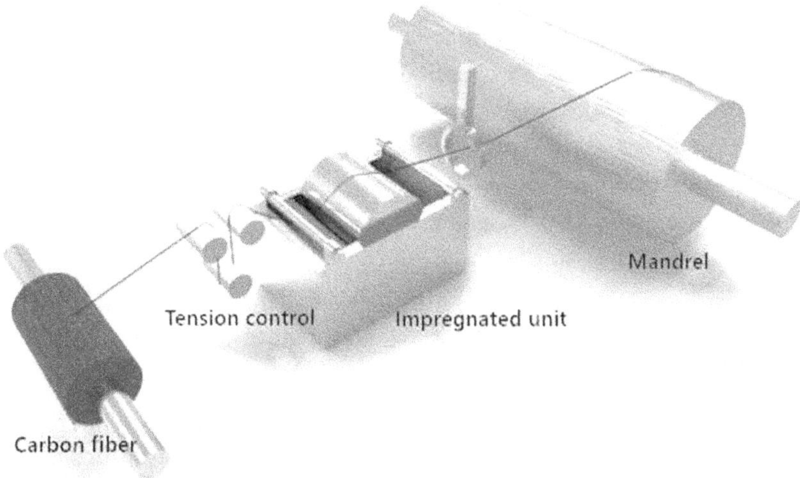

FIGURE 13.8 Capacitors made of carbon fiber fabric.

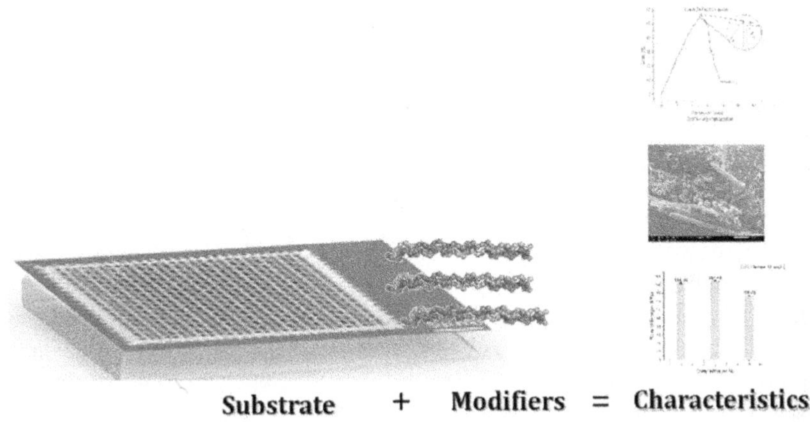

FIGURE 13.9 Surface modification on the substrates.

Low toughened composite applications include the biomedical applications. It requires few biomedical characteristics such as low cytotoxicity, anticorrosive, biocompatible, and biodegradable. There is much work carried out in the scientific community on the use of metals and composite materials as separate entities, and in recent days, it has extended to versatile biomedical applications. Ceramic materials such as aluminum oxide, zirconium dioxide, titanium dioxide, glass ceramics, and carbon are the most widely used forms of ceramics in biomedical industries, which are mostly used as coatings on the substrates to achieve desired properties, as shown in Figure 13.9 (Treccani et al. 2013).

Basic modifications that can be made on a substrate to tailor the properties of the composite is through various approaches, as shown in Figure 13.9. Basic substrate properties required for any base material would include the structural morphology, chemical composition, porosity, solubility, and surface charge. For instance, if roughness of the designed substrate needs to be tailored, by changing the morphology of the microstructure of the composite substrate, by adding few nanoparticles, or by changing the processing parameters, we can achieve the desired mechanical strength. In a similar manner, if surface charge or energy needs to be configured, then the addition of inorganic

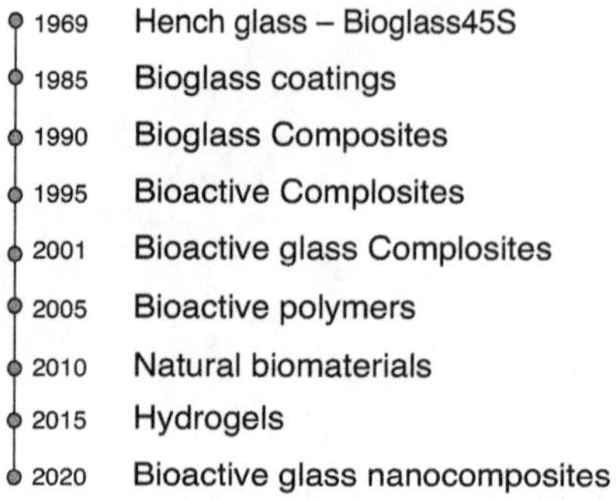

1969 Hench glass – Bioglass45S
1985 Bioglass coatings
1990 Bioglass Composites
1995 Bioactive Complosites
2001 Bioactive glass Complosites
2005 Bioactive polymers
2010 Natural biomaterials
2015 Hydrogels
2020 Bioactive glass nanocomposites

FIGURE 13.10 Timeline of development of bioglass.

components as a dopant on the substrates will enable ion exchange and hence results in better performance; adding organic would likely enhance the reactivity and catalytic activity, which actually require high thermal resistance and when subjected to harsh chemical environments. Also, the addition of biomolecules to the substrate will render properties such as biocompatibility or bioactivity.

There is ample amount of research in this context in the literature on the development of biocompatible tools and materials amenable for biomedical industries. In this context, few works such as Treccani et al. (2013); Kokubo et al. (2000) have reviewed the need for toughened composites for biomedical applications. One such excellent outcome for the biomedical field is bioglass. Bioglass is a form of the composite system with silica, calcium, and phosphate as precursors and other metals such as rare-earth elements, alkaline, and other luminescence materials as dopants.

Case study

It is inferred from the work of Prof. Hench, who introduced bioglass (Hench 2006), that bioglass—a bioactive system combination of silica, phosphate, and calcium as precursors—is processed by a melt quench technique and using a sol–gel process, which is also known as Stober's method. The process includes stirring various materials in the different rpms or specific periodic intervals, after which the powders formed from the sol-gel process were precipitated to form into nanopowders. These nanopowders where then centrifuged at 800 rpm for 3 minutes and then sent for further processing to be amenable for product commercialization. The development of the bioglass in various forms is shown in the timeline in Figure 13.10.

Here, complexities involved in the development of bioglass are mechanical stability and strength. Although they possess good biocompatibility, they lack mechanical characteristics, importantly low fracture toughness. To enhance this property, composites have been combined with the bioglass system, where composites can be natural, synthetic, or hybrid-based polymers with suitable biocompatibility. Recent material advancements led to the development of introducing nanocomposites into the bioglass system to achieve these characteristics (Boccaccini et al. 2010; Kim, Song and Kim 2006; Mozafari et al. 2010; Luz and Mano 2012).

13.5 CLOSING REMARKS

Composites exhibit low fracture toughness, which is one of the major drawbacks for aerospace, automotive, and structural engineering applications. However, their orthotropic nature of desired properties in various directions make them advantageous. In addition to other interesting properties such as high strength-to-weight ratio, good thermal resistance and corrosion resistance make them amenable to various applications in aerospace, automotive, and biomedical industries. However, a more than five decades of research has contributed toward the enhancement of mechanical and tribological properties of composites. Among them fracture toughness is one important characteristic that has specific interest for aerospace applications. Hence many toughening mechanisms have been identified which is specifically reviewed in the current context wherein, processing the material in its own form in a different microstructural treatment, thus tailoring the mechanical and tribological properties. But the enhancement through this approach did not account for significant enhancement, and therefore, new alternative has been identified.

Introducing the elastomer modifiers to the composites have significantly enhanced not only the fracture toughness but also other mechanical characteristics. A short introduction to these elastomer modifiers have been discoursed in the current review. However, it is realized that, over time, the mechanism to enhance the fracture toughness depends on the type of the elastomer modifier and processing methodology adopted, which again depends on the application and type of the composite such as whether a polymer, metal or hybrid configuration. A case studies pertaining to this hierarchy with a suitable example have been discoursed in the current context. However, a sufficient number of articles were cited to have more detailed attention for the reader.

This review identified few literature gaps and scope for future research in terms of development of natural polymer-based biowaste elastomer modifier/nanoparticles for enhancing the mechanical and tribological properties of the toughened composites. This may lead to the use of renewable energy and focus on achieving sustainable development goals.

REFERENCES

Akulichev, A.G., Alcock, B., Tiwari, A. and Echtermeyer, A.T., 2016. Thermomechanical properties of zirconium tungstate/hydrogenated nitrile butadiene rubber (HNBR) composites for low-temperature applications. *Journal of Materials Science*, 51(24): 10714–10726.

Awaji, H., Choi, S.M. and Yagi, E., 2002. Mechanisms of toughening and strengthening in ceramic-based nanocomposites. *Mechanics of Materials*, 34(7): 411–422.

Balakrishnan, S., Start, P.R., Raghavan, D. and Hudson, S.D., 2005. The influence of clay and elastomer concentration on the morphology and fracture energy of preformed acrylic rubber dispersed clay filled epoxy nanocomposites. *Polymer*, 46(25): 11255–11262.

Becher, P.F., 1991. Microstructural design of toughened ceramics. *Journal of the American Ceramic Society*, 74(2): 255–269.

Becher, P.F., Hsueh, C.H., Angelini, P. and Tiegs, T.N., 1988. Toughening behavior in whisker-reinforced ceramic matrix composites. *Journal of the American Ceramic Society*, 71(12): 1050–1061.

Belmonte, M., Nistal, A., Boutbien, P., Román-Manso, B., Osendi, M.I. and Miranzo, P., 2016. Toughened and strengthened silicon carbide ceramics by adding graphene-based fillers. *Scripta Materialia*, 11: 127–130.

Boccaccini, A.R., Erol, M., Stark, W.J., Mohn, D., Hong, Z. and Mano, J.F., 2010. Polymer/bioactive glass nanocomposites for biomedical applications: a review. *Composites Science and Technology*, 70(13): 1764–1776.

Carolan, D., Ivankovic, A., Kinloch, A.J., Sprenger, S. and Taylor, A.C., 2016. Toughening of epoxy-based hybrid nanocomposites. *Polymer*, 97: 179–190.

Chakraborty, S., Sahoo, N.G., Jana, G.K. and Das, C.K., 2004. Self-reinforcing elastomer composites based on ethylene–propylene–diene monomer rubber and liquid-crystalline polymer. *Journal of Applied Polymer Science*, 93(2): 711–718.

Chandrasekaran, S., Sato, N., Tölle, F., Mülhaupt, R., Fiedler, B. and Schulte, K., 2014. Fracture toughness and failure mechanism of graphene based epoxy composites. *Composites Science and Technology*, 97: 90–99.

Chevalier, J., Olagnon, C. and Fantozzi, G., 1999. Crack propagation and fatigue in zirconia-based composites. *Composites Part A: Applied Science and Manufacturing*, 30(4): 525–530.

Coblenz, W.S. and Lewis III, D., 1988. In situ reaction of B_2O_3 with AlN and/or Si_3N_4 to form BN-toughened composites. *Journal of the American Ceramic Society*, 71(12): 1080–1085.

Dong, J., Li, Z., Pasternak, H. and Ciupack, Y., 2020. Reinforcement of fatigue damaged steel structures using CFRP lamellas-Part 3: Numerical simulation. *Bauingenieur*, 95: 362–368.

Drissi-Habti, M., El Assami, Y. and Raman, V., 2021. Multiscale toughening of composites with carbon nanotubes—continuous multiscale reinforcement new concept. *Journal of Composites Science*, 5(5): 135.
Gay, D. and Hoa, S.V., 2007. *Composite Materials: Design and Applications*. CRC Press.
Georgopoulou, A., Kummerlöwe, C. and Clemens, F., 2020. Effect of the elastomer matrix on thermoplastic elastomer-based strain sensor fiber composites. *Sensors*, 20(8): 2399.
Haq, M., Burgueño, R., Mohanty, A.K. and Misra, M., 2008. Hybrid bio-based composites from blends of unsaturated polyester and soybean oil reinforced with nanoclay and natural fibers. *Composites Science and Technology*, 68(15–16): 3344–3351.
Hench, L.L., 2006. The story of bioglass. *Journal of Materials Science: Materials in Medicine*, 17(11): 967–978.
Hoecker, F., and Karger-Kocsis, J. 1996. Surface energetics of carbon fibers and its effects on the mechanical performance of CF/EP composites. *Journal of Applied Polymer Science*, 59(1): 139–153.
Kapgate, B.P., Das, C., Basu, D., Das, A. and Heinrich, G., 2015. Rubber composites based on silane-treated stöber silica and nitrile rubber: Interaction of treated silica with rubber matrix. *Journal of Elastomers & Plastics*, 47(3): 248–261.
Kim, H.W., Song, J.H. and Kim, H.E., 2006. Bioactive glass nanofiber–collagen nanocomposite as a novel bone regeneration matrix. *Journal of Biomedical Materials Research Part A*, 79(3): 698–705.
Kinloch, A.J., Taylor, A.C., Techapaitoon, M., Teo, W.S. and Sprenger, S., 2015. Tough, natural-fibre composites based upon epoxy matrices. *Journal of Materials Science*, 50(21): 6947–6960.
Kokubo, T., Kim, H.M., Kawashita, M. and Nakamura, T., 2000. Novel ceramics for biomedical applications. *Journal of the Australasian Ceramic Society*, 36(1): 37–46.
Komarneni, S., 1992. Nanocomposites. *Journal of Materials Chemistry*, 2(12): 1219–1230.
Kriven, W.M., 1990. Martensitic toughening of ceramics. [MgO-Tb sub 2 O sub 3-B; ZrO sub 2]. *Materials Science and Engineering, A: Structural Materials: Properties, Microstructure and Processing*, 127(2).
Kuo, P.Y., de Assis Barros, L., Yan, N., Sain, M., Qing, Y. and Wu, Y., 2017. Nanocellulose composites with enhanced interfacial compatibility and mechanical properties using a hybrid-toughened epoxy matrix. *Carbohydrate Polymers*, 177: 249–257.
Li, C. and Chou, T.-W., 2006. Multiscale modeling of compressive behavior of carbon nanotube/polymer composites. *Composites Science Technology*, 66: 2409–2414.
Li, V.C. and Maalej, M., 1996. Toughening in cement based composites. Part I: Cement, mortar, and concrete. *Cement and Concrete Composites*, 18(4): 223–237.
Li, Y., Umer, R., Isakovic, A., Samad, Y.A., Zheng, L. and Liao, K., 2013. Synergistic toughening of epoxy with carbon nanotubes and graphene oxide for improved long-term performance. *RSC Advances*, 3(23): 8849–8856.
Luz, G.M. and Mano, J.F., 2012. Chitosan/bioactive glass nanoparticles composites for biomedical applications. *Biomedical Materials*, 7(5): 054104.
Mahaling, R.N., Jana, G.K. and Das, C.K., 2005. Modified nanofiller epichlorohydrin elastomer composite. *Composite Interfaces*, 11(8–9): 701–710.
Mai, Y.W. and Yu, Z.Z., 2006. *Polymer Nanocomposites*. CRC Press.
Markham, M.F. and Dawson, D., 1975. Interlaminar shear strength of fibre-reinforced composites. *Composites*, 6(4): 173–176.
Marouf, B.T., Mai, Y.W., Bagheri, R. and Pearson, R.A., 2016. Toughening of epoxy nanocomposites: Nano and hybrid effects. *Polymer Reviews*, 56(1): 70–112.
Mondal, S., Ganguly, S., Rahaman, M., Aldalbahi, A., Chaki, T.K., Khastgir, D. and Das, N.C., 2016. A strategy to achieve enhanced electromagnetic interference shielding at low concentration with a new generation of conductive carbon black in a chlorinated polyethylene elastomeric matrix. *Physical Chemistry Chemical Physics*, 18(35): 24591–24599.
Mozafari, M., Rabiee, M., Azami, M. and Maleknia, S., 2010. Biomimetic formation of apatite on the surface of porous gelatin/bioactive glass nanocomposite scaffolds. *Applied Surface Science*, 257(5): 1740–1749.
Nair, S.S., Dartiailh, C., Levin, D.B. and Yan, N., 2019. Highly toughened and transparent biobased epoxy composites reinforced with cellulose nanofibrils. *Polymers*, 11(4): 612.
Nasuha, N., Azmi, A.I. and Tan, C.L., 2017. A review on Mode-I interlaminar fracture toughness of fibre reinforced composites. *Journal of Physics: Conference Series*, 908(1): 012024.
Nicolais, L., Meo, M. and Milella, E. 2011. *Composite Materials: A Vision for the Future*. Springer Science & Business Media.
Parker, D.S., 1989. Factors influencing the mode I interlaminar fracture resistance of toughened matrix composites (Doctoral dissertation, University of Michigan).
Paul, D.R. and Robeson, L.M., 2008. Polymer nanotechnology: Nanocomposites. *Polymer*, 49(15): 3187–3204.

Pumera, M., 2009. Electrochemistry of graphene: new horizons for sensing and energy storage. *The Chemical Record*, 9(4): 211–223.

Raghava, R.S., 1989. Toughened thermosetting matrices and their composites. *Reference Book for Composites Technology*, 1: 79.

Rangappa, S.M., Parameswaranpillai, J., Yorseng, K., Pulikkalparambil, H. and Siengchin, S., 2021. Toughened bioepoxy blends and composites based on poly (ethylene glycol)-block-poly (propylene glycol)-block-poly (ethylene glycol) triblock copolymer and sisal fiber fabrics: A new approach. *Construction and Building Materials*, 271: 121843.

Ratna, D., 2008. Toughened FRP composites reinforced with glass and carbon fiber. *Composites Part A: Applied Science and Manufacturing*, 39(3): 462–69.

Rice, R.W., 1985. Mechanisms of toughening in ceramic matrix composites. *Proceedings of 9th Annual Conference on Composites and Advanced Ceramic Materials*, 589: 591.

Rosso, M., 2006. Ceramic and metal matrix composites: Routes and properties. *Journal of Materials Processing Technology*, 175(1–3): 364–375.

Sinnott, S.B., Shenderova, O.A., White, C.T. and Brenner, D.W., 1998. Mechanical properties of nanotubule fibers and composites determined from theoretical calculations and simulations. *Carbon*, 36: 1–9.

Sprenger, S., 2020. Nanosilica-toughened epoxy resins. *Polymers*, 12(8): 1777.

Stefanescu, D.M., Dhindaw, B.K., Kacar, S.A. and Moitra, A., 1988. Behavior of ceramic particles at the solid-liquid metal interface in metal matrix composites. *Metallurgical Transactions A*, 19(11): 2847–2855.

Stinton, D.P., Lackey, W.J., Lauf, R.J. and Besmann, T.M. 1984. Fabrication of ceramic-ceramic composites by chemical vapor deposition. *Ceramic Engineering*: 668–676.

Stoller, M.D., Park, S.J., Zhu, Y.W., An, J.H. and Ruoff, R.S., 2008. Graphene-based ultracapacitors. *Nano Letters*, 8: 3498–502.

Subramaniam, K., Das, A., Häußler, L., Harnisch, C., Stöckelhuber, K.W. and Heinrich, G., 2012. Enhanced thermal stability of polychloroprene rubber composites with ionic liquid modified MWCNTs. *Polymer Degradation and Stability*, 97(5): 776–785.

Tang, L.C., Zhang, H., Sprenger, S., Ye, L. and Zhang, Z., 2012. Fracture mechanisms of epoxy-based ternary composites filled with rigid-soft particles. *Composites Science and Technology*, 72(5): 558–565.

Thostenson, E.T., Li, C. and Chou, T.W., 2005. Nanocomposites in context. *Composites Science and Technology*, 65(3–4): 491–516.

Treccani, L., Klein, T.Y., Meder, F., Pardun, K. and Rezwan, K., 2013. Functionalized ceramics for biomedical, biotechnological and environmental applications. *Acta Biomaterialia*, 9(7): 7115–7150.

Tserpes, K.I., Papanikos, P., Tsirkas, S.A., 2006. A progressive fracture model for carbon nanotubes. *Composites Part B Engineering*, 37: 662–669.

Walker, L.S., Marotto, V.R., Rafiee, M.A., Koratkar, N. and Corral, E.L., 2011. Toughening in graphene ceramic composites. *ACS Nano*, 5(4): 3182–3190.

Wei, J., Jacob, S. and Qiu, J., 2014. Graphene oxide-integrated high-temperature durable fluoroelastomer for petroleum oil sealing. *Composites Science and Technology*, 92: 126–133.

Wong, D.W., Zhang, H., Bilotti, E. and Peijs, T., 2017. Interlaminar toughening of woven fabric carbon/epoxy composite laminates using hybrid aramid/phenoxy interleaves. *Composites Part A: Applied Science and Manufacturing*, 101: 151–159.

Wu, X.F., Zholobko, O., Zhou, Z. and Rahman, A., 2021. Electrospun nanofibers for interfacial toughening and damage self-healing of polymer composites and surface coatings. In *Electrospun Polymers and Composites* Woodhead Publishing: 315–359.

Xia, S., Liu, X., Wang, J., Kan, Z., Chen, H., Fu, W. and Li, Z., 2019. Role of poly (ethylene glycol) grafted silica nanoparticle shape in toughened PLA-matrix nanocomposites. *Composites Part B: Engineering*, 168: 398–405.

Xia, Z., Riester, L., Curtin, W.A., Li, H., Sheldon, B.W., Liang, J., Chang, B. and Xu, J.M., 2004. Direct observation of toughening mechanisms in carbon nanotube ceramic matrix composites. *Acta Materialia*, 52(4): 931–944.

Yoon, B., Kim, J.Y., Hong, U., Oh, M.K., Kim, M., Han, S.B., Nam, J.D. and Suhr, J., 2020. Dynamic viscoelasticity of silica-filled styrene-butadiene rubber/polybutadiene rubber (SBR/BR) elastomer composites. *Composites Part B: Engineering*, 187: 107865.

Zhang, Y., Topping, T.D., Yang, H., Lavernia, E.J., Schoenung, J.M. and Nutt, S.R., 2015. Micro-strain evolution and toughening mechanisms in a trimodal Al-based metal matrix composite. *Metallurgical and Materials Transactions A*, 46(3): 1196–1204.

14 Carbon Nanostructure–Based Composite for Energy-Related Applications

Shrabani Ghosh, Supratim Maity, and K. K. Chattopadhyay
Jadavpur University

CONTENTS

14.1 INTRODUCTION

The accelerating demand for energy and depleting fossil fuel has insisted researchers to explore alternatives which can minimize environmental pollution. For the last few decades, there has been rapid development in the field of clean energy technologies containing water splitting and fuel cells which can be able to compensate for the energy crisis and environmental pollution. To establish renewable energy sources, numerous strategies such as wind, tidal, geothermal, and solar energy have been considered so far but the storing issues have prevented them for being applied widely (Zhang et al. 2017). It is well known that hydrogen is a clean, light weight, non-polluting, and renewable energy carrier with a high caloric value (Zhang et al. 2017). As a renewable energy source, an electrochemical hydrogen fuel cell conducts a hydrogen oxidation reaction (HOR) at the anode and oxygen reduction reaction (ORR) at the cathode producing water and electricity. Hydrogen and oxygen can be evolved at cathode and anode respectively by splitting water (Zhang et al. 2017; Dai 2017; Zheng, Jiao, and Qiao, 2015). These electrochemical reactions are the hearts of renewable energy. In order to realize the system practically, there are requirements for noble metals (Pt, Ru, and Ir) and their oxides as catalysts to enhance the kinetics of the reactions although the scarcity and high cost of noble materials have impeded the actual realization of renewable energy (Wang et al. 2017).

DOI: 10.1201/9780429330575-14

In addition, noble metals often suffer from impurity poisoning, poor durability and selectivity, and fuel crossover effects with harmful consequences on the environment (Dai 2017; Zhao et al. 2019). To eliminate such disadvantages and make the progression of renewable energy faster, there is a high demand to find alternatives. It is highly desirable to promote noble metal-free or metal-free electrocatalysts which are cost efficient, abundant, and stable and also can be a good competitor to noble metals in terms of activity.

In this perspective, non-noble metal chalcogenides, nitride, and carbide have been undergoing rapid development though they suffer from poor stability issues (Lee et al. 2014). Nowadays, nanocarbon-based composites and hybrids are performing exceptionally impressive in the field of electrochemistry due to their composition chemistry, superior conductivity, and negligible environmental impact (Wang et al. 2017). Nanostructural carbon-based composites pursue a porous structure, varying morphologies with high surface areas, resistivity against acid and basic conditions, recyclability, and cheap and easy availability (Hu and Dai 2016). Therefore, it has become a new thirst to investigate electrochemical activity using different allotropes of carbon-based nano-electrocatalysts. In the field of nanotechnology, the well-known carbon allotropes are carbon nanotubes (CNT), reduced graphene oxide (rGO), and carbon quantum dot (CQD) with a different structural orientation. Hence, the composites/hybrids of CNT, rGO, and quantum dot have attracted the special attention of researcher in the field of energy (Chen, Liu, and Huang 2020; Jayaraman et al. 2018).

14.2 THEORETICAL BACKGROUND

14.2.1 Fuel Cell

A fuel cell is an electrochemical device that transforms chemical energy into electrical energy producing zero polluting agents. A hydrogen fuel cell which is introduced as a proton exchange membrane (PEM) cell also uses a PEM. It employs H_2 and O_2 gases as fuels at anode and cathode, respectively, as shown in Figure 14.1. H_2 is oxidized at the anode by HOR and O_2 is reduced at the cathode by ORR, producing water, electricity, and heat (Dave and Gomes 2019; Zheng, Jiao, and Qiao 2015). This is a significant improvement over combustion engines, coal burning, and nuclear plants, all of which leave harmful byproducts for the environment (Zou and Zhang 2015).

Now, these H_2 and O_2 have to be supplied to the fuel cell. Although oxygen is available in a free form in the environment, hydrogen is available in a compound form. Hence, there are technologies such as steam methane reformation, coal gasification, etc., which can generate large amounts of H_2 gas along with harmful CO_2 as a byproduct (Zou and Zhang 2015). In addition, storage, transportation, and distribution are real challenges in the field of hydrogen production. Recently, to store the gas, new materials are getting developed utilizing their high H_2 adsorption–desorption reversibility and stability properties. However, due to the safety and expense of transportation, it is more favorable to produce hydrogen locally. In this perspective, electrochemical H_2 production by water splitting is a well-accepted process. Besides, it can generate pure oxygen with no harmful byproducts. Thus, locally produced H_2 and O_2 can be transferred into the fuel cell for the generation of electricity.

14.2.2 Water Splitting

During the splitting of water, catalysts are utilized as cathode and anode for hydrogen evolution reaction (HER) and oxygen evolution reaction (OER) respectively to speed up the process. By an external voltage, water is decomposed to H_2 and O_2 at cathode and anode by two half-reactions, respectively. Although the reaction path is pH dependent, the thermodynamic voltage of water splitting is maintained to be 1.23 V at 25°C and 1 atm (Yan et al. 2016). Generally, there are intrinsic activation barriers in catalysts with some other resistances which need to be overcome by applying a higher potential than 1.23 V. This extra potential is termed overpotential (Yan et al. 2016).

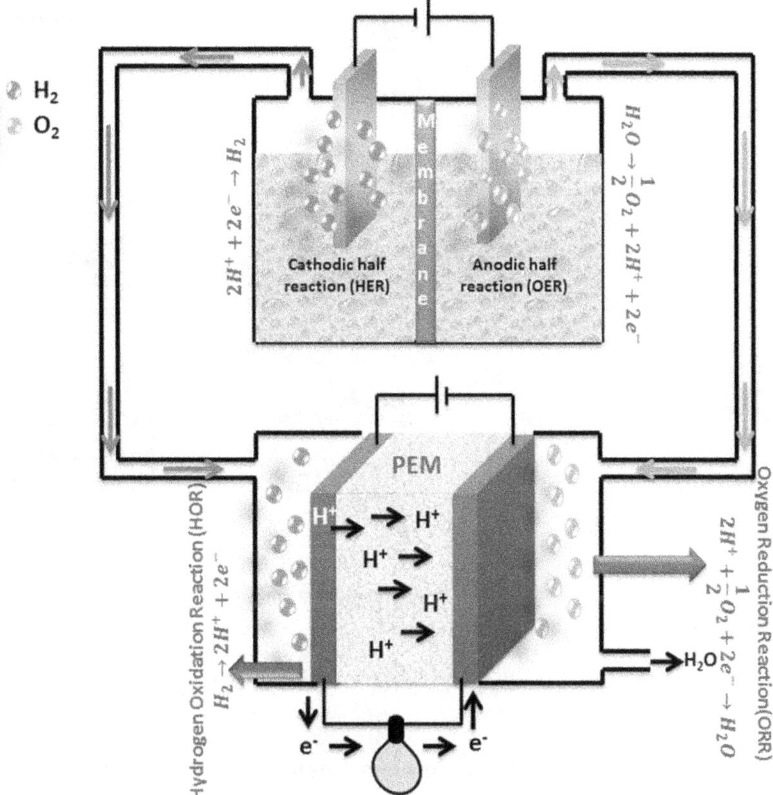

FIGURE 14.1 Interlinked water-splitting device and fuel cell in one frame.

14.2.3 Reaction Mechanism

The reaction paths for HER, OER, and ORR are pH dependent. In general, the water splitting and reverse reaction are sluggish in nature due to the inadequate presence of ions (Yan et al. 2016; Li et al. 2016). Different mediums are considered to increase the conductivity of the electrolyte. Generally, acidic and alkaline solutions are used as electrolytes as they are rich in proton and hydroxyl ions assuring enhancement in the charge transfer phenomenon. The reaction mechanism under acidic and alkaline conditions for HER, OER, and ORR are explained in Figure 14.2. Acidic electrolyte is more preferable for HER than OER/ORR due to the presence of a large number of protons whereas the alkaline electrolyte is more suitable for OER/ORR than HER for the presence of OH^- ions (Yan et al. 2016; Li et al. 2016).

HER is a two-electron transfer process. For HER, hydrogen-binding energy on the surface of the electrocatalyst should be in the optimum range so that H adsorption and desorption can be smooth during the reaction. The reaction proceeds by either Volmer–Heyrovsky or Volmer–Tafel in both acidic and alkaline mediums (Zou and Zhang 2015; Yan et al. 2016; Li et al. 2016). The rate-limiting step is determined by the parameter called Tafel slope. If the coverage of adsorbed hydrogen on the electrocatalyst surface (M) is moderate, the reaction proceeds by Heyrovsky during desorption but if the coverage of adsorbed hydrogen is high enough, two neighbor hydrogen atoms join each other forming H_2 as shown in the Tafel reaction.

OER is a four-electron transfer process indicating a slower reaction than HER. Two different reaction paths are followed for two different mediums as shown in Figure 14.2. In the alkaline medium, hydroxyl ions are adsorbed on the electrocatalyst surface by one electron transfer. In the

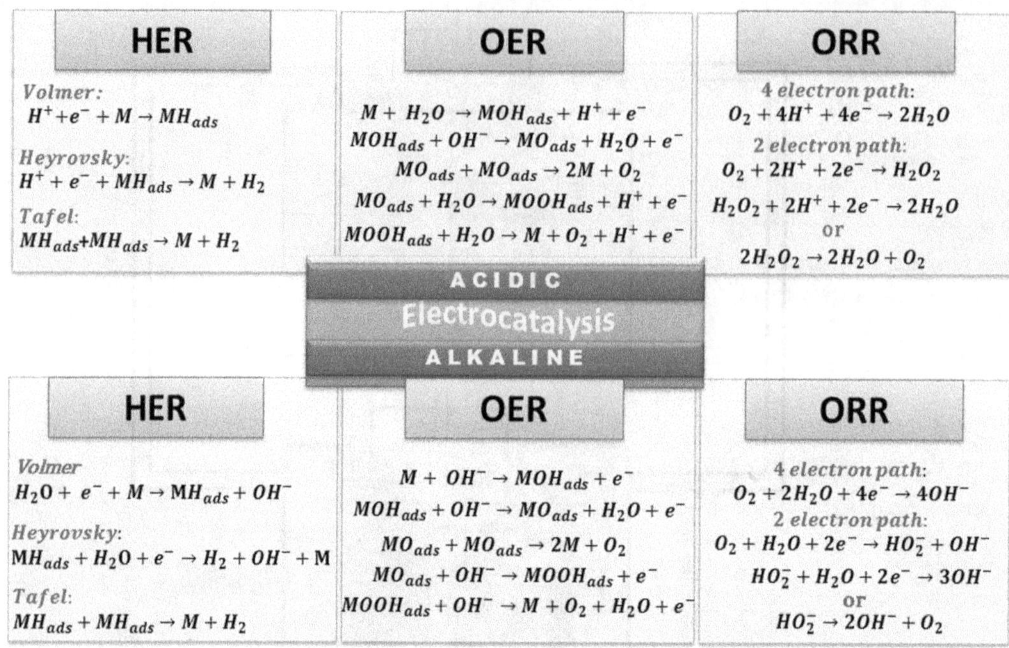

FIGURE 14.2 Reaction mechanism of electrocatalytic HER, OER, and ORR in acidic and alkaline medium.

acidic medium, the adsorbent is H_2O which decomposes to OH ions. This OH is transformed to O_{ads} on the catalyst surface (M) by the second electron transfer. If the catalyst surface is rich in adsorbed oxygen, O_2 is generated by joining two oxygen atoms. For poor coverage of O_{ads}, the reaction follows the path to adsorb OOH ions, and by another two-electron transfer, O_2 is generated (Li et al. 2016; Reier et al. 2017)

ORR by three-electrode experiments in a laboratory setup actually defines the efficiency of a catalyst in a fuel cell. Oxygen can be reduced by either direct four- or two-electron transfers through peroxide formation under a negative potential in a fuel cell under both the mediums as shown in Figure 14.2. In an acidic medium, due to the presence of a large number of protons, the reaction is initiated by H^+, whereas in an alkaline medium, H_2O molecules participate in the reaction (Zhang, Xia, and Dai 2015). Hence, electricity is generated with water as a byproduct. Thus, environmentally friendly electricity can be employed as an energy source as a substitute for the fossil fuel.

14.2.4 Parameters to Evaluate Energy

There are different parameters that have been taken into account so far to evaluate the efficiency of a catalyst (Anantharaj et al. 2018) in an energy application as presented in Figure 14.3.

Polarization Curve and Overpotential: It is one of the most important parameters to judge an electrocatalyst. For HER and OER, the thermodynamic potentials are 0 and 1.23 V (vs. RHE), respectively. Thermodynamically obtained potentials are not experimentally feasible due to the presence of a few other parameters which are not considered at the time of calculation. There is the activation energy of catalyst, ion and gas diffusion, solution concentration, wire and electrode resistance, bubble formation, heat releases, etc. All of them contribute to increasing the potential of water splitting and reverse reaction. Such additional potential is termed "overpotential" which is evaluated from polarization curves such as linear sweep voltammetry (LSV) as displayed in Figure 14.3. The low overpotential signifies the higher activity of the electrocatalyst.

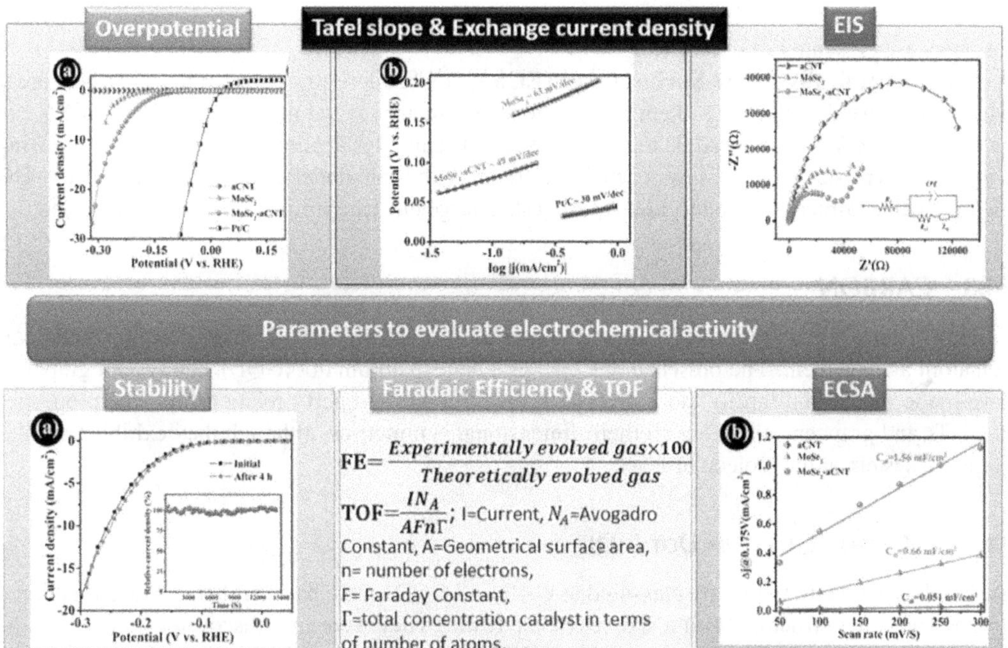

FIGURE 14.3 Parameters to evaluate electrochemical activity. (Reprinted with permission from Maity et al. 2020. Copyright 2020 American Chemical Society.)

In ORR, the onset potential is determined by the polarization curve (LSV). The electron transfer number can be calculated by rotating disc electrode (RDE) measurement. The percentage of produced H_2O_2 and more accurate electron transfer number can be calculated by rotating ring disc electrode (RRDE) measurement. From the cyclic voltammetry (CV) curve, the oxygen reduction potential can be determined for a particular electrocatalyst.

Tafel Slope and Exchange Current Density: Tafel slope exhibits the dependency of a steady-state current density in different overpotentials. In this plot, the logarithmic current density is related to the overpotential by the formula: $\eta = a + b\log|j|$, where, η = overpotential, b = Tafel slope, a = exchange current density at $\eta = 0$

From the graph, the linear portion is considered to calculate the Tafel slope which defines the reaction mechanism of the electrode and a signifies the intrinsic activity of the electrode material under equilibrium conditions. Low b and high a values are highly desirable to obtain an impressive electrocatalytic activity.

Electrochemical Impedance Spectra (EIS): EIS is presented in Figure 14.3 which is the Nyquist plot. It actually defines the series resistance (R_s) from the initial intercept of the circle at a high-frequency region and it assists to eliminate the iR drop from LSV. The second intercept at a low-frequency region or the diameter of the circle confirms the charge transfer resistance (R_{ct}) which is related to the kinetics of HER and OER. Low R_{ct} implies high charge transfer phenomena indicating large current density at an applied potential. In addition, EIS recorded at different potentials can also confirm the Tafel slope which excludes the influence of R_s.

Stability: Stability is an important parameter for any electrocatalyst during long hours of operations. For HER, OER, and ORR, two types of stabilities are investigated such as chronoamperometry (*I* vs. *t*)/chronopotentiometry (*V* vs. *t*) (*cf.* Figure 14.3) and long-hour CV cycle in a defined range.

Faradaic Efficiency (FE) and Turn Over Frequency (TOF): FE reveals selectivity of the electrocatalyst for the corresponding electrochemical reaction. It denotes how efficiently a catalytic system can utilize the applied electrical energy selectively for a particular electrochemical reaction.

TOF is another parameter like the Tafel slope confirming the kinetics of the reaction. The defining formulas are presented in Figure 14.3.

Electrochemically Active Surface Area (ECSA): ECSA determines the exposed surface areas which are active during electrochemical reactions. There are several methods by which ECSA can be confirmed. CV in non-faradaic regions at different scan rates is one of the important measuring ways. By this process, double-layer capacitance (C_{dl}) can be calculated as shown in Figure 14.3. The high C_{dl} value, directly related to ECSA, suggests a large electrochemically active surface area.

14.3 CARBON

Carbon materials (e.g., graphite, diamond) contain different properties which depend on the carbon atom arrangement. The presence of CQD, graphene quantum dot (GQD), CNTs, and graphene have made the carbon family more diverse in the structure–property relationship. The properties of CNTs and graphene change with their dimensional symmetries, although they exhibit a similar graphitic system at the molecular level.

14.3.1 Carbon Quantum Dot (0D)

Carbon-based quantum dots are classified as CQD and GQD on the basis of their surface functional groups and inner structure with a size less than 10 nm. They were first discovered in 2004 at the time of purification of single-walled CNTs (SWCNTs) through preparative electrophoresis and in 2006 by laser ablation of graphite powder and cement. They have got prior attention due to their high solubility and strong luminescence property. In addition, they possess low toxicity, photostability, diminished photobleaching, biocompatibility, electrocatalysis, multiphoton excitation, and upconverted photoluminescence and chemiluminescence. Hence, the quantum dot and its composites with nanostructures have unlocked a new door toward multifunctional applications in the field of science and technology. The unique size insists it to be utilized in drug delivery operations and in the study of kinetics. The mass production of carbon-based quantum dots can be accomplished by different facile synthetic approaches such as hydrothermal, pyrolysis of organic substances, electrochemical, chemical, and microwave-assisted methods, etc. (Wang and Hu 2014; Elvati, Baumeister, and Violi 2017).

14.3.2 Carbon Nanotube (1D)

A CNT can be described as a graphite sheet that is rolled into a nanoscale tubular form. It can be defined as a single-walled carbon nanotube (SWCNT) or a concentric graphene tube. When multiple numbers of SWCNTs are nested inside one another with a regular periodic interlayer spacing located around an ordinary central hollow, such CNT is known as a multiwalled CNT (MWCNT). Nanotubes generally have lengths of several centimeters to submeters and diameters from a few angstroms to tens of nanometers. CNTs can exhibit both semiconducting and metallic nature. On the basis of crystallinity, there are two types of CNTs: crystalline and amorphous nature. Due to the unique structures and extraordinary electrical, mechanical, thermal, and optical properties with a high surface area, CNTs are favorable for various applications in the form of solo to composite materials. Mass production of CNTs is realized by chemical vapor deposition, laser ablation, electric arc discharge, and hydrothermal process (Eatemadi et al. 2014).

14.3.3 Reduced Graphene Oxide (2D)

Since the discovery of graphene in 2004, it has attracted the attention of the researcher. It is an atom-thick planar sheet of sp^2-bonded carbon in a honeycomb network. Graphene owns excellent features which include exceptional mechanical properties resulting from strong planar σ bonds,

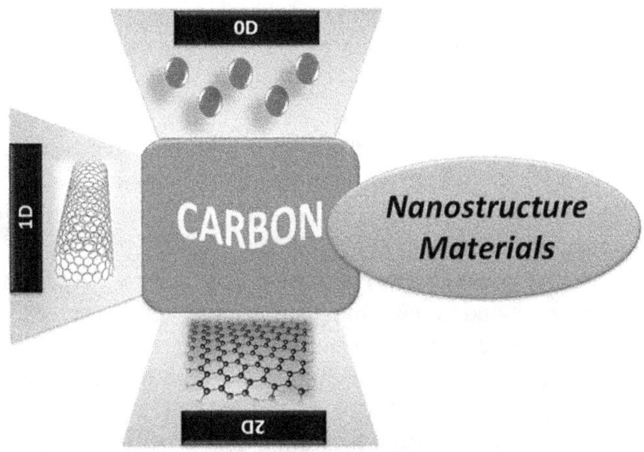

FIGURE 14.4 Carbon-based composites.

high specific surface area, outstanding thermal conductivity, high Young's modulus, impressive electrical conductivity from π bonds, and efficient charge mobility. In addition, transparent single-layer graphene can be employed as a transparent electrode also. Unlike graphene, graphene oxide (GO) holds the single layer of carbon atoms that are covalently functionalized by oxygen-containing groups (hydroxyl, epoxide, carbonyl, etc.) on the basal plane and edges. The properties of GO are greatly hampered due to the synthesis process but GO can recover its unique graphene-like features through additional reductive exfoliation treatments. Thus, it can be transformed into rGO that has relatively good conductivity. The easiest way to synthesize rGO is from a desirable amount of cheap GO by using a variety of (electro) chemical, microwave, and photo-assisted thermal methods. In addition, it is also possible to produce RGO by employing serigraph-guided, radiation-induced, and solar-mediated reduction processes. Therefore, different qualities of rGO can be manufactured depending on its degree of reduction (Tarcan et al. 2020). Since rGO possesses similar mechanical, optical, and conductive assets like pristine graphene with additional structural defects and oxidized functional groups, it has diversified its applications as composites with nanostructured materials in a wide range of science and technology.

The following discussion is based on the schematic of carbon-based composites in Figure 14.4.

14.4 CARBON COMPOSITE IN ENERGY APPLICATIONS

14.4.1 Hydrogen Evolution Reaction

Carbon Quantum Dot (0D): Carbon-based quantum dots and their composites with nanostructure have shown superior electrochemical HER activity in acidic and alkaline mediums. GQD and CQD both contain a high surface area with a large number of active sites which participate in an efficient charge transfer mechanism. Heteroatom doping (B, N, S, etc.) in carbon-based quantum dots enhances the charge transfer process by the generation of greater amounts of active sites and defects facilitating optimum adsorption/desorption of H atoms with low charge transfer resistance. There are many reports on GQD-based composites such as N-GQD on graphite foam, CoP/GQD/graphene, $NiCo_2P_2$/GQD, MoS_2@GQDs, Ni_3S_2/N-doped GQDs/Ni foam, etc., exhibiting excellent HER activity (Dave and Gomes 2019).

In general, HER is less active in an alkaline medium due to the inadequate existence of protons even in the presence of Pt-based electrocatalysts whereas CQD-Ru nanoparticles are able to show exceptional catalytic activity with the overpotentials of –10 mV to draw the current density of 10 mA/cm^2 (Dave and Gomes 2019). In addition, CQDs have improved the electrocatalytic HER activity of various pristine nanostructured catalysts by developing composites like N, S co-doped CQDs/CoS, N, P co-doped CQDs/CoS_2, CQD/MoP, etc. (Dave and Gomes 2019; Chen et al. 2020). Here, the heteroatom-doped CQD prohibits the agglomeration or dissolution of pristine materials in high pH conditions.

Carbon Nanotube (1D): One-dimensional pristine carbon nanotubes (CNTs) are inactive/less active as electrocatalysts in HER without any modification. The activity of CNTs depends mainly on the number of walls in the tubes. Two- to three-walled CNTs are established to be better HER catalysts than single- and multi-walled nanotubes (Cheng and Zhang 2015). It is proposed by Cheng and Zhang (2015) that the charge transfer between outer and inner walls of CNTs is dependent on the distribution of driving force and potential applied to inner tubes; thereby, MWCNT is unsuitable for the electrochemical charge transfer phenomena (Cheng and Zhang 2015). CNTs assist any metal or non-metal active catalysts as a support system to enhance electrochemical HER. The heteroatom doping by any nonmetal or metal elements is one of the synthetic ways to improve the catalytic activity of CNTs (Zhang et al. 2017; Jayaraman et al. 2018). The density functional theory (DFT) calculation assures that Gibb's free energy of hydrogen adsorption on nanocarbon catalysts is reduced by heteroatom doping resulting in efficient HER (Wang et al. 2017). Even by co-doping with transition metal or metal elements in CNTs, the HER activity can be enriched (Jayaraman et al. 2018). The doping in CNT increases the electron mobility, electrical conductivity, pseudocapacitance, etc (Lee et al. 2014). This doping phenomenon also assists to reduce the work function of the CNT (Lee et al. 2014). Hence, the heteroatom-doped CNTs and their hybrids have been proposed as an efficient cathodic electrode material. In addition, metal or metal compounds can be embedded inside CNTs so that the mechanical and chemical stabilities can be achieved with an increment of electrical conductivity. Such coating or enclosure prevents the metal or metal-based nanoparticle compounds from leaching or aggregating (Zhang et al. 2017; Wang et al. 2017). Moreover, the numbers of electrochemically active sites are increased after doping and encapsulation which also raise the quality of electrochemical performance of the catalysts. Besides, low-dimensional surface defects can cause edges/corners which act as catalytic centers during the reaction (Wang et al. 2017). There are numerous reports on doped CNT composites which have performed actively in hydrogen evolution. Such composites are MoS_x/CNT, MoS_2/CNT, P-$NiCo_2S_4$@CNT/carbon nanofibers (CNFs), FeCo alloy in N-CNT, Co in N rich CNTs, Co-doped FeS_2 with CNTs, MoS_x/N-CNT, Mo_2C–CNT, $Fe_{0.9}Co_{0.1}S_2$/CNT, CoP/MoS_2–CNTs, etc (Zhang et al. 2017; Jayaraman et al. 2018). In one of our previous works of $MoSe_2$/aCNT, amorphous CNT (aCNT) has assisted to reduce the overpotential of HER in $MoSe_2$ composite. As a backbone of $MoSe_2$, aCNTs own a lot of defect sites and dangling bonds in the walls of the carbon network, thereby eliminating the cost of functionalization or doping (Maity et al. 2020). Thus, the synergism and intimate electronic interplay between both constituent materials in the composites play a significant role in the improvement of electrocatalytic HER.

Reduced Graphene Oxide (2D): Transition metal and its compound are well known as alternatives of Pt-based catalysts for HER but they also suffer from poor durability and stability issues under high pH conditions. To eliminate this drawback, graphene-encapsulated core-shell structure can be an impressive way out. It prevents the metal core from leaching or dissolution during a reaction (Zhang et al. 2017). Additionally, it can increase the graphitization degree of carbon materials. RGO possesses a large surface area with uncountable protrudes acting as active sites in an electrochemical study. Like CNTs, heteroatom doping in rGO improves its electrocatalytic HER activity depending on the doping parameters. The metal and N doping on rGO synergistically can moderate the hydrogen adsorption Gibb's free energy, thereby increasing the catalytic activity (Wang et al. 2017). As per the DFT study, the electron transfer between metal core and doped rGO can modulate the electronic configuration of the rGO layer. Therefore, the binding energy of the reaction intermediates and electronic conductivities are altered on superficial rGO layers resulting in a better catalytic activity. rGO-based composites such as g-C_3N_4 with N-doped graphene, atomic Co in N-doped graphene, CoNi alloy in N-doped graphene, MoO_2 with P-doped porous carbon and rGO, MoC_x,

FeCo in N-doped graphene, WS_2 nanolayers with P/N/O-doped graphene, FeP nanoparticles in the graphene sheet, SnS on N-reduced graphene sheets, etc., have shown excellent HER activity due to the synergistic properties in composites (Zhang et al. 2017; Wang et al. 2017; Hu and Dai 2016).

14.4.2 Oxygen Evolution Reaction

Carbon Quantum Dot (0D): Carbon-based quantum dots and their composites have acted as efficient electrocatalysts for OER same as HER due to the increment in the surface area and active sites with a large number of defects (Chen, Liu, and Huang 2020). There are various composite-based electrocatalysts made up of GQDs and CQDs that include nickel–iron layer dihydroxide (NiFe-LDH/CQD), CQDs/SnO_2–Co_3O_4, CQD-CoP, B, N, S-CQD/RGO/CoO, N-GQDs/Co_3O_4, Ni_3S_2/N-doped GQDs/Ni foam, etc (Dave and Gomes 2019). Having the strong interaction between the carbon dot and nanostructured materials along with high electrical conductivity, the composites have been revealed to be improved catalysts with a significant enhancement in OER. The synergistic effects of the constituent materials in composite assist to outperform the pristine samples.

Carbon Nanotube (1D): As discussed before, CNTs with optimum numbers of walls are active for OER as well as HER. Besides, CNTs are structurally and chemically stable in an alkaline medium during an electrochemical reaction. Hence, they are more efficient as positive electrodes in OER than HER and ORR due to the presence of a large number of hydroxyl ions (Cheng and Zhang 2015). Same as other electrochemical catalyses, heteroatom doping and encapsulation of metal/metal compounds by CNTs increase the number of active sites participating in the electrochemical OER. Although co-doping of transition metals in CNTs does not affect much in improving the OER activity unlike HER (Zhang et al. 2017), doping types, sites, and levels can be tuned to tailor the OER performance (Zhang et al. 2017). The enhanced electrical conductivity and nanostructure-induced large surface area are important factors to emphasize for enrichment in OER. CNT-based composites as OER catalysts are highly reported such as N-doped mesoporous carbon nanosheet/CNT, N/S co-doped CNTs, S-doped CNTs, Co embedded in N-CNTs, Co/N embedded in CNTs, FeNi alloy nanoparticles in CNTs, Co_3O_4 nanoparticles embedded in N-doped carbon/CNTs, $CoFe_2O_4$/PANI-MWCNT, Fe_2O_3/CNT, etc (Zhang et al. 2017; Liu et al. 2019). Thus, the electrochemical OER activity is enriched by employing the synergistic effect and electronic interplay between the constituent materials in the composites.

Reduced Graphene Oxide (2D): Like HER, rGO plays a significant role in improving the OER activity by its large surface area, high electrical conductivity, large numbers of protrudes acting as active sites, good electronic conductivity, moderate adsorption capability of reaction intermediates, etc. By heteroatom doping on rGO and encapsulation keeping the metal/metal compound as a core, the properties are substantially improved. Therefore, impressive OER activity is observed. There are various reports on rGO-based OER composites like N-doped porous carbon@graphene, Fe-N co-doped graphene, Ni nanoparticles encapsulated in N-doped graphene, CoP/rGO, etc., have been reported to be efficient OER catalysts (Zhang et al. 2017; Liu et al. 2019; Jayaraman et al. 2018).

14.4.3 Oxygen Reduction Reaction

Carbon-Based Quantum Dot (0D): In a carbon quantum dot, the interface between the carbogenic core and functional groups controls the electron transfer properties. This property can be influenced by heteroatom doping. The presence of profuse edges and high specific surface area of GQDs are attributed to large numbers of electron transfers (Dave and Gomes 2019). The size of GQD is responsible for electron trapping between the surface coating and GQD. By heteroatom doping of S, N, B, etc., in GQD, electrochemical ORR can be amended. Due to different electronegativities with sp^2 carbon, heteroatoms can polarize the adjacent C atoms and ease the adsorption capacity of oxygen molecules by lowering the oxygen dissociative energy barrier (Dave and Gomes 2019). There are several reports on GQD-based composites such as graphene hydrogel embedded in GQD,

N-doped GQD assembled with GO, S-doped GQD composited with rGO, B, N co-doped GQD assembled with graphene, GQDs on graphene nanoribbons, Pt/GQD, S-doped g-C_3N_4/GQD, etc (Chen, Liu, and Huang 2020; Dave and Gomes 2019) which have shown excellent ORR activity comparable to noble Pt-based catalysts. All these hybrids contain a huge number of active sites with defects that facilitate the electron transfer during a reaction. In addition, the synergistic effects of the constituent materials present in hybrids attribute to highly efficient ORR activities.

CQD/GQD both are enriched with functional groups on the surface and these groups facilitate functionalization for designing composites to enhance the electrochemical activities. Thus, intermolecular electron transfer becomes easier between the components. There are various reports on CQD-based composites which include PO_4–CQD/Au, N-doped CQD/carbon nanosheets, Pt/N-doped CQD/graphene, Co_3O_4-CQD/g-C_3N_4, N-CQDs/graphene, N-CQD/nanosheets, N-CQD/nanospheres, N-CQDs@WN, etc. (Chen, Liu, and Huang 2020; Dave and Gomes 2019), showing an outstanding ORR performance. Like GQD, heteroatom doping enhances the electrocatalytic ORR activity in CQD by charge or spin redistributions.

Carbon dot–based composites are prepared by physical mixing, chemical bonding, and in situ growth. The synthesis procedures are safe, simple, and efficient. Moreover, the size of the carbon-based quantum dot is responsible to enlarge the surface area of the composites as a whole facilitating electrochemical catalysis.

Carbon Nanotube (1D): For the last few years, different design and engineering principles are explored which can cause modification of electronic and doping configuration in nanocarbons to achieve high ORR activity. In general, pristine carbon is inactive for ORR. As already discussed, a few ways are discovered such as heteroatom doping, insertion of structural defects such as zigzag edges, pentagon rings, etc. N-doping has the ability to modify the sp^2 carbon for ORR due to the similar atomic radius of N and C. Since ORR is a purely surface phenomenon, concentrating the dopants on the surface enhances the ORR activity rather than the N-doping in bulk reduces the conductivity due to the scattering effect. Apart from N, different nonmetals such as S, P, and B can be doped, and co-doping in a carbon matrix can boost the ORR activity significantly by improving the adsorption capability of reaction intermediates (Yang et al. 2019). Different combinations of doping may lead to dissimilar ORR activities. Although ORR is active mainly in alkaline pH, co-doping in carbon can tune it to be activated in the acidic medium. Transition metal suffers from instability in high pH during reaction with peroxide in the Fenton process producing hydroxyl ions which can oxidize the carbon support, thereby destroying the active structure. Hence, encapsulation of metal and metal compounds with CNTs can prevent degradation to some extent. Researchers have formed several CNT-based electrocatalysts for ORR such as Co supported on N-doped CNT, Co nanoparticles (NPs) encapsulated in N-doped CNTs, Co- N embedded in CNTs, CoFe NPs embedded in N-doped bamboo-like CNTs tangled with rGO nanosheets, FeCo NPs encapsulated in in situ grown N-doped graphitic CNTs with bamboo-like structure, core-shell Co@Co_3O_4 NPs embedded in CNT-grafted N-doped carbon-polyhedra, $NiCo_2S_4$ NCs anchored on N-doped CNT, etc (Liu et al. 2019; Yang et al. 2019; Hu et al. 2018).

Reduced Graphene Oxide (2D): rGO can produce a large number of active sites by encapsulation on metal or metal compounds which can be protected from dissolution in high pH during a reaction. Same as HER and OER, by heteroatom doping of non-metal and metal in graphene matrix, ORR activity can be tuned to betterment. In edge sites of graphene, dangling bonds are more reactive for ORR than the covalent bonds in the basal plane. Nitrogen and sulfur doping are done on the graphene nanosheets for enhanced ORR activity (Yang et al. 2019). As per theoretical analysis, doping in graphene affects the charge and spin distribution in the graphitic matrix. Dai and co-workers prepared edge selectively halogenated graphene nanoplate by a ball milling process, which has shown remarkably improved ORR activity (Yang et al. 2019). As per the previously reported data, N-doped graphene, N-doped nanoporous graphene, S-doped rGO, N, S-doped graphene, etc., have outperformed pristine graphene in ORR. Besides, different composites are synthesized by researchers such as CuPc/rGO, C_3N_4@N-doped G, g-C_3N_4 nanoribbon-G (Zhao et al. 2019), mesoporous

CoNC NP-coated graphene framework, atomically thin mesoporous Co_3O_4 layers strongly coupled with N-rGO nanosheets, etc (Yang et al. 2019). Therefore, the introduction of such rGO-based composites in ORR has increased the electrochemically active sites remarkably.

14.5 CONCLUSION

In the field of energy, carbon materials such as graphene, CNTs, CNFs, carbon nanospheres, carbon nanodots, etc., are considered as advanced electrocatalysts. By doping metal/non-metal atoms and employing a metal or its compound nanoparticles as core, electronic configuration and density of states of carbon materials can be tuned to generate active sites for electrochemical reactions. The carbon-based materials have been fascinated as superior, durable, and long-life energy conversion device electrocatalysts due to their high availability, high endurance in corrosive environments such as extreme pH conditions, high electronic conduction, capability to create large surface area for the active components of the catalyst to be deposited, good electronic coupling between carbon and hetero atoms, stability, and insolubility in water. Therefore, carbon-based composites are obtained by the incorporation of non-noble metals, heteroatoms, inexpensive earth-abundant materials, etc., to introduce next-generation catalysts for energy storage and conversion applications. Carbon materials may just play the role of support to the active component of the catalyst. As observed, carbon skeletons such as CNT, rGO, and quantum dot can act as a backbone or an outer protective layer and they have the ability to improve the performance of the active component of the catalyst. Moreover, pristine carbon materials can be established as an efficient electrocatalyst for HER, OER, and ORR by careful modulation of the carbon skeleton, which is still under research.

ACKNOWLEDGMENT

One of the authors (SG) wishes to thank the Council for Scientific and Industrial Research (CSIR) (File no: 09/096(0926)/2018-EMR-I), the Government of India, for providing her with a senior research fellowship through "CSIR-SRF." The authors wish to acknowledge the University Grants Commission (UGC), the Govt. of India for the support under the "University with Potential for Excellence (UPE-II)" scheme.

REFERENCES

Anantharaj, S., S. R. Ede, K. Karthick, S. Sam Sankar, K. Sangeetha, P. E. Karthik, and S. Kundu 2018. Precision and correctness in the evaluation of electrocatalytic water splitting: revisiting activity parameters with a critical assessment. *Energy & Environmental Science* 11(4): 744–771.

Chen, B. B., M. L. Liu, and C. Z. Huang 2020. Carbon dot-based composites for catalytic applications. *Green Chemistry* 22(13): 4034–4054.

Cheng, Y., and J. Zhang 2015. Are metal-free pristine carbon nanotubes electrocatalytically active?. *Chemical Communications* 51(72): 13764–13767.

Dai, L. 2017. Carbon-based catalysts for metal-free electrocatalysis. *Current Opinion in Electrochemistry* 4(1): 18–25.

Dave, K., and V. G. Gomes 2019. Carbon quantum dot-based composites for energy storage and electrocatalysis: mechanism, applications and future prospects. *Nano Energy* 66: 104093.

Eatemadi, A., H. Daraee, H. Karimkhanloo, M. Kouhi, N. Zarghami, A. Akbarzadeh, M. Abasi, Y. Hanifehpour, and S. W. Joo 2014. Carbon nanotubes: properties, synthesis, purification, and medical applications. *Nanoscale Research Letters* 9(1): 1–13.

Elvati, P., E. Baumeister, and A. Violi 2017. Graphene quantum dots: effect of size, composition and curvature on their assembly. *RSC Advances* 7(29): 17704–17710.

Hu, C., and L. Dai 2016. Carbon-based metal-free catalysts for electrocatalysis beyond the ORR. *Angewandte Chemie International Edition* 55(39): 11736–11758.

Hu, C., Y. Xiao, Y. Zou, and L. Dai, 2018. Carbon-based metal-free electrocatalysis for energy conversion, energy storage, and environmental protection. *Electrochemical Energy Reviews* 1(1): 84–112.

Jayaraman, T., A. P. Murthy, V. Elakkiya, S. Chandrasekaran, P. Nithyadharseni, Z. Khan, R. A. Senthil, R. Shanker, M. Raghavender, and P. Kuppusami 2018. Recent development on carbon based heterostructures for their applications in energy and environment: a review. *Journal of Industrial and Engineering Chemistry* 64: 16–59.

Lee, W. J., U. N. Maiti, J. M. Lee, J. Lim, T. H. Han, and S. O. Kim 2014. Nitrogen-doped carbon nanotubes and graphene composite structures for energy and catalytic applications. *Chemical Communications* 50(52): 6818–6830.

Li, X., X. Hao, A. Abudula, and G. Guan 2016. Nanostructured catalysts for electrochemical water splitting: current state and prospects. *Journal of Materials Chemistry A* 4(31): 11973–12000.

Liu, D., Y. Tong, X. Yan, J. Liang, S. X. Dou 2019. Recent advances in carbon-based bifunctional oxygen catalysts for zinc-air batteries. *Batteries & Supercaps* 2: 743–765.

Maity, S., B. Das, M. Samanta, B. K. Das, S. Ghosh, and K. K. Chattopadhyay 2020. MoSe2-amorphous CNT hierarchical hybrid core–shell structure for efficient hydrogen evolution reaction. *ACS Applied Energy Materials* 3: 5067–5076.

Reier, T., H. N. Nong, D. Teschner, R. Schlögl, and P. Strasser 2017. Electrocatalytic oxygen evolution reaction in acidic environments–reaction mechanisms and catalysts. *Advanced Energy Materials* 7(1): 1601275.

Tarcan, R., O. Todor-Boer, I. Petrovai, C. Leordean, S. Astilean, and I. Botiz 2020. Reduced graphene oxide today. *Journal of Materials Chemistry C* 8(4): 1198–1224.

Wang, J., F. Xu, H. Jin, Y. Chen, and Y. Wang., 2017. Non-noble metal-based carbon composites in hydrogen evolution reaction: fundamentals to applications. *Advanced Materials* 29(14): 1605838.

Wang, Y. and A. Hu 2014. Carbon quantum dots: synthesis, properties and applications. *Journal of Materials Chemistry C* 2(34): 6921–6939.

Yan, Y., B. Y. Xia, B. Zhao, and X. Wang 2016. A review on noble-metal-free bifunctional heterogeneous catalysts for overall electrochemical water splitting. *Journal of Materials Chemistry A* 4(45): 17587–17603.

Yang, L., J. Shui, L. Du, Y. Shao, J. Liu, L. Dai, and Z. Hu 2019. Carbon-based metal-free ORR electrocatalysts for fuel cells: past, present, and future. *Advanced Materials* 31: 1804799.

Zhang, J., Z. Xia, and L. Dai 2015. Carbon-based electrocatalysts for advanced energy conversion and storage. *Science Advances* 1(7): e1500564.

Zhang, L., J. Xiao, H. Wang, and M. Shao 2017. Carbon-based electrocatalysts for hydrogen and oxygen evolution reactions. *ACS Catalysis* 7(11): 7855–7865.

Zhao, S., D.-W. Wang, R. Amal, and L. Dai 2019. Carbon-based metal-free catalysts for key reactions involved in energy conversion and storage. *Advanced Materials* 31(9): 1801526.

Zheng, Y., Y. Jiao, and S. Z. Qiao 2015. Engineering of carbon-based electrocatalysts for emerging energy conversion: from fundamentality to functionality. *Advanced Materials* 27(36): 5372–5378.

Zou, X., and Y. Zhang 2015. Noble metal-free hydrogen evolution catalysts for water splitting. *Chemical Society Reviews* 44(15): 5148–5180.

15 Self-Healing Composites
Capsule- and Vascular-Based Extrinsic Self-Healing Systems

Bhavya Parameswaran and Nikhil K. Singha
Indian Institute of Technology Kharagpur

CONTENTS

15.1 INTRODUCTION

Innovations in the field of self-healing polymer composites are very important to humankind since they are the most promising materials in a variety of engineering fields. The perception of self-healing composites started gaining popularity with the pioneering work of White et al. (2001). Extensive pieces of research in the field of self-healing composites have been carried out over the past two decades on account of the insight that focusing on the mechanical properties alone will not improve the performance and lifetime of the product. Failures or damages are bound to happen in any product, so besides mechanical strength, the main factor to be considered while designing a polymer composite is its ability to repair autonomously (Nosonovsky Michael et al. 2009).

DOI: 10.1201/9780429330575-15

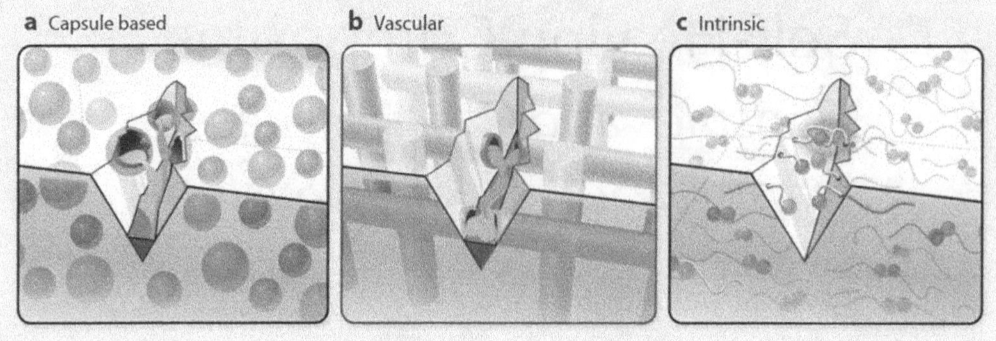

FIGURE 15.1 Different approaches to self-healing composite (a) capsule based, (b) vascular network based, (c) intrinsic self-healing (Blaiszik et al. 2010).

The progress in the field of self-healing of inorganic materials is still in its preliminary stage. Apart from polymers, various other materials like ceramics, cement, and metals have also been targeted for self-healing studies recently. Healing in cementitious materials is carried out by two strategies namely autogenous and autonomous healing. Autonomous healing is a naturally occurring process in which microcracks are healed due to continuous hydration of residual clinkers or carbonation of dissolved [$Ca(OH)_2$]. Deliberate incorporation of materials like polymer resins or adhesive agents, bacteria, or minerals into the cement matrix for accomplishing the self-healing process can be termed autonomous healing (Li et al. 2018). Self-healing may be induced in metals by mixing a high melting point alloy and a low melting point solder, circulating a liquid sealant through the metal surface and releasing it whenever the crack occurs, or adding materials like shape memory alloys (Aouadi, Gu, and Berman 2020). The extreme inert nature of ceramic materials which keeps them stable at elevated temperatures makes the healing process difficult as the triggering temperature is beyond 800°C. One of the main strategies employed in this situation is the incorporation of self-healing centers at the time of fabrication of a ceramic composite either through oxidation or precipitation method (Mauldin and Kessler 2010). The wide range of applications in various fields like biomedical, coating, automobile industries, etc., makes self-healing composites the most promising smart materials.

Based on the self-healing chemistries that the polymer composites exhibit in response to a fracture, they can be grouped into intrinsic and extrinsic self-healing mechanisms. Different approaches to self-healing composites are depicted in Figure 15.1.

15.2 INTRINSIC SELF-HEALING MECHANISM

In the intrinsic self-healing mechanism, the polymer composite is healed by employing a reversible molecular bond in the structure of the material, which may be triggered by forces like heat, light, UV rays, pH, etc (Zhong and Post 2015). At a molecular level, healing is accomplished owing to the low T_g of the polymer, which in the heating and cooling process will lead to molecular diffusion, randomization, recombination of the chain end, etc. It does not employ any external healing agent or catalyst for accompanying the healing process. Based on the triggering forces, the repairing mechanism can be sorted into (i) heat triggered, (ii) photochemical triggered, (iii) electrically triggered, and (iv) moisture triggered self-healing system (Kavitha, Amalin, and Singha 2009). Some of the prominent intrinsic self-healing chemistries based on covalent bonds and supramolecular bonding are Diels–Alder chemistry (Sarkar et al. 2021; Raut et al. 2021; Mondal et al. 2019), di-sulfide chemistry (Nie et al. 2020; Behera et al. 2021), hydrogen bonding (Banerjee et al. 2018), ionic interactions (Mandal et al. 2021; Das et al. 2015), metal coordination (Kanu et al. 2019), etc.

15.3 EXTRINSIC SELF-HEALING MECHANISM

This is an autonomous healing mechanism in which the sealants are loaded in different types of vessels as an isolated phase and are embedded in the matrix. It uses an external healing agent as an additive. Based on the type of vessels, the self-healing process can be categorized into (i) microcapsule-based and (ii) vascular network-based. Certain eminent extrinsic or autonomous self-healing mechanisms are dicyclopentadiene (DCPD)/5-thylidene-2-norbornene (ENB)-based, siloxane-based, epoxy-based, amine-epoxy-based, thiol-epoxy-based, azide-alkyne-based, maleimide-based healing system, etc (Hillewaere and Du Prez 2015).

Certain limitations of extrinsic self-healing structures are that the conviction of healing depends mainly on the response after or at the onset of damage. Also, the points which are once damaged are highly prone to further damage. Furthermore, the healing process can be completely ceased when the vessel containing the healing agent becomes empty.

15.4 MICROCAPSULE-BASED SELF-HEALING SYSTEMS

Microcapsule-based self-healing composites are the most potential candidate among the extrinsic self-healing in automobile and coating industries. In this method, a healing agent in the liquid phase is dispersed in the matrix inside microcapsules along with a catalyst which may either be in the liquid or solid phase. When a crack or failure propagates along the plane, the microcapsules get ruptured. The healing agent combines with the catalyst and flows toward the crack by capillary action and fills the crack as shown in Figure 15.2.

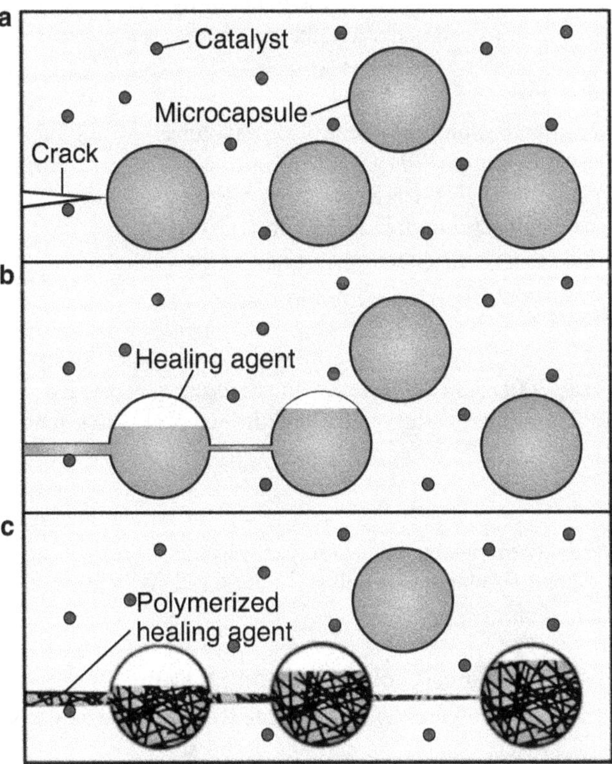

FIGURE 15.2 Prototype of capsule-based self-healing. (a) External damage which results in the formation of crack. (b) Crack ruptures the microcapsules and releases the healing agent. (c) The healing agent and catalyst comes in to contact which initiates the polymerisation process and closes the crack. (Reprinted by permission from White et al. 2001.)

Polymerization occurs when the healing agent reacts with the catalyst which leads to adherence of the two cracked planes (Banerjee at al. 2018).

15.4.1 Fabrication

Capsule-based self-healing systems are often preferred over vascular networks due to their ease of fabrication. This process involves normally two steps; encapsulation of the healing agent in the capsule and incorporation of the healing agent–loaded capsule into the composite matrix. The most popular method of preparing capsule-based systems is in situ polymerization in the oil–water emulsion. Blaiszik et al. (2010) reported encapsulation of DCPD healing agents in urea–formaldehyde shells or capsules. In situ polymerization of the healing agent and capsule was carried out in oil–water emulsion.

Encapsulation of healing agents and/or catalysts in the microcapsules can be achieved using structures like hollow fibers, nanotubes, and microspheres (Scheiner et al. 2016).

15.4.1.1 Hollow Fibers

In this technique, the healing agent is loaded inside the hollow fibers, which is feasible in the use of both cementitious and fiber-reinforced polymer composites. This method of incorporating healing agents is also used in vascular networks with the key difference that here the hollow fibers cannot be filled externally in a continuous fashion. The majority of researches on hollow fibers is based on hollow glass fibers (HGFs). Dry (1996) showed self-healing in the reinforced polymer composite with cyanoacrylate resin as a healing agent filled in hollow glass pipettes. Trask and Bond (2006) developed a two-part healing system contained in hollow fibers. Studies indicated that inclusion of hollow fibers reduced the initial strength of the material by 16%. After healing, the self-healing laminate could regain 87% of its original strength.

15.4.1.2 Nanotubes

SiO_2–polymer hybrid nanotubes and polyelectrolyte nanotubes are the most successful nanocapsules used in anticorrosive coatings (Li et al. 2013; Kopeć et al. 2015). Lanzara et al. (2009) studied utilization of carbon nanotubes (CNTs) as a nano reservoir for the healing agent. As CNTs are noted for their spectacular mechanical properties and reinforcing capabilities, their rupturing was difficult as expected, which limits their use as nanocapsules for encapsulation.

15.4.1.3 Microspheres

White et al. developed the first polymer composite with microspheres loaded with the healing agent and catalysts (White et al. 2001). Later, a great amount of research is devoted to encapsulating healing agents and catalysts in microspheres (Rule et al. 2005; Brown et al. 2004; Jung 1997; Liu et al. 2017).

15.4.2 Mechanisms

Researchers have used different mechanisms to impart capsule-based healing in a polymer composite, among which ring-opening metathesis polymerization (ROMP), polycondensation, and epoxy-based systems are prominent.

15.4.2.1 Ring-Opening Metathesis Polymerization (ROMP)

ROMP attracted the attention of researchers with its long shelf life, low monomer viscosity, volatility, rapid polymerization at ambient conditions, and low shrinkage upon polymerization (Kovačič and Slugovc 2020). ROMP generally employs the usage of a transition metal catalyst which can be dormant to a broad range of functionalities, water, oxygen, etc., and also shows good metathesis activities.

White et al. (2001) adopted a living ROMP reaction with DCPD as the healing agent and a transition metal catalyst called Grubbs catalyst for polymerization. Polymerization with unterminated chain ends facilitated multiple healing events. DCPD was encapsulated in urea–formaldehyde

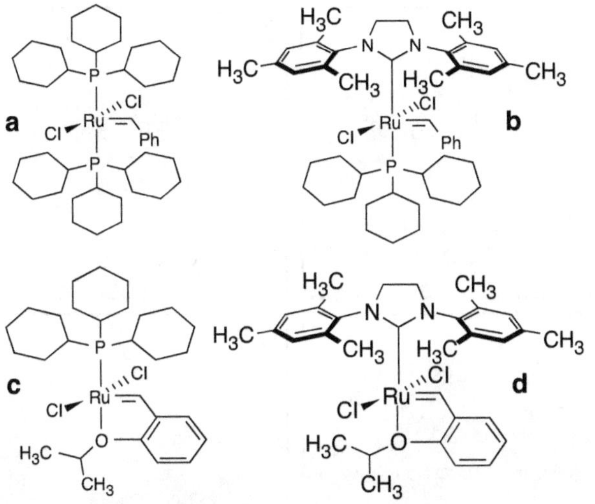

FIGURE 15.3 Structures of different Grubbs catalysts (a) first generation, (b) second generation, (c) Hoveyda–Grubbs first generation, and (d) Hoveyda–Grubbs second generation. (Reprinted by permission from Ullah et al. 2016.)

microcapsules (50–200 μm) in an epoxy matrix. Damage to the epoxy matrix initiates a crack on the capsule shells, which causes a living ROMP between the healing agent and catalyst, which in turn results in polymerization leading to the structural continuity of the matrix. Despite its wide acceptance, DCPD–Grubbs catalyst healing system had limitations like incomplete healing due to poor dispersion. The epoxy curing agent diethylenetriamine can also deactivate Grubbs catalyst. Rule et al. (2005) introduced an extra wax microsphere protection layer to the Grubbs catalyst which could not only reduce the catalyst's concentration to a significant extent but also provided superior healing efficiency to the former.

ENB along with norborene-based ROMP crosslinker was used as the two-component healing system embedded in melamine-urea-formaldehyde (MUF) microspheres with shell thickness around 700–900 nm and a microcapsule size of 120 μm. The two-component healing system exhibited faster polymerization and improved crosslinking density and thermal, mechanical, and adhesive properties than DCPD-Grubbs catalyst-based healing system. Also, MUF microcapsule shells contributed higher thermal stability and improved handling and shelf life properties than UF microcapsule shells (Liu et al. 2009; Yuan et al. 2008; Cosco et al. 2006; Kovačič and Slugovc 2020).

Raimondo et al. (2015) studied the ROMP of ENB/DCPD blend activated by Hoveyda–Grubbs' first-generation catalyst, which occurs at very low temperatures (–50°C) in aeronautical applications. Structures of different generations of Grubbs catalyst are depicted in Figure 15.3.

15.4.2.2 Polycondensation

Self-healing property can be imparted to functionalized polydimethylsiloxane with the help of a variety of catalysts through a polycondensation reaction. One of the classic examples of self-healing via polycondensation reactions is based on siloxane systems. The shortcomings of Grubbs catalyst, mainly poor dispersion and reaction toward amine functionalities, inspired Cho et al. for developing organotin catalysts. It consists of phase-separated globules of hydroxyl end-functionalized polydimethylsiloxane (HOPDMS), poly-diethoxy siloxane (PDES) as a crosslinker, and organotin catalyst embedded inside polyurethane microcapsules. When a crack appears, there occurs a polycondensation reaction between the phase-separated healing agent and crosslinker along with the help of a catalyst, which results in healing. Dibutyltin laurate -based catalysts have good dispersion and they are more stable than the Grubbs catalyst (Cho et al. 2006).

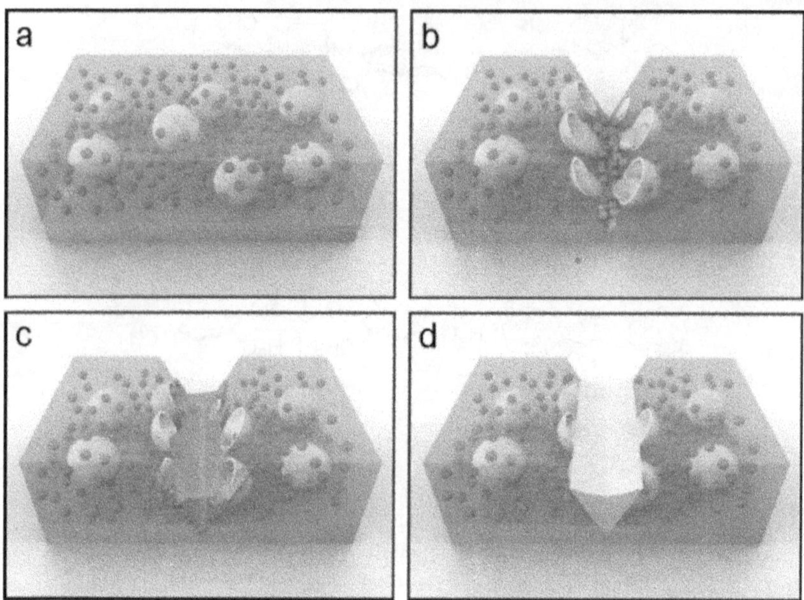

FIGURE 15.4 Schematic representation of self-healing process by polycondensation. (Reprinted by permission from Cho, White, and Braun 2009.)

Another tin-based catalyst, dimethyltin dineodecanoate, showed significant healing (approximately 40%), and the healing efficiency could increase with higher temperatures. Tetrakis (acetoxydi- butyltinoxy) silane is also a successful tin-based catalyst that could heal at room temperature (Figure 15.4). Thes catalyst does not require any moisture for initiating the healing activity and hence it is highly suitable for aerospace applications (Rule et al. 2005).

15.4.2.3 Epoxy-Based Systems

Epoxy-based healing mechanism is the most popular microcapsule approach while comparing to ROMP and polycondensation due to their easy accessibility. The system consists of epoxy resin and a hardener encapsulated separately and embedded inside the composite matrix. When a crack occurs, the epoxy resin and the hardener get mixed up and flow toward the crack to fulfill the healing process (Caruso et al. 2007; Blaiszik et al. 2009), developing a microcapsule containing a reactive resin and solvent as the core material. The system could exhibit significant healing with epoxy films visible on the crack plane after healing. The regions of the epoxy films deposited are clearly shown in Figure 15.5.

There are both positive and negative sides to microcapsule-based self-healing composites. The healing process can be autonomously activated and also the catalytic action does not require any stochiometric mixing of the healing agent. This method is limited since it imparts complications to the base design and also causes a significant reduction in the strength of the product. Because microcapsule vessels could not be dispersed all over the material, the healing process is confined only to a limited volume. There are possibilities for the leaking of the healing agent and also this is a single-time usage healing system (Blaiszik et al. 2009).

15.5 SELF-HEALING SYSTEMS BASED ON VASCULAR NETWORKS

While capsule-based healing structures resemble the natural healing of living organisms on a microscopic level, the vascular network–based healing systems mimic the circulatory system (Cohades et al. 2018). Vascular network–based self-healing systems consist of channels all over the matrix

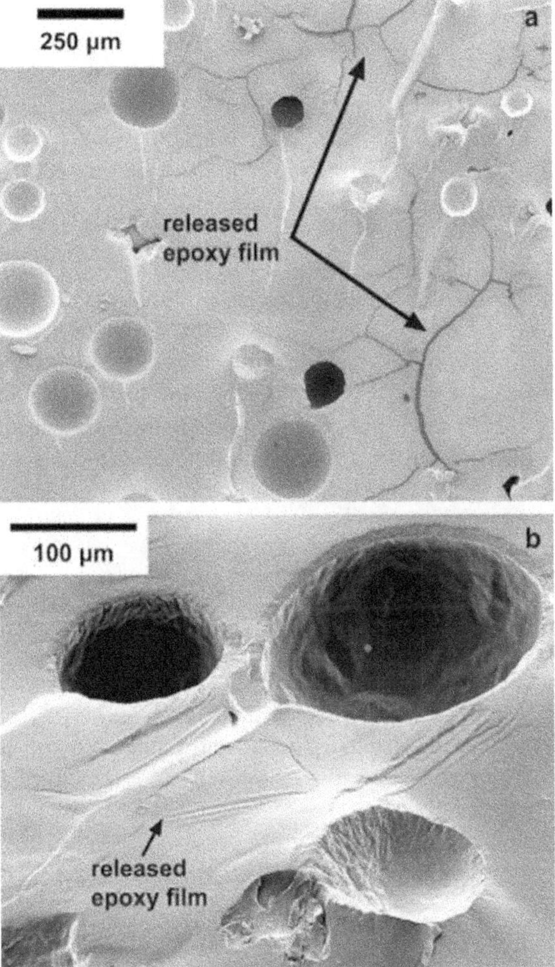

FIGURE 15.5 Microcapsules and regions of the deposited epoxy film. (Reprinted from Blaiszik et al. 2009 with permission from Elsevier.)

which are facilitated by a continuous flow of healing agents through it. The channels may either be composed of hollow fibers, CNTs, sacrificial materials, or shape memory polymers depending on several factors like the type of healing agent used, fabrication method, interaction between matrix and healing agent, etc. Also, the channels are interconnected, enabling the healing agent to be supplied across the network, hence promoting multiple local healing events (Wu et al. 2008).

15.5.1 Fabrication

The fabrication process plays an important role in the self-healing capacity since it affects the interaction between the composite materials, healing agent, and network components. The design of the network should meet certain parameters like precise network diameter, broader network coverage, adequate strength, and good connectivity between the networks for a better healing process to occur. Fabrication of a vascular network is indeed a more difficult process than developing a microcapsule shell. Recently, researchers are concentrating on improving the network design and fabrication process of the vascular network, as it plays an equally important role as that of the healing agent or catalyst (Bekas et al. 2016).

15.5.1.1 Hollow Glass Fibers

In this method of fabrication, tubes containing the healing agent are installed inside the composite matrix. Any of the conventional glass drawing techniques can be used for the fabrication of HGFs (Hia et al. 2016). Keeping apart the fact that the activity of HGFs is restricted to a one-dimensional plane they are advantageous because of their better compatibility with usual polymer matrices and their inert behavior with the healing agents. Pang et al. have reported 95% healing efficiency using HGFs in an epoxy-based system under ambient conditions for 24 hours (Pang and Bond 2005).

15.5.1.2 Scaffolds

Scaffolds or sacrificial fibers filled with healing agents are infused within the composite matrix and are made up of materials that are easy to remove or destroy. The scaffold material should be able to survive the cure temperature, and once the curing process is over, the scaffolds are destroyed either by increasing the temperature or pH, etc., leaving the hollow channels inside. Scaffolds can either be simple one-dimensional hollow structures or complicated three-dimensional structures. For constructing complex three-dimensional structures, the following methods are adopted (Blaiszik et al. 2009):

15.5.1.3 Soft Lithography

Soft lithography is a potential method used in developing microfluidic devices and designing three-dimensional vascular networks. It works by replicating the existing patterns through molding and embossing an elastomer on a mold. Golden and Tien (2007) (Figure 15.6) fabricated a vascular network arrangement on a glass substrate as a duplicate, which is filled with gelatin and solidified. These scaffolds were incorporated with hydrogels and heated to 37°C which in turn could melt the gelatin and keep the hydrogel in the channel. Though soft lithography is a cost-effective method, it is limited to producing two-dimensional networks since it needs additional stacking steps to obtain a multidimensional network.

Bellan et al. (2012) combined the soft lithography technique with the melt spinning technique by fabricating primary vessels with the former and tertiary small vessels with the latter technique.

15.5.1.4 Electrospinning and Melt Spinning

Electrospinning has been used to fabricate vascular networks ranging in size between 10 nm and 10 μm. This technique employs an electric charge to molten polymer solutions and it is extruded through a nozzle forming fibers or threads. Two immiscible liquids are chosen as a healing agent and network material, which are spun coaxially to form the inner and outer layers respectively (Bellan et al. 2012). Gualandi et al. (2013) and Wu et al. (2013) developed core shell structures that could act as vascular networks through this technique. During the spinning process, the DCPD healing agent has directly contained polyacrylonitrile shells, and when a crack occurs in shells, the healing agent could directly flow to the damaged site. In this work, the step of removing or destroying the scaffold is avoided. Melt spinning is very much similar to electrospinning. Patrick et al. (2014) developed microvascular scaffolds with polylactic acid fibers (300 μm) infused with tin oxalate through melt spinning.

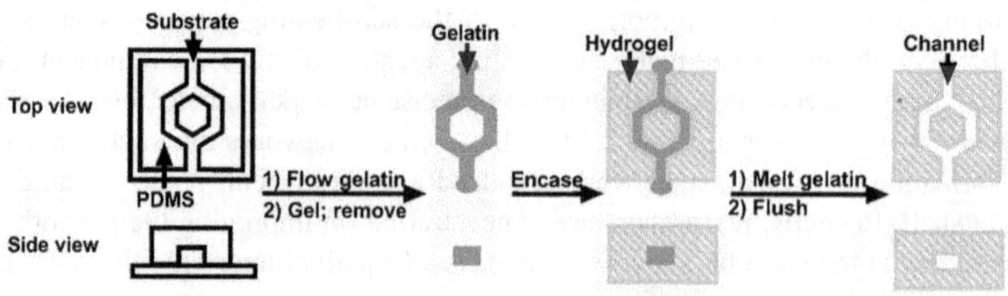

FIGURE 15.6 Schematic representation of soft lithography process. (Golden Andrew and Tien 2007.)

The main advantages of the spinning processes are their simplicity and lower time consumption for fabrication. However, this technique can be used only to produce single fibers, and their incorporation into the composite matrix has to be done manually.

15.5.1.5 Electrostatic Discharge

This is the most rapid fabrication technique, which is suitable for developing a vascular network of any size. Huang et al. (2009) explored this technique to fabricate a highly branched tree-like microvascular network pattern through an instantaneous process (Figure 15.7). However, it is difficult to obtain accurate patterns using this method since the process is not controllable.

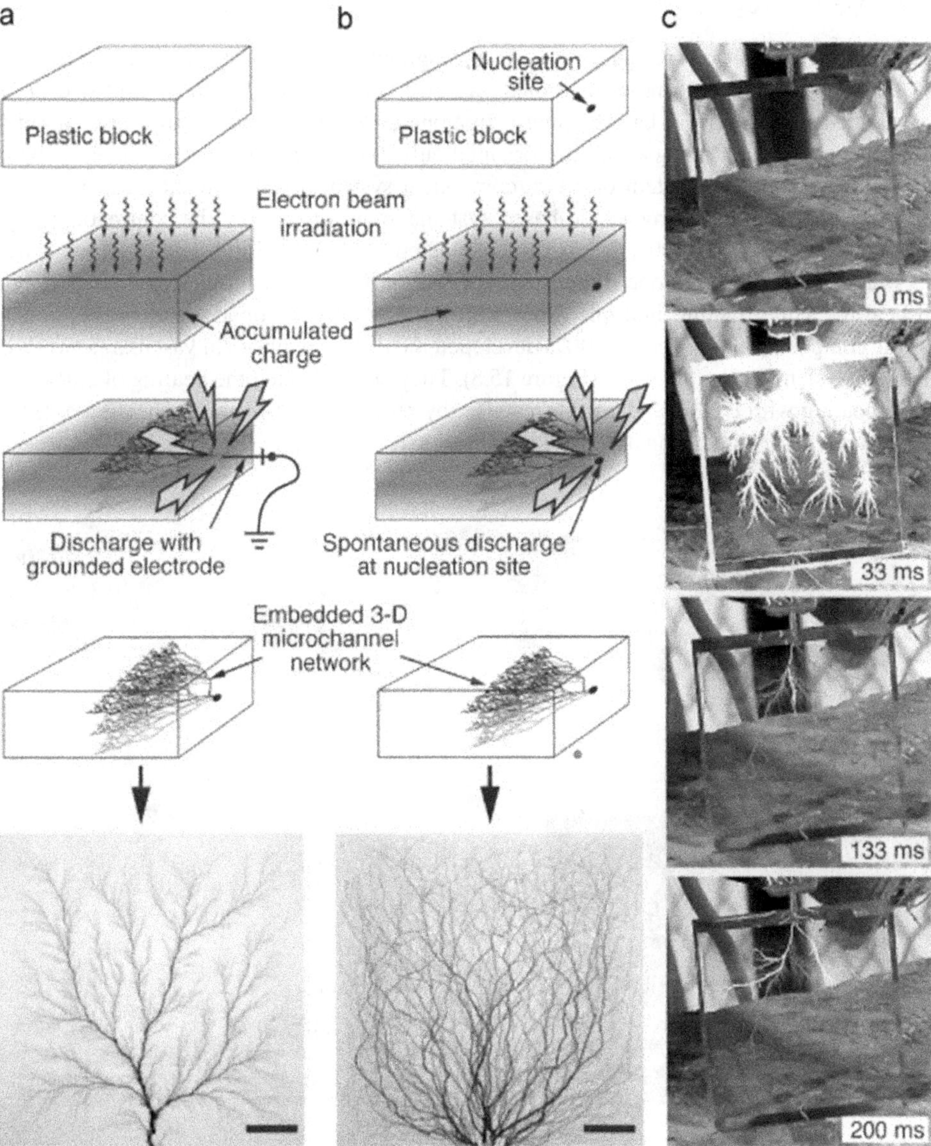

FIGURE 15.7 Highly branched tree-like vascular network electrostatic discharge (Huang et al. 2009). (a) Initiating a sudden energy release by bringing the substrate in contact with a grounded electrode, which in turn vaporizes the surrounding material leaving a tree-like branched microchannel network. (b) Initially a defect is created on the substrate. Then by exposure to irradiation the internal electrical charge of the surface exceeds a critical level which initiates the nucleation from the defected and for a micro network structure. (c) Stagewise images of the first process (part a.). (Reprinted by permission from John Wiley and Sons protected by Copyright (2009).)

15.5.1.6 Laser Machining

Laser machining technique has been proved to be one of the best methods to produce two-dimensional network patterns. Mostly, polycaprolactone and polyamide are the powder materials used in this technique which are sintered by laser to form a solid layer of agglomerated powder (Huang et al. 2009; Wang et al. 2017). Here, the laser solidifies the powder in a space predefined by a model vascular network and binds the material in that specific portion to create the network patterns. The method is comparatively costlier than the methods discussed earlier. Also, it needs additional polishing steps due to its poor surface finishing.

15.5.2 Mechanisms

Based on the arrangement of vascular networks throughout the composite, their self-healing mechanism can be classified into one-dimensional vascular network-based systems and two or three-dimensional vascular network-based systems. In a one-dimensional system, the vascular networks or channels containing healing agents are oriented only in one direction which can either be in the form of a one-component system or a two-component system. This mechanism is similar to the hollow fiber approach of microcapsule-based systems, contrasting only the continuous supply of healing agents. During the 1990s, the research pioneered by Dry and Sottos (1993) on epoxy healing agents in HGFs is considered one-dimensional networks. The higher dimensional healing systems could heal the same fracture multiple times with the help of interconnected networks connected across the composite. Toohey et al. (2007) developed the three-dimensional vascular network-based self-healing system for the first time (Figure 15.8). They demonstrated the healing of crack damage in epoxy coating with DCPD and Grubbs catalyst by repeated delivery of the healing agent to the crack site via a three-dimensional network arrangement.

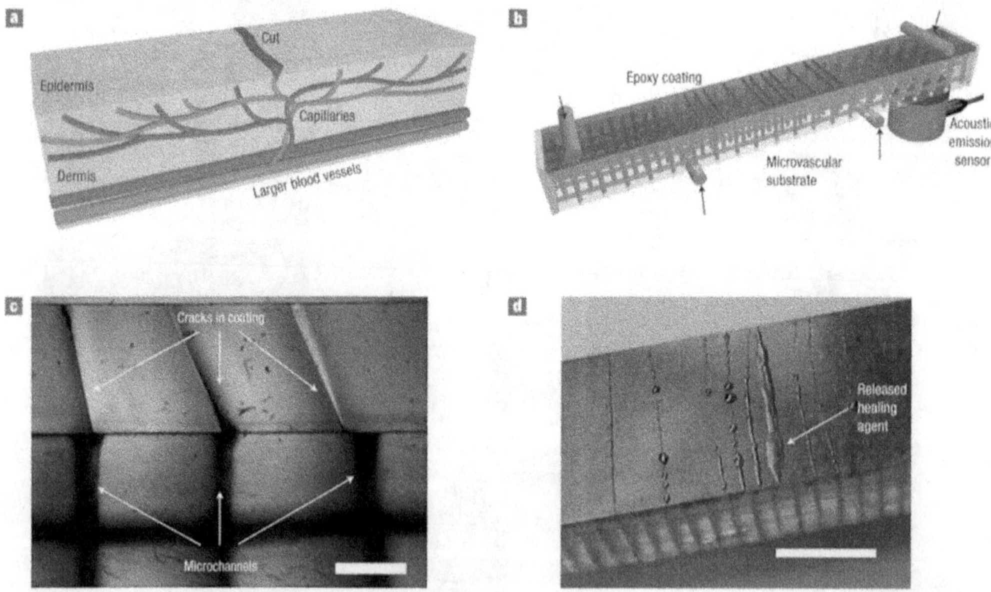

FIGURE 15.8 Self-healing materials with 3D microvascular networks (Toohey et al. 2007). (a) Blood vessels under the epidermal layer of the skin with a cut. (b) Microvascular self-healing network and brittle epoxy coating embedded in catalyst. (c) Cross section image of the coating showing the crack which is initiated at the surface, propagates towards the microchannel opening. (d) Optical images of self-healing network after cracks are formed in the coating. (Reprinted by permission from Springer nature; Toohey, K.S., Sottos, N.R., Lewis, J.A., Moore, J.S., and White, S.R 2007. Self-healing materials with microvascular networks. Nature Materials 6(8): 581–585; ©2007.)

15.6 COMPARISON OF HEALING PERFORMANCE

The healing performance of a self-healing composite is mainly evaluated by its self-healing efficiency which can be defined as the percent recovery of the virgin material. The selection of a healing agent plays an important role in the self-healing efficiency of a composite. Most of the healing agents have either enhancing or degrading effects on the composite. Typically, with the higher loading of a healing agent, the original mechanical properties of the composite gets diminished while the healability of the composite is enhanced. Apart from underlying mechanisms, healing time and temperature play an important role in self-healing efficiency (Toohey et al. 2007).

In the microcapsule approach, small- and medium-sized fractures can be healed in a single cycle. Self-healing based on the basic ROMP reaction with DCPD and Grubbs catalyst yielded 75% of healing efficiency in 48 hours while the later generation of Grubbs catalyst showed higher healing efficiency with higher temperature saving the healing time and enhancing the mechanical properties (White et al. 2001; Rule et al. 2005; Brown et al. 2004). Self-healing by polycondensation with HOPDMS and PDES exhibited complete recovery of the original properties within 48 hours at 150°C in epoxy- and fiber-reinforced composite structures (Cho et al. 2006, 2009). Epoxy-based healing systems gained greater popularity as self-healing composites because of their healing capability in milder conditions. Epoxy-based healing systems with healing efficiency greater than 100% and ultrafast epoxy systems that could heal medium-sized fractures within seconds have contributed significant diversity and new opportunities to this area of research. A wide range of structural and dynamic factors influence the healing efficiency of capsule-based healing structures. Structural features like capsule size, thickness, dispersion of the capsules, and ratio of healing agent and catalyst should be given greater importance while designing a capsule-based healing system. Apart from dynamic factors like healing time, temperature, and pressure, aging and fatigue characteristics also could control the healing efficiency. A couple of results were reported by Neuser and Michaud (Neuser and Michaud 2013) regarding the inconsistency of self-healing efficiency with aging. The self-healing efficiency of a freshly prepared system was reduced from 77% to 13% within 77 days of aging.

Vascular-based healing systems have both advantages and disadvantages. The most beneficial qualities of vascular-based healing systems are they could implement healing in fiber-reinforced composites and also they could initiate multiple healing events within a single system. Regardless of their merits, these types of systems have less healing efficiency (generally up to 80%) as compared to capsule-based healing structures. The healing efficiency of vascular-based healing systems is undoubtedly affected by architectural and dynamic factors as well as the composition of the healing agent. Architectural factors include the channel size, surface, and pattern. A large-sized and smooth-surfaced channel could facilitate an easy flow of healing agents through the channels and hence could achieve better healing efficiency. The network pattern with a larger circulation area across the matrix could improve the healing efficiency. Dynamic factors like healing time, temperature, hydraulic pressure, and mixing process affect the healing efficiency to a greater extent. Longer time along with higher healing temperature improves the healing efficiency. A better pumping strategy with the precise amount of pressure through the network could facilitate an easy flow of healing agents (Neuser, Michaud, and White 2012).

15.7 CONCLUSION AND FUTURE OUTLOOKS

In this section, healing in polymer composites has been demonstrated emphasizing microcapsule and vascular network-based self-healing systems. Developments in the field of self-healing composites have been highly exponential over the past decade. Autonomous repair of self-healing composites has been mainly carried out by capsule- and vascular network-based structures. Capsule-based structures are highly efficient for healing small cracks while the vascular network-based self-healing structures give the best results on large cracks. Capsule-based healing systems are noted for their simple healing concept, easy integration inside the polymer composite matrix, and high healing efficiency, but they could perform only a single healing event in a cycle and the early exhaustion of the healing agent can cause incomplete healing. The vascular approach is very much promising for large cracks and could perform multiple healing events, but their integration in the composite matrix is highly complicated and also a small block in any of the network of a system could disrupt healing. Either through improving the healing mechanism or by altering the fabrication techniques, there are so many possibilities for producing a new-generation self-healing composite. Though extrinsic systems are widely used in self-healing composites, they have a few disadvantages like multiple healing which is difficult to achieve. Dispersion of the vascular and capsular systems in the composite matrix is not always homogeneous, so the healing efficiency is not uniform in the entire system. During the last decade, a tremendous amount of research is focused on the intrinsic self-healing system.

REFERENCES

Aouadi, S.M., J. Gu, and D. Berman 2020. Self-healing ceramic coatings that operate in extreme environments: A review. *Journal of Vacuum Science & Technology A: Vacuum, Surfaces, and Films* 38(5): 050802.

Banerjee, S.L., R. Hoskins, T. Swift, S. Rimmer, and N.K. Singha 2018. A self-healable fluorescence active hydrogel based on ionic block copolymers prepared via ring opening polymerization and xanthate mediated RAFT polymerization. *Polymer Chemistry* 9(10): 1190–1205.

Behera, P.K., S.K. Raut, P. Mondal, S. Sarkar, and N.K. Singha 2021. Self-healable polyurethane elastomer based on dual dynamic covalent chemistry using Diels–Alder “click” and disulfide metathesis reactions. *ACS Applied Polymer Materials* 3(2): 847–856.

Bekas, D.G., K. Tsirka, D. Baltzis, and A.S. Paipetis 2016. Self-healing materials: A review of advances in materials, evaluation, characterization and monitoring techniques. *Composites Part B: Engineering* 87: 92–119.

Bellan, L.L., T. Kniazeva, E.E. Kim, A.A. Epshteyn, D.D. Cropek, R. Langer, and J.J. Borenstein 2012. Fabrication of a hybrid microfluidic system incorporating both lithographically patterned microchannels and a 3D fiber-formed microfluidic network. *Advanced Healthcare Materials* 1(2): 164–167.

Blaiszik, B.J., M.M. Caruso, D.A. McIlroy, J.S. Moore, S.R. White, and N.R. Sottos 2009. Microcapsules filled with reactive solutions for self-healing materials. *Polymer* 50(4): 990–997.

Blaiszik, B.J., S.L.B Kramer, S.C. Olugebefola, J.S. Moore, N.R. Sottos, and S.R. White 2010. Self-healing polymers and composites. *Annual Review of Materials Research* 40: 179–211.

Brown, E.N., S.R. White, and N.R. Sottos 2004. Microcapsule induced toughening in a self-healing polymer composite. *Journal of Materials Science* 39(5): 1703–1710.

Caruso, M.M., D.A. Delafuente, V. Ho, N.R. Sottos, J.S. Moore, and S.R. White 2007. Solvent-promoted self-healing epoxy materials. *Macromolecules* 40(25): 8830–8832.

Cho, S.H., H.M. Andersson, S.R. White, N.R. Sottos, and P.V. Braun 2006. Polydiniethylsiloxane-based self-healing materials. *Advanced Materials* 18(8): 997–1000.

Cho, S.H., S.R. White, and P.V. Braun 2009. Self-healing polymer coatings. *Advanced Materials* 21(6): 645–649.

Cohades, A., C. Branfoot, S. Rae, I. Bond, and V. Michaud 2018. Progress in self-healing fiber-reinforced polymer composites. *Advanced Materials Interfaces* 5(17): 1800177.

Cosco, S., V. Ambrogi, P. Musto, and C. Carfagna 2006. Urea-formaldehyde microcapsules containing an epoxy resin: Influence of reaction parameters on the encapsulation yield. *Macromolecular Symposia* 234(1): 184–192.

Das, A., A. Sallat, F. Böhme, M. Suckow, D. Basu, S. Wießner, K.W. Stöckelhuber, B. Voit, and G. Heinrich 2015. Ionic modification turns commercial rubber into a self-healing material. *ACS Applied Materials & Interfaces* 7(37): 20623–20630.

Dizon, J.R.C., A.H. Espera, Q. Chen, and R.C. Advincula 2018. Mechanical characterization of 3D-printed polymers. *Additive Manufacturing* 20: 44–67.

Dry, C. 1996. Procedures developed for self-repair of polymer matrix composite materials. *Composite Structures* 35(3): 263–269.

Dry, C.M., and N.R. Sottos 1993. Passive smart self-repair in polymer matrix composite materials. In *Smart Structures and Materials 1993: Smart Materials, 1916, International Society for Optics and Photonics*, SPIE: 438–444.

Golden, A.P., and J. Tien 2007. Fabrication of microfluidic hydrogels using molded gelatin as a sacrificial element. *Lab on a Chip* 7(6): 720–725.

Gualandi, C., A. Zucchelli, M. Fernández Osorio, J. Belcari, M.L. Focarete 2013. Nanovascularization of polymer matrix: Generation of nanochannels and nanotubes by sacrificial electrospun fibers. *Nano Letters* 13(11): 5385–5390.

Hia, I.L., V. Vahedi, and P. Pasbakhsh 2016. Self-healing polymer composites: Prospects, challenges, and applications. *Polymer Reviews* 56(2): 225–261.

Hillewaere, X.K.D., and F.E. Du Prez 2015. Fifteen chemistries for autonomous external self-healing polymers and composites. *Progress in Polymer Science* 49: 121–153.

Huang, J.H., J. Kim, N. Agrawal, A.P. Sudarsan, J.E. Maxim, A. Jayaraman, and V.M. Ugaz 2009. Rapid fabrication of bio-inspired 3D microfluidic vascular networks. *Advanced Materials* 21(35): 3567–3571.

Jung, D. 1997. *Performance and Properties of Embedded Microspheres for Self-Repairing Applications.* Master's thesis, University of Illinois at Urbana-Champaign: 176.

Kanu, N.J., E. Gupta, U.K. Vates, and G.K. Singh 2019. Self-healing composites: A state-of-the-art review. *Composites Part A: Applied Science and Manufacturing* 121: 474–486.

Kavitha, A.A., and N.K. Singha 2009. "Click chemistry" in tailor-made polymethacrylates bearing reactive furfuryl functionality: A new class of self-healing polymeric material. *ACS Applied Materials & Interfaces* 1(7): 1427–1436.

Kopeć, M., K. Szczepanowicz, G. Mordarski, K. Podgórna, R. P. Socha, P. Nowak, P. Warszyński, and T. Hack 2015. Self-healing epoxy coatings loaded with inhibitor-containing polyelectrolyte nanocapsules. *Progress in Organic Coatings* 84: 97–106.

Kovačič, S., and C. Slugovc 2020. Ring-opening Metathesis Polymerisation derived poly (dicyclopentadiene) based materials. *Materials Chemistry Frontiers* 4(8): 2235–2255.

Lanzara, G., Y. Yoon, H. Liu, S. Peng, and W.I. Lee 2009. Carbon nanotube reservoirs for self-healing materials. *Nanotechnology* 20(33): 335704.

Li, G.L., Z. Zheng, H. Möhwald, and D.G. Shchukin 2013. Double-walled hybrid nanotubes: Synthesis and application as stimuli-responsive nanocontainers in self-healing coatings. *ACS Nano* 7(3): 2470–2478.

Li, W., B. Dong, Z. Yang, J. Xu, Q. Chen, H. Li, F. Xing, and Z. Jiang 2018. Recent advances in intrinsic self-healing cementitious materials. *Advanced Materials* 30(17): 1705679.

Liu, S., Y. Lin, Y. Wei, S. Chen, J. Zhu, and L. Liu 2017. A high performance self-healing strain sensor with synergetic networks of poly (ε-caprolactone) microspheres, graphene and silver nanowires. *Composites Science and Technology* 146: 110–118.

Liu, X., X. Sheng, J.K. Lee, and M.R. Kessler 2009. Synthesis and characterization of melamine- urea-formaldehyde microcapsules containing ENB-based self-healing agents. *Macromolecular Materials and Engineering* 294(6–7): 389–395.

Mandal, S., F. Simon, S. Shankar Banerjee, L.B. Tunnicliffe, C. Nakason, C. Das, M. Das et al. 2021. Controlled release of metal ion cross-linkers and development of self-healable epoxidized natural rubber. *ACS Applied Polymer Materials* 3(2): 1190–1202.

Mauldin and M.R. Kessler 2010. Self-healing polymers and composites. *International Materials Reviews* 55(6): 317–346.

Mondal, P., G. Jana, P.K. Behera, P.K. Chattaraj, and N.K. Singha 2019. A new healable polymer material based on ultrafast Diels–Alder 'click' chemistry using triazolinedione and fluorescent anthracyl derivatives: A mechanistic approach. *Polymer Chemistry* 10(37): 5070–5079.

Neuser, S., and V Michaud 2013. Effect of aging on the performance of solvent-based self-healing materials. *Polymer Chemistry* 4(18): 4993–4999.

Neuser, S., V. Michaud, and S.R White 2012. Improving solvent-based self-healing materials through shape memory alloys. *Polymer* 53(2): 370–378.

Nie, J., J. Huang, J. Fan, L. Cao, C. Xu, and Y. Chen 2020. Strengthened, self-healing, and conductive ENR-based composites based on multiple hydrogen bonding interactions. *ACS Sustainable Chemistry & Engineering* 8(36): 13724–13733.

Nosonovsky, M., R. Amano, J.M. Lucci, and P.K. Rohatgi 2009. Physical chemistry of self-organization and self-healing in metals. *Physical Chemistry Chemical Physics* 11(41): 9530–9536.

Pang, J.W.C., and I.P. Bond 2005. A hollow fibre reinforced polymer composite encompassing self-healing and enhanced damage visibility. *Composites Science and Technology* 65(11–12): 1791–1799.

Patrick, J.F., K.R. Hart, B.P. Krull, C.E. Diesendruck, J.S. Moore, S.R. White, and N.R. Sottos 2014. Continuous self-healing life cycle in vascularized structural composites. *Advanced Materials* 26(25): 4302–4308.

Patrick, J.F., M.J. Robb, N.R. Sottos, J.S. Moore, and S.R. White 2016. Polymers with autonomous life-cycle control. *Nature* 540(7633): 363–370.

Raimondo, M., P. Longo, A. Mariconda, and L. Guadagno 2015. Healing agent for the activation of self-healing function at low temperature. *Advanced Composite Materials* 24(6): 519–529.

Raut, S.K., P.K. Behera, T.S. Pal, P. Mondal, K. Naskar, and N.K. Singha 2021. Self-healable hydrophobic polymer material having urethane linkages via a non-isocyanate route and dynamic Diels–Alder 'click' reaction. *Chemical Communications* 57(9): 1149–1152.

Rule, J.D., Brown, E.N., Sottos, N.R., White, S.R., and Moore, J.S. 2005. Wax-protected catalyst microspheres for efficient self-healing materials. *Advanced Materials* 17(2): 205–208.

Sarkar, S., S.L. Banerjee, and N.K. Singha 2021. Dual-responsive self-healable carboxylated acrylonitrile butadiene rubber based on dynamic Diels–Alder "Click Chemistry" and disulfide metathesis reaction. *Macromolecular Materials and Engineering* 306(13): 2000626.

Scheiner, M., T.J. Dickens, and O. Okoli 2016. Progress towards self-healing polymers for composite structural applications. *Polymer* 83: 260–282.

Shields, Y., N. De Belie, A. Jefferson, and K. Van Tittelboom 2021. A review of vascular networks for self-healing applications. *Smart Materials and Structures* 30(6), 063001.

Thakur, V.K., and M.R. Kessler 2015. Self-healing polymer nanocomposite materials: A review. Polymer 69: 369–383.

Toohey, K.S., Sottos, N.R., Lewis, J.A., Moore, J.S., and White, S.R 2007. Self-healing materials with microvascular networks. *Nature Materials* 6(8): 581–585.

Trask, R.S., and I.P. Bond 2006. Biomimetic self-healing of advanced composite structures using hollow glass fibres. *Smart Materials and Structures* 15(3): 704–710.

Ullah, H., K.A.M. Azizli, Z.B. Man, M.B. Che Ismail, and M. Irfan Khan 2016. The potential of microencapsulated self-healing materials for microcracks recovery in self-healing composite systems: A review. *Polymer Reviews* 56(3): 429–485.

Wang, X., M. Jiang, Z. Zhou, J. Gou, and D. Hui 2017. 3D Printing of polymer matrix composites: A review and prospective. *Composites Part B Engineering* 110: 442–458.

Wang, Y., D.T. Pham, and C. Ji 2015. Self-healing composites: A review. *Cogent Engineering* 2(1): 1075686.

White, S.R., N.R. Sottos, J. Moore, P. Geubelle, M. Kessler, and E. Brown 2001. *Nature* 409(6822): 794–797.

Wu, D.Y., S. Meure, and D. Solomon 2008. Self-healing polymeric materials: A review of recent developments. *Progress in Polymer Science* 33(5): 479–522.

Wu, X.F., Rahman, A., Zhou, Z., Pelot, D.D., Sinha-Ray, S., Chen, B., Payne, S., and Yarin, A.L., 2013. Electrospinning core-shell nanofibers for interfacial toughening and self-healing of carbon-fiber/epoxy composites. *Journal of Applied Polymer Science* 129(3): 1383–1393.

Yuan, L., A. Gu, and G. Liang 2008. Preparation and properties of poly (urea–formaldehyde) microcapsules filled with epoxy resins. *Materials Chemistry and Physics* 110(2–3): 417–425.

Zhong, N., and W. Post 2015. Self-repair of structural and functional composites with intrinsically self-healing polymer matrices: A review. *Composites Part A: Applied Science and Manufacturing* 69: 226–239.

16 Role of Nanoreinforcement onto Intra- and Interlaminar Fracture of Hybrid Laminated Composites

Subhankar Das
University of Petroleum and Energy Studies

Parnab Das
The University of Alabama

Sudipta Halder and Pannalal Choudhury
National Institute of Technology Silchar

CONTENTS

16.1 INTRODUCTION

Fiber-reinforced plastics (FRPs) are composite materials consisting of high strength and modulus fibers reinforced in a polymer matrix. Usually, glass, carbon, and aramid fibers are used as reinforcement, whereas epoxy, vinyl ester, or polyester is used as matrix materials. These continuous fibers are twisted together to form a yarn and used by themselves or woven into a fabric. Yarns or fabrics are stacked together and impregnated with the polymer matrices to form various laminated composite structures of different shapes and sizes. Both thermoplastics and thermosets are considered potential polymer matrices for FRPs. However, thermosetting polymers are mostly preferred over thermoplastics owing to their beneficial processing which consists of prepreg tack and lower viscosity during the process. Epoxy resins are a very attractive class of thermosetting polymer due to their high adhesive property, strength and hardness, thermal stability, and solvent resistance and ease of processing (Zamanian et al. 2013).

Continuous fiber/epoxy matrix laminated composites are considered the key FRP material for aeronautical and aerospace applications as they are characterized by high strength and stiffness at low weight (Warrier et al. 2010). Even though continuous fiber/epoxy matrix laminated composites

DOI: 10.1201/9780429330575-16

possess high strength in the fiber direction, they have poor resistance to failure in the transverse direction due to the low strength and brittleness of the polymer matrix (Yokozeki et al. 2007). These irregularities result in a complex damage development and finally lead to unavoidable intra- and interlaminar failure in polymer matrix composite laminates (De Greef et al. 2011a,b; Lubineau and Rahaman 2012). In addition to that, strength in the through-thickness direction is also limited in FRPs due to the absence of fibers oriented in the thickness direction for sustaining the transverse load. To mitigate this kind of damage, a different ply orientation is used to prepare multidirectional laminates. However, for multidirectional laminates, the first damage which appears under mechanical loading is often microcracking of the matrix at the juncture of the off-axis plies (De Greef et al. 2011a,b; Lubineau and Rahaman 2012). This microcracking leads to separation or debonding of the adjacent brittle matrix within the interlaminar as well as the intralaminar zone in laminates (Lubineau and Rahaman 2012). Therefore, laminated composite structures are extremely susceptible to crack initiation and propagation site along with laminar interfaces that finally lead to delamination (Lubineau and Rahaman 2012). Delamination thus is the most preventing life-limiting growth mode in laminated composite structures. Delamination often results in significant loss of strength, stiffness, and fatigue life of composite laminates. Consequently, attainment of improved delamination resistance and, in turn, more damage-tolerant composite structures has been a major goal of continuous fiber/epoxy matrix laminated composite development activities (Soliman 2011). Transverse properties of laminated composites can be enhanced by improving the strength and stiffness and also by increasing the ductility of the matrix material (Hui et al. 2008; Kamar et al. 2015; Wang et al. 2016; Fan, Santare, and Advani 2008). This is generally done by incorporating second-phase micron- or nanofillers. Various types of micron- (Hussain, Nakahira, and Niihara 1996) and nanofillers (Kamae and Drzal 2012; Srinivasan et al. 2010) have been introduced at the constituent level to enhance the mechanical behaviors of continuous fiber/epoxy matrix laminated composites by introducing an additional reinforcement scale using different reinforcement strategies as schematically shown in Figure 16.1.

Indeed, fibers can be modified by growing carbon-based nanostructured materials on their surfaces by gaseous phase deposition (chemical vapor decomposition) (Koissin, Warnet, and Akkerman 2013) or by nanodeposition in the liquid phase (electrophoresis or mechanical coating) (Zhang et al. 2012). Another simple approach is to enrich the matrix separately with nanostructured materials. Though it has been identified that each of these strategies influences the mechanical response of the continuous fiber/epoxy matrix laminated composites differently, matrix modification can be said the most technoeconomically feasible strategy for industries (Pathak et al. 2016; Hui et al. 2008). At this point, the matrix modification technique has been largely employed in the framework of micro- or nanoreinforced polymers (bulk reinforced matrix); however, less work can be found on hybridizing laminated structures with functionalized nanoreinforcement that can modulate the matrix stiffness at the interfaces (including both micron and nanofillers and classical fibers). Therefore, in this present chapter, the degradation mechanism of classical laminated composites is presented, and a strategy is discussed to design more damage-tolerant composite structures with improved delamination resistance.

16.2 MECHANISMS OF DEGRADATION OF LAMINATED COMPOSITES

Though FRPs are widely used in several engineering applications nowadays, their strength analysis for any specific application is found very complex. The strength analysis becomes further complicated and nonlinear for multilayer laminated composites (Das et al. 2020). The strengthening mechanism or failure pattern of FRPs mainly depends on its fabrication quality, ply orientation, sequence, and nature of loading (Lubineau and Rahaman 2012). At first, the ultimate failure and degradation patterns of FRPs (Figure 16.2a) appear complex at the macroscopic scale; however, the microscopic analysis discloses that several well-identified mechanisms coalesce together before a final fracture (Lubineau and Rahaman 2012; Das et al. 2020).

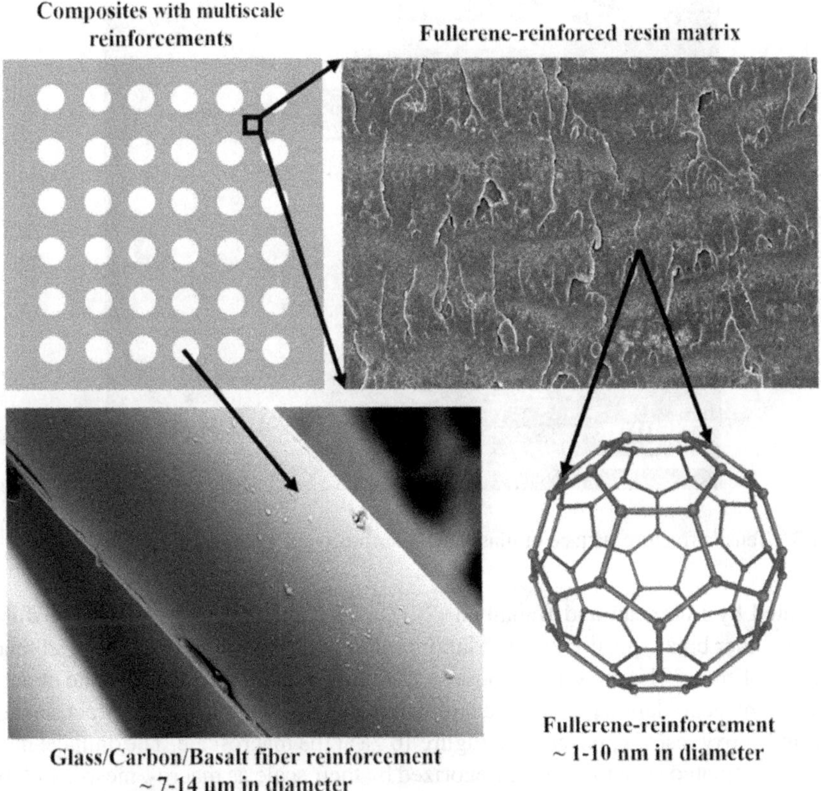

FIGURE 16.1 Schematic of hybrid laminated composites (Jiang et al. 2008; Koissin, Warnet, and Akkerman 2013).

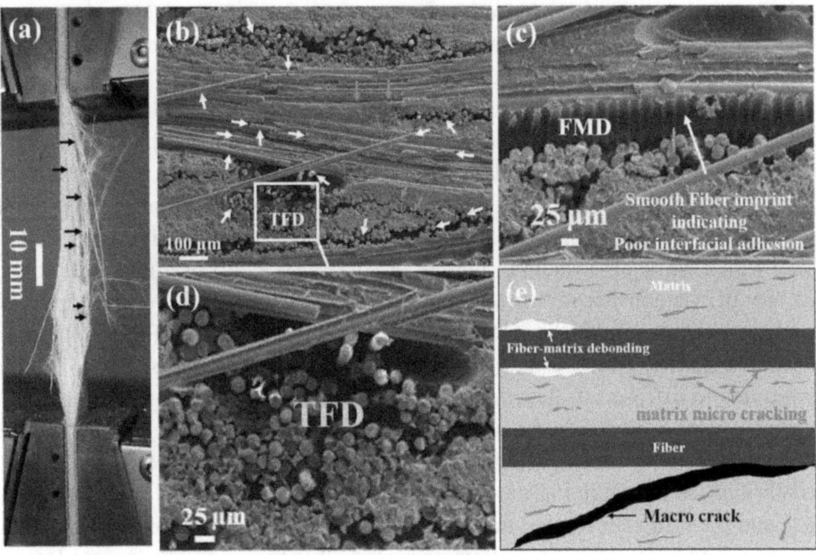

FIGURE 16.2 Ultimate failure and degradation patterns of FRPs at various scales. (a) Complex tensile failure patter of bidirectional GF-epoxy laminated composite at a macroscopic scale where ply to ply separation (indicate with black arrowhead) is visible with the splitting of fibers like a broom, (b) FESEM observations of the mechanisms of degradation of a bidirectional GF-epoxy laminated composite at micro-scale showing (b) diffused intralaminar damage, (c) Fiber-matrix debonding, and (d) Transverse fiber debundling, and (e) Scheme representing the mechanisms of degradation at micro-scale.

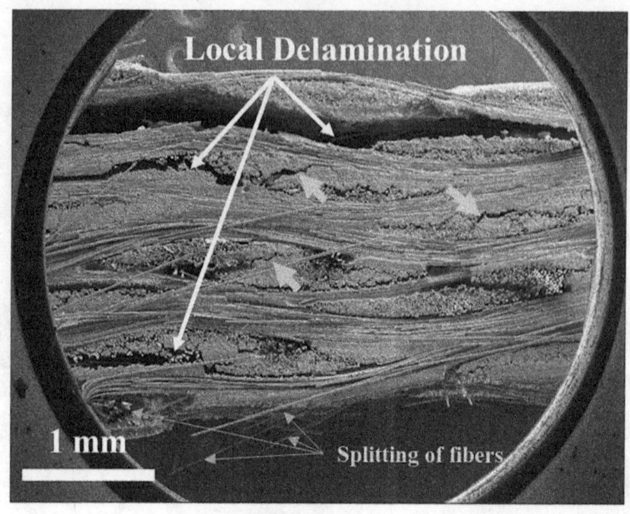

FIGURE 16.3 Tensile fracture surface of glass fiber-reinforced epoxy composites at mesoscale.

As advocated by Lubineau and Rahaman (2012), simple mechanisms such as diffuse intralaminar damage, fiber breakage, diffuse interlaminar damage, transverse cracking, and macroscopic delamination (Soliman 2011) are superimposed together resulting in a complex state of degradation. As an example, the degradation pattern of a laminated composite made with bidirectional glass fiber (GF) fabric and epoxy resin is presented in Figure 16.2a at the macroscale. The damage mechanisms of such classic laminated structures are categorized by their scale as macro-, meso-, and microscale (Lubineau and Rahaman 2012). The brittle nature of matrix material initiates matrix microcracking (indicated with red arrowheads in Figure 16.2b) which is visible at low/microscale under an electron microscope which otherwise could not be visible through naked eyes. Moreover, insufficient adhesion between fiber and matrix also resulted in fiber–matrix debonding [(FMD), indicated with yellow arrowheads in Figure 16.2b] visible at the microscale. The enlarged view of the FMD region as shown in Figure 16.3c portrays the appearance of a smooth fiber imprint on the matrix without any visible sign of matrix tearing resulting from the poor interfacial adhesion between fiber and matrix (Halder et al. 2016). A similar failure mode is also observed by Greef et al. (2010) and Prusty et al. (2017) for control specimen made with plain weave carbon and GF fabrics respectively. The matrix microcracking is added with FMD and finally leads to transverse fiber debundling (area enclosed with a yellow square in Figure 16.2a) as shown in the enlarged view in Figure 16.3d. Fiber breakage (indicated with a white arrowhead in Figure 16.3b) is also a part of microscale damage. These microcracks are diffuse consistently throughout the thickness of the ply resulting in diffuse intralaminar damages which resulted in a substantial reduction of stiffness of FRPs without any visible sign of transverse cracking. Moreover, a high amount of shear-induced plasticity in laminar materials is observed due to diffuse intralaminar damages that cause frictions at the juncture of widely debonded surfaces (Das et al. 2020). Another important failure mode in FRPs is transverse cracking (indicated with a green arrowhead) where cracks grow and spread completely across the thickness of the elementary ply as shown in Figure 16.3 at the mesoscale. They usually initiate at the edges of the laminates and then transfer toward the core through a stable or unstable way depending on the stacking sequence and type of loading (static or cyclic). Transverse cracks initiate local delamination at the interfaces between plies due to induced overloading at the crack tips. Transverse cracking and local delamination are visible and can be characterized under a simple microscope. All these meso- and microscale failure modes are merged and propagate to ply boundaries as schematically shown in Figure 16.2e leading to a ply-to-ply separation with splitting of fibers visible at macroscale through naked eyes as shown in Figure 16.2a.

16.3 HYBRID LAMINATED COMPOSITES WITH BULK MATRIX REINFORCEMENT

It has been reported that limited wettability and poor adsorption of sizing agent present on GFs or carbon fibers are unable to provide required interfacial adhesion at the juncture of fiber and epoxy matrix (Prusty, Rathore, and Ray 2017; Chen et al. 2014; Sánchez et al. 2013). This difficulty is a big challenge for their use in advanced next-generation applications and due to the unwanted diffuse intralaminar damages at the microscale (De Greef et al. 2011; Lubineau and Rahaman 2012). Therefore, various authors studied the effect of matrix modification on FRPs. They added various inorganic and organic fillers (TiO_2, Al_2O_3, SiO_2, $CaCO_3$, boron nitride, and clay nanoparticles) (Halder et al. 2016; Nayak, Dash, and Ray 2014; Eskizeybek et al. 2018; Koricho et al. 2015; Kwon et al. 2017) and carbon-based fillers (CNTs, MWCNTs, fullerene, graphene nanoplatelets) to uplift the toughness as well as interlaminar properties of the epoxy network (Punetha et al. 2017; Goyat et al. 2015) and subsequently to their FRPs (Fan, Santare, and Advani 2008; Hui et al. 2008; Kamar et al. 2015; Wang et al. 2016). They found that matrix modification has improved the wettability with the reinforcement fibers and established an improved interfacial bonding that greatly assists or uplifts the transfer of stress from the matrix to the reinforcing fibers (Pathak et al. 2016; Hui et al. 2008). FRP laminates made with such a modified matrix are generally referred to as multiscale composites or hybrid FRP laminated composites. Though various strategies are available to improve the delamination resistance of laminated composites, structures made with different forms and types of fibers and modified matrices are reinforced with various nanoreinforcement, but here in this present chapter, the discussion is restricted to hybrid GF/epoxy laminated composites reinforced with silanized milled graphite nanoparticles only.

16.4 SILANIZATION OF MILLED GRAPHITE NANOPARTICLES

Silanization is a process to cover mineral surfaces with organofunctional alkoxysilane molecules (Guan et al. 2016). The organofunctional groups may vary from amino, epoxide, vinyl, or methacryloxy groups according to the need and applications (Wang et al. 2012). A few examples of silane coupling agents based on organofunctional groups are (3-aminopropyl)triethoxysilane, (3-mercaptopropyl)trimethoxysilane, (3-glycidyloxypropyl)trimethoxysilane, and triethoxysilane. These silane coupling agents are mostly used to modify the surface property of glass and metal oxide surfaces. Recently, the surface functionality of nanoparticles is also improved by these silane coupling agents (Halder et al. 2016; Khan et al. 2021; Vennerberg, Rueger, and Kessler 2014). Formation of a covalent -Si-O-Si- bond with an end functional group onto nanoparticles helps them form a better interface with organic components present in epoxy resins, paints, adhesives, etc., which could be difficult otherwise (Halder et al. 2016).

(3-Aminopropyl)triethoxysilane abbreviated as APTES is an industrial-friendly silane coupling agent mostly used to graft the amino functional group (NH_2) onto organic and carbon fillers (Halder et al. 2016; Das et al. 2016, 2017, 2018a,b). APTES is a low-cost aminosilane that is easily available and often used in the process of silanization (Zhu et al. 2010; Lee, Rhee, and Park 2011). The general physical and chemical properties of APTES are provided in Table 16.1 with the chemical structure as shown in Figure 16.4.

Carbon-based reinforcements are needed to be oxidized before silanization to produce oxygenated functional groups into carbon fillers. Oxidant, oxidant concentration, and oxidation times are varying based on the reinforcement physicochemical properties. Oxidation of pristine milled graphite nanoparticles (PGr) is given here as an example where 0.3 g of PGr was dispersed in 20 mL of HNO_3 and the solution was kept in a 150 mL conical flask. The solution was mechanically stirred at 600 rpm using a magnetic stirrer for 90 minutes at 80°C under an oil bath as shown in Figure 16.5a. After oxidation, the solution was washed and filtered using double-distilled water, ethanol, and acetone in a sequence to pH 7 and dried in the oven at 80°C for 8 hours.

TABLE 16.1
Physical and Chemical Properties of APTES

Physical and Chemical Properties	APTES
Appearance	Clear liquid
Purity	98%
Formula weight	221.37
Boiling point	217°C
Flash point	104°C (219°F)
Density	0.948
Solubility	Miscible with toluene, acetone, chloroform, and ethanol

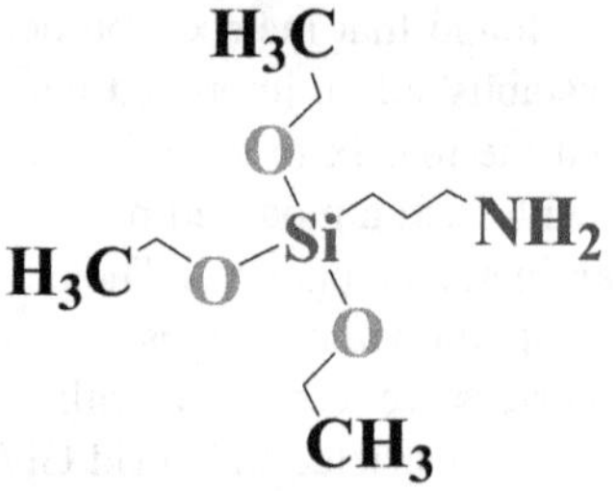

FIGURE 16.4 Chemical structure of (3-aminopropyl)triethoxysilane.

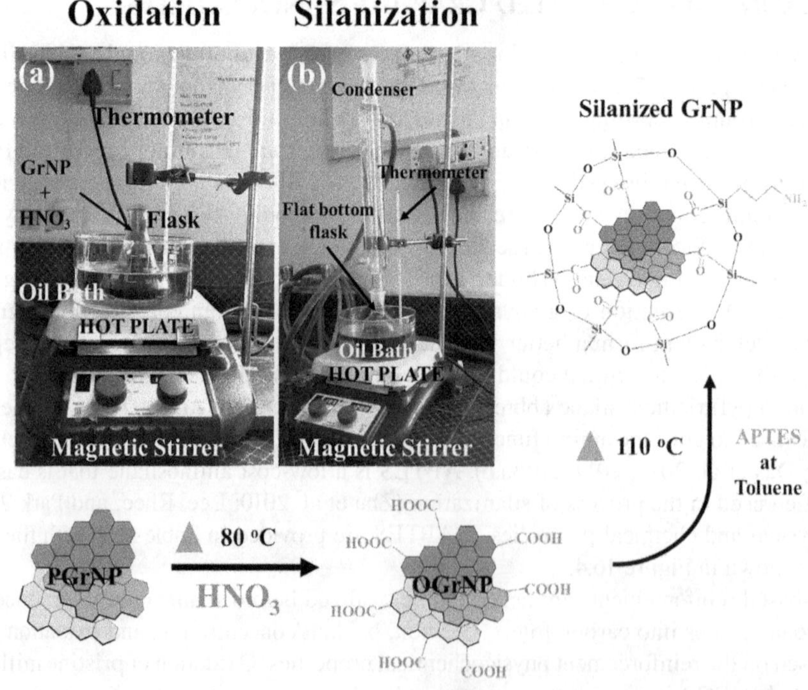

FIGURE 16.5 Schematic representation of Oxidation and silanization process of GrNPs, where (a) Representing the oxidation setup consists of magnetic stirrer with hot plate, oil bath, and 150 ml conical flask containing a mixture of 0.3 g of GrNP and 20 ml of conc. HNO3, (b) Representing the silanization setup that consists of magnetic stirrer with hot plate, oil bath, and 150 ml flat bottom flask containing a mixture of 30 ml toluene, 0.3 g of oxidized GrNPs, and 1 wt% APTES.

Later, oxidized carbon fillers were subjected to a silanization reaction in the presence of APTES. 0.3 g of oxidized graphite nanoparticles (OGr) was poured into a round bottom flask having 25 mL of toluene and bath sonicated for 1 hour. During sonication, a mixture of 5 mL toluene with APTES (1 wt% of oxidized carbon fillers) was dropwise added. The resulting solution was refluxed for 8 hours at a constant temperature of 110°C and simultaneously stirred at 600 rpm as shown in Figure 16.5b. The refluxed solution was filtered and washed several times with double-distilled water, ethanol, and acetone in a sequence. Silanized carbon fillers were obtained after the oven drying at 80°C for 8 hours. After successful silanization, amine groups (NH_2) will form at the surface of OGr as schematically shown in Figure 16.5. Silanized graphite nanoparticles (SGr) have the capability to link with the epoxide group of epoxy resins, thereby enhancing the performance of the matrix material. Silanization of various other organic and carbon-based fillers is covered in the following articles (Halder et al. 2016; Guan 2016; Wang et al. 2012; Khanet al. 2021; Das et al. 2016, 2017, 2018a,b; Khan et al. 2019).

16.5 FABRICATION OF HYBRID GLASS FIBER-REINFORCED PLASTIC (GFRP) ADDED WITH SILANIZED GRAPHITE NANOPARTICLES

Epoxy resin containing PGr was prepared by blending 0.5 wt% of PGr in resin using a high-speed impeller at 1,500 rpm using a Remi Lab stirrer. The mixture was then degassed to remove the entrapped air using a vacuum desiccator equipped with a vacuum pump (~10–3 Torr). Later, a curing agent was added to the mixture with a proportion of 12:1 (resin:hardener) by weight according to the technical data sheet provided by Atul Limited, India. The mixture is mechanically mixed for another 5 minutes at room temperature followed by further degassing. A similar process was followed for modifying epoxy resin containing PGr and SGr.

Fabrication of hybrid GFRP laminates containing eight plies and modified epoxy resin was completed by a simple hand lay-up process. The hand lay-up process has been chosen due to its simplicity in composite processing. The hand lay-up process involves less infrastructural requirement and the processing steps are easy to adopt. First of all, a stainless steel plate as shown in Figure 16.6a is used here as a mold plate which was cleaned with a mold cleaner (TAG Chemicals, GmbH).

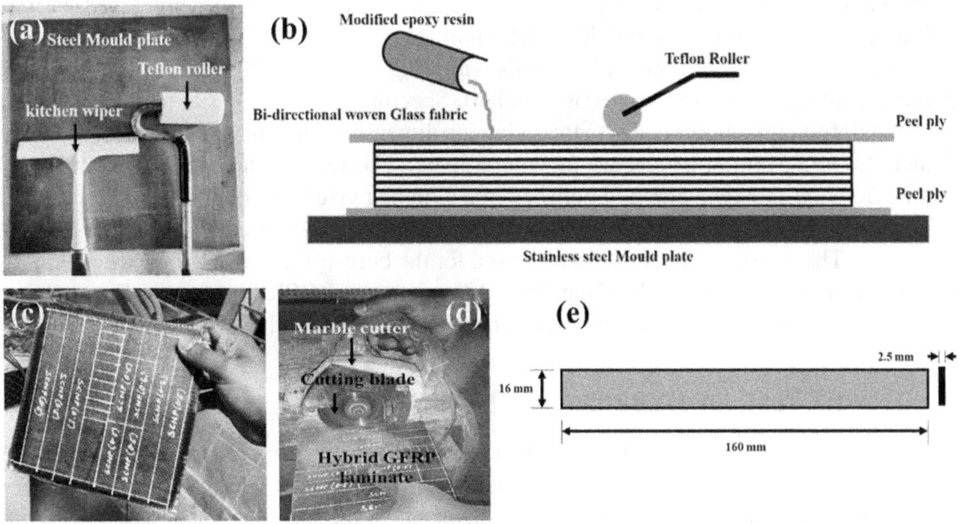

FIGURE 16.6 (a) Tools used for making GFRP laminated structures, (b) hand lay-up technique, (c) cured hybrid GFRP laminate, (d) cutting of hybrid GFRP laminate, and (e) schematic diagram of a tensile specimen of hybrid GFRP laminate.

TABLE 16.2
Designations of Hybrid GFRP Laminated Composites and Their Corresponding Fiber Vol.%

S. No.	Fillers	Fillers wt %	Code of the Hybrid GFRP Laminated Composites	Fiber Volume %	σ_u (MPa)
1.	–	–	NE/GF	50.24±0.43	358.25±11.11
2.	PGr	0.5	PGr/GF	50.75±1.32	417.72±10.61
3.	SGr	0.5	SGr/GF	50.80±1.43	475.86±12.22

The cleaned mold plate is then coated with a mold sealer followed by a mold release agent (TAG Chemicals, GmbH). The mold release agent is coated on the mold surface to avoid sticking of epoxy to the mold surface and easy removal of the cured GFRP laminates.

A total of eight plies of size 180 mm × 180 mm were cut from the bulk bidirectional woven glass fabric. The plies are placed at the surface of mold over peel ply as schematically presented in Figure 16.6b. Peel ply being a synthetic perforated cloth made up of nylon is used here to drape over the epoxide surface of the laminates. Once the laminated composites are cured, peel plies were peeled off to get perfectly flat and smooth surfaces. The modified epoxy resin is poured onto the surface of the peel ply already placed in the mold. The epoxy resin is uniformly spread with the help of a kitchen wiper (Figure 16.6a). The first layer of woven GF fabric is then placed on the epoxy resin surface, and a Teflon roller (Figure 16.6b) is rolled with mild pressure to remove any trapped air as well as the excess resin present as schematically presented in Figure 16.6b. The process is repeated for each layer of modified epoxy resin and GF fabric till the required number of layers is stacked. After placing the last fabric, another peel ply is placed and rolled with a Teflon roller. A top mold steel plate weighing ~3.7 kg is kept on the stacked layers to create uniform pressure for preventing distortion of the laminates during curing. The system is then placed in a woven at room temperature for 24 hours according to the technical data sheet provided by Atul Limited, India. After curing at room temperature, the mold is opened and the developed composite plate is taken out as shown in Figure 16.6c. The control GFRP specimen fabricated without any nanoreinforcement is coded as NE/GF, whereas hybrid GFRP made with modified epoxy resin containing PGr and SGr is coded as PGr/GF and SGr/GF, respectively. The laminate plate is further processed to prepare the required test specimens. The hybrid GFRP laminate plate was machined using a water-cooled marble cutter (model: Planet Power EC4) as shown in Figure 16.6d. Static tensile test samples are prepared according to the ASTM D3039 standard. Specimens for the tensile test were machined (Figure 16.6d) from the sample plates to the following dimensions: length = 160 mm, width = 16 mm, and thickness = 2.5 mm as presented in Figure 16.6e. The rectangular tensile specimens were machined from the sample plates and polished in a twin belt grinder to obtain the required dimension. Edges of specimens were further polished with different grades of emery paper to remove any surface flaws. These test specimens were also used for the burn-off test to calculate the weight fraction as well as the volume fraction of the fiber of the respective GFRP laminates. Five specimens of respective samples were used for the burn-off test. Respective results are shown in Table 16.2.

16.6 TENSILE TEST OF HYBRID GFRP LAMINATES

A minimum of five replicate specimens are needed for testing each sample as shown in Figure 16.7a for obtaining standard deviation. Here, the tensile testing of hybrid GFRP laminated composite specimens was conducted on an Instron: 5969 universal testing machine under a 50 kN load cell with a crosshead speed of 1 mm/minute as shown in Figure 16.7b. The stress–strain curves of tensile test specimens are used for the determination of tensile strength (σ_u), elastic modulus (E), and maximum extension (ε). The results are reported after averaging at least five measurements.

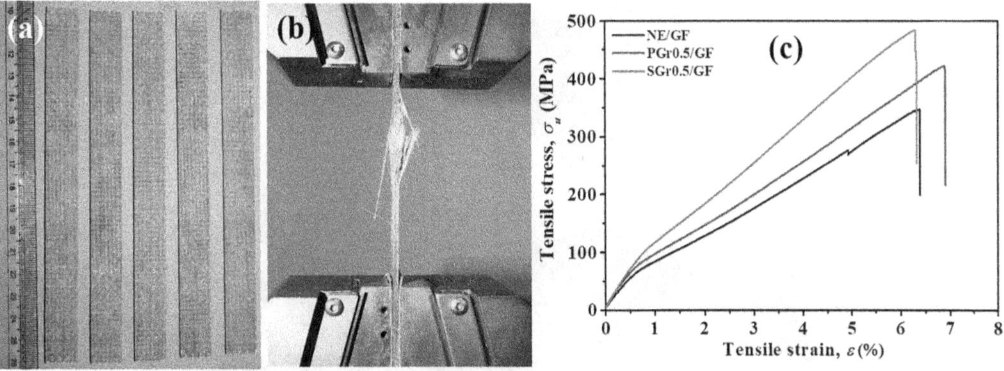

FIGURE 16.7 (a) Machined and polished tensile specimens, (b) tensile testing of hybrid GFRP specimen up to fracture, and (c) stress–strain plot of the specimens.

The effectiveness of PGr and SGr on the strength of the hybrid GFRP is depicted by the investigation of in-plane tensile properties (Das et al. 2020). Uniaxial tensile tests were performed on hybrid GFRP added with 0.5 wt% of PGr and SGr respectively, and respective tensile stress–strain curve is presented in Figure 16.7c. The tensile stress–strain curves of all samples showed almost linear elongation (~0.65% of strain) followed by a second nonlinear zone as visible in Figure 16.7c. FMD, matrix microcracking, and fiber fracture at the crimped region are the reasons for the nonlinear behavior at a higher strain value (Das et al. 2020). Fibers at the crimped region were tried to fragment at a higher load causing FMD and generate small cracks within the matrix. These microscopic and macroscopic damages were assembled at higher strain leading to a nonlinear behavior before final ruptures at the maximum load (Das et al. 2020; Lubineau and Rahaman 2012). σ_u, E, and ε of NE/GF were reported as 358.25±11.11 MPa, 9.64±0.12 GPa, and 6.34%±0.18%, respectively. It was seen that σ_u (417.72±10.61 MPa), E (10.86±0.62 GPa), and ε (6.88%±0.40%) were improved by ~16%, ~13%, and ~8.5%, respectively for PGr/GF compared to that of NE/GF. It has been reported that the higher content (1.0 and 1.5 wt%) of PGr was unable to improve the tensile performance of hybrid GFRP further (Das et al. 2020). The reduction in tensile properties at higher wt% of PGr was attributed to their intensification of agglomeration. The self-aggregated PGr are susceptible to generating microcracks within the matrices as well as initiating crack sites at the fiber/matrix interfaces (Halder et al. 2016). Moreover, aggregated PGr promote stress concentration at the PGr/epoxy interfaces and result in faster propagation of microcracks (Godara et al. 2009). On the other hand, a significant rise in the tensile properties was observed for SGr/GF: incorporation of SGr at 0.5 wt.% led to a ~33%, ~21%, and ~5% increase in σ_u, E, and ε, respectively, to that of NE/GF. It has been reported that ε of hybrid GFRP (for all variations of SGr content) was slightly lesser than that of hybrid GFRP fabricated with PGr. Strength enhancement in SGr/GF is mainly attributed to the formation of a better interface between the SGr and epoxy matrix, resulting in better stress transfer and inclusion of a more effective strengthening mechanism. Addition of 0.5 wt% of SGr in hybrid GFRP also restricts FMD and delays crack propagation, hence improving the tensile properties. In addition, a well-formed fiber/matrix interface may restrict additional elongation of crimped bundles, thereby enhancing the stiffness and reducing the total elongation of laminate for SGr/GF compared to that of PGr/GF.

16.7 MORPHOLOGICAL CHARACTERISTICS OF TENSILE FRACTURE SURFACES OF HYBRID GFRPs

It has been reported that homogeneous dispersion of nanoreinforcement in epoxy resin enhances its strength and toughness (Das et al. 2020), hence improving the delamination resistance of hybrid laminated composites made with such modified epoxy resins (Halder et al. 2016). As an example,

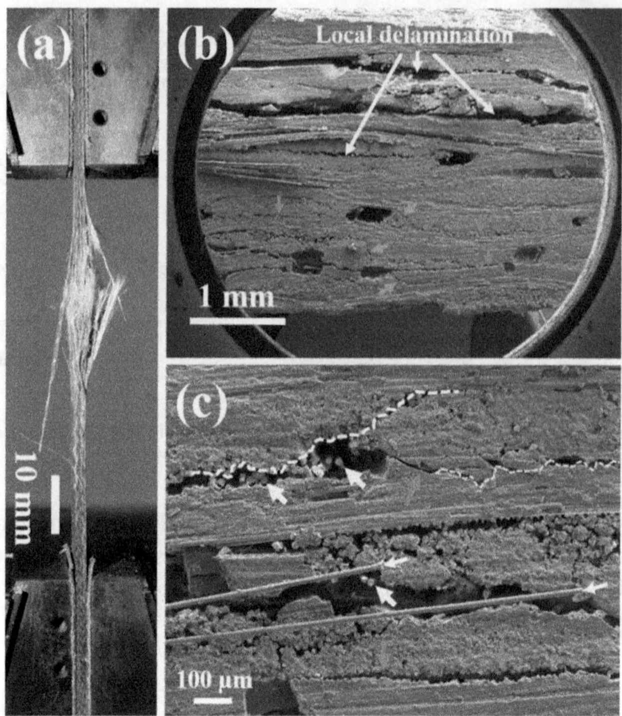

FIGURE 16.8 (a) Macroscale image of tensile fracture surface, (b) mesoscale fracture surface, and (c) microscale fracture surface of hybrid GFRP sample added with PGr.

at the microscale (Figure 16.8a), the tensile fracture surface of hybrid GFRP reinforced with PGr depicts reduced ply-to-ply separation with the reduced blooming effect of longitudinal fibers. Such behavior has resulted from improved interfacial bonding between GFs and epoxy matrix after the addition of PGr.

Field emission scanning electron microscopy at the mesoscale made it easy to study other features that help restrict the ply-to-ply separation throughout the length. The presence of matrix microcracking and transverse microcracking is noticeable at meso- and microscale; however, such cracks are seized after the addition of PGr. The cracks are arrested and forced to move their patch in a serpentine way (indicated by yellow dotted lines) owing to the presence of PGr. The modified matrix provided good interfacial property, hence adhering to GF with partial fiber breakage (indicated by white arrowhead) as shown in Figure 16.8c. Hence, it can be said that the addition of PGr improves the fiber/matrix interface and restricts the crack propagation within the matrices, thereby improving the tensile properties of PGr/GF compared to NE/GF as shown in Figure 16.8c. Despite the restriction of matrix microcracking, the presence of FMD is prominent at the microscale (Figure 16.8c). It is reported that the presence of aggregated PGr causes stress concentration and catalyzes localized delamination resulting in FMD (Das et al. 2020).

The effect of SGr on hybrid GFRP laminates is visible in Figure 16.9a at the microscale, where unlike NE/GF the delamination is restricted at a localized region, and the plies are found held together for SGr/GF. The damage prevention is also noticeable at the mesoscale as shown in Figure 16.9b. Most of the fibers and matrix are found intact with the presence of shortened microcracks and reduced local delamination. Though matrix microcracking, transverse microcracking, and fiber breakage are visible in SGr/GF at the microscale, their presence is very minimal compared to NE/GF and PGr/GF as evident in Figure 16.9c. Therefore, it can be said that the addition of functionalized nanoreinforcement is an effective solution to restrict the intralaminar as well as

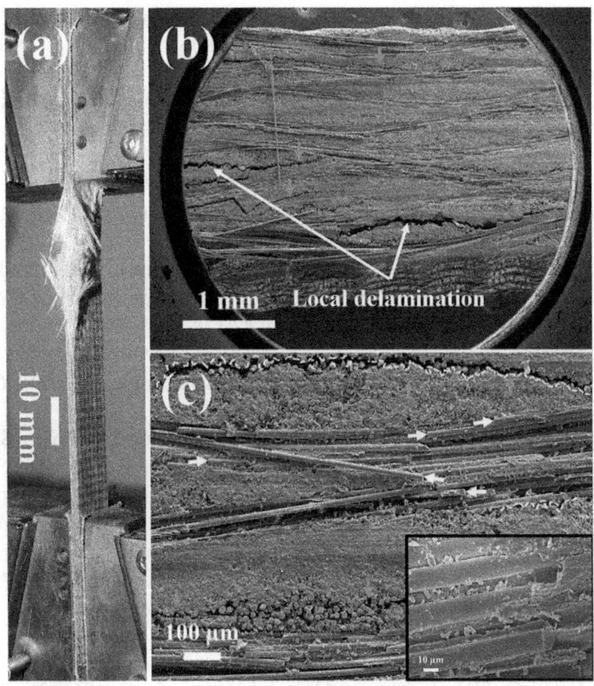

FIGURE 16.9 (a) Macroscale image of tensile fracture surface, (b) mesoscale fracture surface, and (c) microscale fracture surface of hybrid GFRP sample added with SGr.

interlaminar diffuse damages. In the present event, the amine group present on SGr (Figure 16.5) reacts with the epoxide group of diglycidyl ether of bisphenol-A resin forming a stable interfacial region, hence being able to hold the fiber together as shown in the inset of Figure 16.9c. Thus, the performance of hybrid laminated composites can be achieved by suppressing the microcracks under homogeneous dispersion of nanoreinforcement in the matrix and their settlement on fibers as well as in the matrix (Das et al. 2020).

REFERENCES

Chen, J., D. Zhao, X. Jin, C. Wang, D. Wang, and H. Ge 2014. Modifying glass fibers with graphene oxide: Towards high-performance polymer composites. *Composites Science and Technology* 97: 41–45.

Das, S., N. Islam Khan, and S. Halder 2018a. Thermo-mechanical stability of epoxy composites induced with surface silanized recycled carbon fibers. *IOP Conference Series: Materials Science and Engineering* 377(1): 012172.

Das, S., S. Halder, A. Sinha, M.A. Imam, and N.I. Khan 2018b. Assessing nanoscratch behavior of epoxy nanocomposite toughened with silanized fullerene. *ACS Applied Nano Materials* 1(7): 3653–3662.

Das, S., S. Halder, J. Wang, M.S. Goyat, A. Anil Kumar, and Y. Fang, 2017. Amending the thermo-mechanical response and mechanical properties of epoxy composites with silanized chopped carbon fibers. *Composites Part A: Applied Science and Manufacturing* 102: 347–356.

Das, S., S. Halder, and K. Kumar 2016. A comprehensive study on step-wise surface modification of C60: Effect of oxidation and silanization on dynamic mechanical and thermal stability of epoxy nanocomposite. *Materials Chemistry and Physics* 179: 120–128.

Das, S., S. Halder, N.I. Khan, B. Paul, and M.S. Goyat 2020. Assessing damage mitigation by silanized milled graphite nanoparticles in hybrid GFRP laminated composites. *Composites Part A: Applied Science and Manufacturing* 132: 105784.

De Greef, N., L. Gorbatikh, A. Godara, L. Mezzo, S.V. Lomov, and I. Verpoest 2011a. The effect of carbon nanotubes on the damage development in carbon fiber/epoxy composites. *Carbon* 49(14): 4650–4664.

De Greef, N., L. Gorbatikh, S.V. Lomov, and I. Verpoest, 2011b. Damage development in woven carbon fiber/epoxy composites modified with carbon nanotubes under tension in the bias direction. *Composites Part A: Applied Science and Manufacturing* 42(11): 1635–1644.

Eskizeybek, V., H. Ulus, H.B. Kaybal, Ö.S. Şahin, and A. Avcı 2018. Static and dynamic mechanical responses of $CaCO_3$ nanoparticle modified epoxy/carbon fiber nanocomposites. *Composites Part B: Engineering* 140: 223–231.

Fan, Z., M.H. Santare, and S.G. Advani 2008. Interlaminar shear strength of glass fiber reinforced epoxy composites enhanced with multi-walled carbon nanotubes. *Composites Part A: Applied science and manufacturing* 39(3): 540–554.

Godara, A., L. Mezzo, F. Luizi, A. Warrier, S.V. Lomov, A.W. Van Vuure, L. Gorbatikh, P. Moldenaers, and I. Verpoest 2009. Influence of carbon nanotube reinforcement on the processing and the mechanical behaviour of carbon fiber/epoxy composites. *Carbon* 47(12): 2914–2923.

Goyat, M.S., S. Suresh, S. Bahl, S. Halder, and P.K. Ghosh 2015. Thermomechanical response and toughening mechanisms of a carbon nano bead reinforced epoxy composite. *Materials Chemistry and Physics* 166: 144–152.

Guan, L.-Z., J.-F. Gao, Y.-B. Pei, L. Zhao, L.-X. Gong, Y.-J. Wan, H. Zhou, et al. 2016. Silane bonded graphene aerogels with tunable functionality and reversible compressibility. *Carbon* 107: 573–582.

Halder, S., S. Ahemad, S. Das, and J. Wang 2016. Epoxy/glass fiber laminated composites integrated with amino functionalized ZrO2 for advanced structural applications. *ACS Applied Materials & Interfaces* 8(3): 1695–1706.

Hui, Z., Z. Zhang, H. Murayama, and K. Okamoto 2008. Improved bonding between PAN-based carbon fibers and fullerene-modified epoxy matrix. *Composites Part A: Applied Science and Manufacturing* 39(11): 1762–1767.

Hussain, M., A. Nakahira, and K. Niihara 1996. Mechanical property improvement of carbon fiber reinforced epoxy composites by Al_2O_3 filler dispersion. *Materials Letters* 26(3): 185–191.

Kamae, T., and L.T. Drzal 2012. Carbon fiber/epoxy composite property enhancement through incorporation of carbon nanotubes at the fiber–matrix interphase–Part I: The development of carbon nanotube coated carbon fibers and the evaluation of their adhesion. *Composites Part A: Applied Science and Manufacturing* 43(9): 1569–1577.

Kamar, N.T., M.M. Hossain, A. Khomenko, M. Haq, L.T. Drzal, and A. Loos 2015. Interlaminar reinforcement of glass fiber/epoxy composites with graphene nanoplatelets. *Composites Part A: Applied Science and Manufacturing* 70: 82–92.

Khan, N.I., S. Halder, N. Talukdar, S. Das, and M.S. Goyat 2021. Surface oxidized/silanized graphite nanoplatelets for reinforcing an epoxy matrix. *Materials Chemistry and Physics* 258: 123851.

Khan, N.I., S. Halder, S. Das, and J. Wang 2019. Exfoliation level of aggregated graphitic nanoplatelets by oxidation followed by silanization on controlling mechanical and nanomechanical performance of hybrid CFRP composites. *Composites Part B: Engineering* 173(2019): 106855–67.

Koissin, V., L.L. Warnet, and R. Akkerman 2013. Delamination in carbon-fibre composites improved with in situ grown nanofibres. *Engineering Fracture Mechanics* 101: 140–148.

Koricho, E.G., A. Khomenko, M. Haq, L.T. Drzal, G. Belingardi, and B. Martorana 2015. Effect of hybrid (micro-and nano-) fillers on impact response of GFRP composite. *Composite Structures* 134: 789–798.

Kwon, D.-J., P.-S. Shin, J.-H. Kim, Y.-M. Baek, H.-S. Park, K.L. DeVries, and J.-M. Park 2017. Interfacial properties and thermal aging of glass fiber/epoxy composites reinforced with SiC and SiO_2 nanoparticles. *Composites Part B: Engineering* 130: 46–53.

Lee, J.-H., K.Y. Rhee, and S.J. Park 2011. Silane modification of carbon nanotubes and its effects on the material properties of carbon/CNT/epoxy three-phase composites. *Composites Part A: Applied Science and Manufacturing* 42(5): 478–483.

Lubineau, G., and A. Rahaman 2012. A review of strategies for improving the degradation properties of laminated continuous-fiber/epoxy composites with carbon-based nanoreinforcements. *Carbon* 50(7): 2377–2395.

Nayak, R.K., A. Dash, and B.C. Ray 2014. Effect of epoxy modifiers (Al_2O_3/SiO_2/TiO_2) on mechanical performance of epoxy/glass fiber hybrid composites. *Procedia Materials Science* 6: 1359–1364.

Pathak, A.K., M. Borah, A. Gupta, T. Yokozeki, and S.R. Dhakate 2016. Improved mechanical properties of carbon fiber/graphene oxide-epoxy hybrid composites. *Composites Science and Technology* 135: 28–38.

Prusty, R.K., D.K. Rathore, and B.C. Ray 2017. Evaluation of the role of functionalized CNT in glass fiber/epoxy composite at above-and sub-zero temperatures: Emphasizing interfacial microstructures. *Composites Part A: Applied Science and Manufacturing* 101: 215–226.

Punetha, V.D., S. Rana, H.J. Yoo, A. Chaurasia, J.T. McLeskey Jr, M. Sekkarapatti Ramasamy, N.G. Sahoo, and J.W. Cho 2017. Functionalization of carbon nanomaterials for advanced polymer nanocomposites: A comparison study between CNT and graphene. *Progress in Polymer Science* 67: 1–47.

Sánchez, M., M. Campo, A. Jiménez-Suárez, and A. Ureña 2013. Effect of the carbon nanotube functionalization on flexural properties of multiscale carbon fiber/epoxy composites manufactured by VARIM. *Composites Part B: Engineering* 45(1): 1613–1619.

Soliman, E. 2011. *New Generation Fiber Reinforced Polymer Composites Incorporating Carbon Nanotubes*. PhD diss., The University of New Mexico.

Srinivasan, V., N. Mohamad Raffi, R. Karthikeyan, and V. Kalaichelvi 2010. Characteristics of Al_2O_3 nanoparticle filled GFRP composites using wear maps. *Journal of Reinforced Plastics and Composites* 29(19): 3006–3015.

Vennerberg, D., Z. Rueger, and M.R. Kessler 2014. Effect of silane structure on the properties of silanized multiwalled carbon nanotube-epoxy nanocomposites. *Polymer* 55(7): 1854–1865.

Wang, F., L.T. Drzal, Y. Qin, and Z. Huang 2016. Size effect of graphene nanoplatelets on the morphology and mechanical behavior of glass fiber/epoxy composites. *Journal of Materials Science* 51(7): 3337–3348.

Wang, X., W. Xing, P. Zhang, L. Song, H. Yang, and Y. Hu 2012. Covalent functionalization of graphene with organosilane and its use as a reinforcement in epoxy composites. *Composites Science and Technology* 72(6): 737–743.

Warrier, A., A. Godara, O. Rochez, L. Mezzo, F. Luizi, L. Gorbatikh, S.V. Lomov, A. Willem Van Vuure, and I. Verpoest 2010. The effect of adding carbon nanotubes to glass/epoxy composites in the fibre sizing and/or the matrix. *Composites Part A: Applied Science and Manufacturing* 41(4): 532–538.

Yokozeki, T., Y. Iwahori, and S. Ishiwata 2007. Matrix cracking behaviors in carbon fiber/epoxy laminates filled with cup-stacked carbon nanotubes (CSCNTs). *Composites Part A: Applied Science and Manufacturing* 38(3): 917–924.

Zamanian, M., M. Mortezaei, B. Salehnia, and J.E. Jam 2013. Fracture toughness of epoxy polymer modified with nanosilica particles: Particle size effect. *Engineering Fracture Mechanics* 97: 193–206.

Zhang, X., X. Fan, C. Yan, H. Li, Y. Zhu, X. Li, and L. Yu 2012. Interfacial microstructure and properties of carbon fiber composites modified with graphene oxide. *ACS Applied Materials & Interfaces* 4(3): 1543–1552.

Zhu, J., S. Wei, J. Ryu, M. Budhathoki, G. Liang, and Z. Guo 2010. In situ stabilized carbon nanofiber (CNF) reinforced epoxy nanocomposites. *Journal of Materials Chemistry* 20(23): 4937–4948.

17 Electrochemical and Microscopic Studies on Controlling of Cracking Behavior of SiC_p, Al_2O_3-Reinforced Aluminum Metal Matrix Composites

Chinmoy Bhattacharya, Biswanath Samanta, Paramita Hajra, Harahari Mandal, Sanjib Shyamal, and Debasis Sariket
Indian Institute of Engineering Science & Technology (IIEST), Shibpur

Sri Bandyopadhyay
UNSW Sydney
Aus Defence DSTO MRL

CONTENTS

DOI: 10.1201/9780429330575-17

17.1 INTRODUCTION

Al-Si-Mg alloy (6061 aluminum) is an important material with wide ranges of industrial applications and marine technology; however, this alloy has some drawbacks, like low hardness, poor tribological properties and crevice corrosion resistance which still need to be improved. The demand for materials of superior mechanical, thermal and electrical properties has focused attention on development of aluminum metal matrix composites (MMCs). MMCs have tremendous potential for the future. Though addition provides strength, the heterogeneous microstructure makes them to susceptible to localized corrosion.

17.1.1 Corrosion

Corrosion is the deterioration of materials by chemical interaction with their environment. The term corrosion is sometimes also applied to the degradation of plastics, concrete and wood, but generally refers to metals. The present chapter details the corrosion behavior of Al-6061 alloy and its SiC_p, Al_2O_3-reinforced MMCs and the influence of some inorganic anions on controlling the degradation behavior of these materials.

17.1.2 Aluminum Alloys and Localized Corrosion

Corrosion is a serious and ongoing problem in reinforced concrete structures, especially in marine environments or regions where deicing salts are used (Koch et al. 2002). When left uncorrected, corrosion degrades the strength of a structure and extensive repairs are often required. Early detection of corrosion can significantly decrease the repair costs. Many methods have been developed to detect corrosion but they are often unreliable, difficult to implement or very expensive (Kitowski 1993).

The corrosion of aluminum and its alloy is the subject of critical technological importance due to the increasing industrial application of these materials. Aluminum and its alloy represent an important category of materials for high technological value and wide range of industrial applications particularly due to their high strength/weight ratio, good formability, good corrosion resistance and recyclability potential in vehicles, household items, infrastructures, constructions, aerospace, transportation and marine technology (Pardo et al. 2005; Costa, Velasco, and Toralba 2000; Pardo et al. 2003). In Al alloys, Si, Fe, Mn, Cr, Cu and Mg are introduced at various levels, mainly to improve the mechanical strength. Si and Fe are normally present as unavoidable impurities in commercially pure Al up to a total of 0.5 wt.%, but may also be introduced at higher levels. During the manufacturing process stages, these elemental additives may create various kinds of insoluble intermetallic particles (IMPs) in the alloys and, to a lesser extent, precipitate from soluble alloying compounds and influence the final product properties (Vargel, Jacques, and Schmidt 2004). It has been calculated that for each kilogram of weight saved in a vehicle, a saving of ca20 kg of CO_2 emissions can be achieved. Therefore, the use of aluminum in the automotive industry in order to produce more fuel-efficient vehicles and to reduce the energy consumption and air pollution has increased greatly over the last few decades, from 20 kg in 1960 to a predicted level of more than 160 kg per vehicle in 2010 and 250–340 kg by 2015 (Miller et al. 2000). One of the increasing applications of Al alloys in vehicles is in heat exchangers (with tube and fin components) such as radiators, evaporators, engine cooling and air-conditioning systems. In the past, aluminum heat exchangers were assembled mechanically, but nowadays tubes and fins are joined together by a brazing process using a brazing Al-Si alloy layer that has a lower eutectic temperature than the tube or fin core alloy.

Aluminum is a very active metal but it naturally creates a passive layer, and its corrosion resistance depends on the passivity produced by this protective oxide layer (Winston Revie 2011). In the pH range of 4–9, the amorphous protective oxide (with an external side of bayerite, Al_2O_3

$3H_2O$ hydroxide gel) layer is formed in water or atmosphere with 2–4 nm thickness. The dissolution potential of aluminum in most aqueous media is in the order of −500 mV with respect to hydrogen electrode (SHE), while its E^0 is −1,660 mV with respect to SHE. Because of this highly electronegative potential, Al is one of the easiest metals to oxidize. However, due to the presence of the naturally passive layer, aluminum behaves as a very stable metal, especially in oxidizing media such as air and water. The few existing defects in the protective oxide layer, which are inevitable even for the purest aluminum alloys, will cause the corrosion initiation. In the alloyed aluminum, the second phases are either cathodic or anodic compared to the aluminum matrix and they give rise to galvanic cell formation because of the potential difference between them. Chloride-containing solutions are the most harmful ones as regards localized corrosion of aluminum alloys.

17.1.3 Metal Matrix Composites

Artificial combination of matrix phase (alloy) and dispersed phase (reinforcement). The demand for lighter weight, cost-effective and high-performance materials for use in a spectrum of structural and nonstructural applications has resulted in the need for fabrication of MMCs of various types. In recent years, the aluminum alloy-based MMCs have offered designers many added benefits as they are particularly suited for applications requiring good strength at high temperatures, good structural rigidity, dimensional stability, light weight and low thermal expansion (Seah et al. 2006). The major advantages of Al-based MMCs include greater strength, improved stiffness, reduced density, improved high-temperature properties, controlled thermal expansion coefficient, thermal/heat management, enhanced and tailored electrical performance, improved abrasion and wear resistance and improved damping capabilities (Bishop and Kinar 1995).

One of the main disadvantages of particulate-reinforced MMCs is the influence of reinforcement on the corrosion resistance. This is of particular importance in aluminum alloy-based composites where corrosion resistance is imparted by a protective oxide film. The addition of a reinforcing phase could lead to further discontinuities or flaws in the protective film, increasing the sites for corrosion initiation and rendering the composite liable to severe corrosion attack. The study on the corrosion behavior of MMCs in different aggressive environments has continued to attract considerable attention because of the several important applications of these materials. These composites frequently come in contact with acids or bases during the process like cleaning, pickling and descaling. It is known that aluminum alloys and their composites exhibit high corrosion rate in HCl- or NaCl-containing solutions. Therefore, studying the corrosion behavior of aluminum alloys and their composites in these environments is of prime importance. The present chapter will provide a detailed study on the corrosion behavior of a typical high-strength Al-Si-Mg alloy (Al-6061 alloy) and its 20% SiC_P, 10% and 20% Al_2O_3-reinforced MMC in aqueous sodium chloride and hydrochloric acid solution.

17.1.4 Corrosion Behavior of Aluminum Samples in Different Environmental Conditions

Although pitting is the most common form of aluminum corrosion (Bishop and Kinar 1995), in practice, different localized corrosion attack morphologies have been observed on aluminum alloys in different solutions (Ambat et al. 2006). For pure Al with a crystallized structure, pitting develops along closely packed (100) planes, resulting in crystallographic corrosion. Alloying generally leads to the initiation and development of pit formation that is less sensitive to the microstructure of the alloy matrix, but more related to the secondary phases. Corrosion of low-alloyed Al alloys is slightly localized along grain boundaries. A further increase in the content of alloying elements enhances the localization of pitting, as the amount of IMPs grows substantially. High-alloyed Al alloys are often susceptible to intergranular corrosion (IGC). One reason is that, in this case, the structure of the grain boundaries becomes more complicated due to the appearance of precipitates, the zones depleted of alloying elements and the zones enriched in certain alloying elements and dislocation piles. Tunnel-like pitting and narrow long channels have been observed on some commercial Al alloys, due to microstructure, solution composition, electrode potential and temperature. The reason for this behavior was discussed in terms of coupling of dissolution and mass transport. Some studies suggested that the alkalinity developed at cathodic IMPs on Al alloys in aerated solutions can dissolve the adjacent alloy matrix, creating grooves or pit-like clusters. Later on, these cavities may switch to an acid-pitting mechanism. Other authors, however, referred to the alkaline attack as pitting or treat the problem as galvanic corrosion between particles and matrix, or self-regulating cathodic reaction occurring on the particles. Electrochemical behavior of Al_3Fe phase in Al-Mn-Fe-Si system in high pH NaOH solution revealed that near the corrosion potential, Al_3Fe phase undergoes a selective dissolution of Al and the surface of Al_3Fe crystals becomes richer in Fe, which is detrimental to the cathodic behavior of this type of IMPs. The presence of Mn or Si in the Al_3Fe, such as α-Al(Fe, Mn)Si and δ-AlFeSi phases, reduces the effects of Fe as regards both anodic and cathodic reaction rates. The positive effect of an increase of Mn in Al-Mn alloy in a solid solution leads to a shift of the potential of matrix to the cathodic direction, while an increase in the Mn/Fe ratio of the IMPs shifts their potential to the anodic direction.

It can be concluded that generally, in commercial engineering aluminum alloys, localized corrosion is often triggered by microscopic defects, such as nonmetallic inclusions or IMPs having a size range of several microns to nanometer. Literature reports on pitting corrosion of aluminum alloys were reviewed by Szklarska-Smialowska (1999) and Frankel (1998), who pointed out the need to further explore the formation of pits on cathodic IMPs and the influence of the second phase. Therefore, the understanding of the corrosion resistance and electrochemical behavior of aluminum for the future industrial applications and development is vital. One example of aggressive environments is seawater. Seawater systems are used by many industries such as shipping, offshore oil and gas production, power plants and coastal industrial plants. Exposure to these structures in marine environments will cause corrosion that finally leads to total damage. Therefore, it is very important to study on corrosion prevention in this environment. The fact of corrosion represent a tremendous economic loss and much can be done to control it. There are many ways to reduce corrosion rate and one of the most popular and acceptable practices is the use of inhibitors. Large numbers of organic compounds were studied and are being studied to investigate their corrosion inhibition potential. All these studies revealed that organic compounds especially those with N, S and O showed significant inhibition efficiency.

17.1.5 Role of Inhibitors in Controlling Corrosion Damage of Aluminum Samples

An inhibitor is a substance that, when added in small concentrations to an environment, decreases the corrosion rate. In a sense, an inhibitor can be considered as a retarding catalyst. There are numerous inhibitor types and compositions. Most inhibitors have been developed by empirical experimentation, and many inhibitors are proprietary in nature and thus their composition is not disclosed. Inhibition is not completely understood because of these reasons, but it is possible to classify inhibitors according to their mechanism and composition. The safety and environmental issues of corrosion inhibitors arisen in industries have always been a global concern. In recent days, many alternative eco-friendly corrosion inhibitors have been developed. The growing needs for the corrosion inhibition become increasingly necessary to delay or stop the attack of metal in aggressive solutions. Many efforts made to find suitable natural sources to be used as corrosion inhibitors in various corrosion media (Iannuzzi 2010; Danilidis, Davenport, and Sykes 2007). This study considered this particular issue when applying selected 6061 aluminum alloy to its application which would be suitable for our natural environment for instance tropical seawater. Nitrate, chromate, molybdate and metavanadate offer interesting possibilities for corrosion inhibitor due to its safe use, low cost and availability.

One of the important characteristics required for a material for structural applications is its inherent ability to resist environmental degradation. Although the incorporation of the second phase into a matrix material can enhance the physical and mechanical properties of the material, but also segregate to form IMPs. The heterogeneous microstructure of high-strength Al alloys and MMCs makes them particularly susceptible to localize attack, such as pitting, crevice, intergranular and exfoliation corrosion (Despic, Parkhutik, and Bockris 1989). Most of the high-strength Al alloys and composites are susceptible to general and localized corrosion in aqueous environment containing Cl^- ions. Many important applications of Al alloys and MMCs have resulted in research into its electrochemical behavior and corrosion resistance in a wide variety of media, including investigations of the properties of the surface oxide film, formed naturally or by anodization. Generally, this consists of a thin barrier film adjacent to the metal (~25 nm thick) covered by a thicker porous oxide layer. In aggressive media such as Cl^--ion-containing solution, localized corrosion can occur leading to the breakdown of the passive layer and pit formation (Staab and Krause-Rehberg 2001). The electrochemistry of aluminum in chloride solution at a variety of pH values has been widely investigated (Cai and Cheng 2007). The effect of the addition of other specific anions to the Cl^--ion-containing solution in terms of electrochemical behavior inhibiting or enhancing corrosion has yet to be investigated thoroughly. Pit formation can occur by migration (absorption) of aggressive ions within the porous oxide layer due to the high electric field, via film cracking due to anion adsorption causing local mechanical stress and through the formation of soluble complex ions by chemical reactions. It is evident that these processes will occur most easily at defects in the oxide film, or where the film is thinnest. Thus, useful corrosion inhibition must protect against the attack of this kind.

It is well known that chromate in solution inhibits the metal dissolution reactions as well as blocks the counter-reduction reactions, say, oxygen reduction reaction for neutral or alkaline medium and hydrogen evolution reaction for acidic environment. Recently, the mechanism of corrosion protection by soluble inhibitors like CrO_4^{2-} has been the subject of active research, which has attempted to understand and replicate its inhibiting functions with less toxic chemical substances. In this work, the inhibition of Al alloys and MMCs' corrosion by two soluble inhibitor anions, CrO_4^{2-} and NO_3^-, was studied. Generally, the pitting corrosion of Al depends greatly on anion composition of the electrolyte. In this respect, inhibition effects of various anion additives on corrosion behavior of Al-6061 alloy, its 20% SiC_p, its 10% and 20% Al_2O_3-reinforced MMC have been investigated in aqueous HCl solutions. Among the inhibitive anions CrO_4^{2-} and NO_3^- ion retard the pit initiation in aqueous hydrochloric acid solutions effectively. The inhibition mechanism of CrO_4^{2-} and NO_3^- corrosion on the samples has been studied mainly in terms of adsorption of the anions. In this study it is observed that pitting corrosion of Al-6061 alloy its SiC_p and Al_2O_3-reinforced MMC in NO_3^- and CrO_4^{2-} ion containing HCl solution by the adsorption of CrO_4^{2-} ion and incorporation of NO_3^- ion, respectively (Uzuo et al. 2002).

For more than a few decades, chromate has been extensively used in many surface pre-treatments and in coatings to provide long-term corrosion protection. In relatively high concentrations, it is a very much effective anodic corrosion inhibitor for iron, zinc and aluminum alloys. However, the use of chromate is restricted in recent times to a few applications where there are no alternatives or where close environmental control is possible. Examples of the use of chromate species in organic coatings include aerospace applications, for inhibition of corrosion of aluminum alloys, and profiled organic-coated galvanized steel used in the construction industry for cladding of buildings (i.e., roofs and sides). In this latter application, chromate species are both used for pre-treatment of the galvanized layer (to enhance coating adhesion) and, as strontium chromate additions, used as an inhibitor by incorporation into organic primer coatings. The toxicity and carcinogenic nature of hexavalent chromium have led to international efforts to develop alternative inhibitive pigments for coating use. However, a nontoxic, low-cost and effective corrosion inhibitor, as an alternative to chromate, has yet to be successfully developed (Costa 2006). In this context, inhibitive action of ammonium metavanadate and sodium molybdate may be considered particularly in chloride-containing solutions.

17.2 ROLE OF CrO_4^{2-} AND NO_3^- ON PROTECTING Al ALLOY AND MMC IN ACIDIC CONDITIONS

Figure 17.1a and b represents typical polarization curves (Tafel plots) recorded for the Al-6061, its 20% SiC_p, 10% and 20% Al_2O_3-reinforced MMCs in 0.01(M) HCl along with 0.01(M) CrO_4^{2-} and 0.01(M) NO_3^--containing solutions, respectively. The polarization curves exhibit typical Tafel behavior (Baorong, Jinglei, Yanxu, and Fangying 2001). The point of intersection of the tangents to the cathodic and anodic branches of the Tafel plot gives corrosion potential (E_{corr}). At the corrosion potential, the rate of the metal dissolution process (i.e., the oxidation process; $Al \rightarrow Al^{3+} + 3e$) is equal to that of the rate of the hydrogen evolution process (i.e., the reduction process; $2H^+ + 2e \rightarrow H_2$) and this point corresponds to the corrosion rate of the system expressed in terms of corrosion current density (I_{corr} in A/cm²). The variations of E_{corr} and I_{corr} for all the materials in all of these three environments, as calculated from Tafel analysis, are presented in Table 17.1. From the polarization experiments, a maximum value of I_{corr} is obtained for all the materials in 0.01(M) HCl-containing solution, and the cathodic branch of the Tafel plot intersects the anodic curve in the active region, leading to rapid corrosion of the specimen. When NO_3^- ions were added to the Cl^--containing solution, cathodic branch remains almost unchanged but the anodic branch current density suppressed significantly leading to E_{corr} value shifted toward less negative side, i.e., more noble potential region, as presented in Table 17.1. A broad passive region was observed for both the materials in NO_3^--containing solutions, which is extended up to −100 and −50 mV vs SCE, for the Al-6061 and its SiC_p-reinforced MMC, respectively. The passivity of the materials in the presence of NO_3^- ions may be explained by the formation of a resistive metastable compound $Al(NO_3)_3$, by the incorporation of NO_3 ions onto the surface oxide film, as soon as it generates over the Al metal.

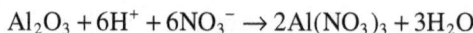

$$Al_2O_3 + 6H^+ + 6NO_3^- \rightarrow 2Al(NO_3)_3 + 3H_2O$$

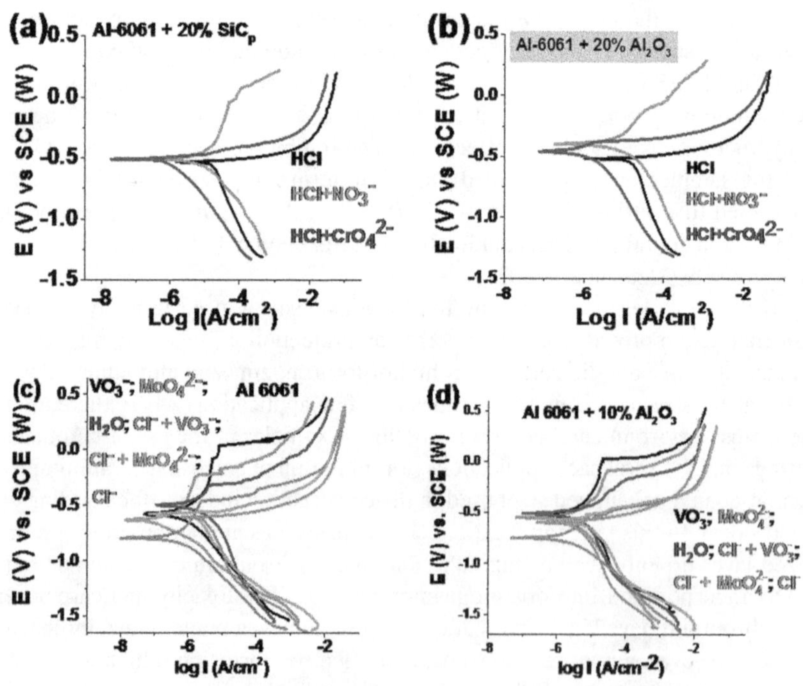

FIGURE 17.1 (a–d) Potentiodynamic polarization (Tafel) plots of different samples in the presence of various electrolytic media

TABLE 17.1
Tafel Parameters of Different Samples in the Presence of NO_3^- and CrO_4^{2-} Inhibitors in Acidic Environments

Solution	0.01(M) HCl	0.01(M) HCl+0.01(M) NO_3^-	0.01(M) HCl+0.01(M) CrO_4^{2-}
		Sample Al-6061+20% SiC_p	
I_{corr} (μA/cm²)	61.00	20.50	4.20
E_{corr} (V)	−0.495	−0.492	−0.47
C.R. (mm/year)	0.62	0.21	0.041
		Sample Al-6061	
I_{corr} (μA/cm²)	25.00	7.30	2.10
E_{corr} (V)	−0.512	−0.498	−0.494
C.R. (mm/year)	0.28	0.075	0.019
		Al-6061+10% Al_2O_3	
I_{corr} (μA/cm²)	29.00	9.20	3.50
E_{corr} (V)	−0.612	−0.607	−0.62
C.R. (mm/year)	0.32	0.091	0.020
		Al-6061+20% Al_2O_3	
I_{corr} (μA/cm²)	45.0	2.60	1.30
E_{corr} (V)	−0.521	−0.421	−0.45
C.R. (mm/year)	0.49	0.043	0.012

The transitory compound is able to control the anodic dissolution of Al underneath the native film before the pitting potential is attained. However under very high overpotential (i.e., beyond the pitting potential), the material loses its transpassive character and the film gets disintegrated due to predominant chemical reaction.

$$Al(NO_3)_3 + NO_3^- \rightarrow Al(NO_3)_4^-.$$

On the other hand, when the materials were polarized in HCl solution containing CrO_4^{2-} ions, the hydrogen evolution reaction ($2H^+ + 2e \rightarrow H_2$) rate decreases (cathodic inhibition) leading to reduction in cathodic branch current density of the Tafel plot, and the E_{corr} values appeared in the positive side to that of the pure HCl solution. A significant decrease in the cathodic current density (cathodic inhibition) accompanied by a nominal passivity was also observed for all the samples in the CrO_4^{2-} -containing environment. It was evident that I_{corr} reaches the maximum level for both the materials in 0.01(M) HCl solution which is decreased significantly in the presence of NO_3^- ions, almost 1/3rd order to that of pure HCl solution, whereas, in the presence of CrO_4^{2-} ions, the I_{corr} values decreased further and reaches to the minimum level, about 15–20th order less in magnitude as compared to that for pure HCl solution. In the presence of CrO_4^{2-} ions, the metal surface exhibits significant inhibition property toward Cl^--containing environment due to the factors like (i) formation of surface chelates; (ii) incorporation into the film of the oxidizing inhibitor (here, CrO_4^{2-}), which itself is reduced to Cr^{3+}; (iii) incorporation of the inhibitor into the barrier oxide film or into its pores; and (iv) formation of a precipitate on the oxide film. Furthermore, chromate ions are mobile in solution and migrate easily to the exposed areas on the Al alloy surface. Chromate adsorbs on active corrosion sites of the surface and is reduced to form a monolayer of a Cr^{3+} species. This monolayer inhibits further reduction of CrO_4^{2-} species, but may adsorb a second layer of CrO_4^{2-} to form mixed oxides. This layer is effective at reducing the activity of both cathodic and anodic sites in the matrix. Inhibition in either or both cases is effected by blocking the active sites and reducing the tunneling rate of electrons through the inhibiting chromate film. Anodic inhibition involves the reduction of localized corrosion initiation sites since chromate ions are mobile in solution and migrate easily to the exposed areas on the Al alloy surface. The higher order of I_{corr} values for the MMC as compared to that of the alloy sample in all of the three electrolytic media may be explained due to the addition of 20% SiC_p into the Al-6061 matrix leading to the incorporation of surface heterogeneity, which may act as the potential sites for further attack by the corrosive environments.

17.3 PROTECTION OF AL SAMPLES BY MoO_4^{2-} AND VO_3^- IN NEUTRAL CHLORIDE ENVIRONMENTS

Figure 17.1c and d represents the typical polarization curves recorded for the materials in 0.05(M) NaCl, 0.05(M) NaCl+0.01(M) Na_2MoO_4 and 0.05 (M) NaCl+0.01(M) NH_4VO_3 solutions. The polarization curves exhibit Tafel-type behavior, and the corresponding corrosion potential (E_{corr}) and corrosion current density (I_{corr} in A/cm²) were evaluated. The corrosion parameters derived from Tafel analysis are given in Table 17.2. From Tafel plot, a maximum corrosive behavior of all the materials is observed in 0.05 (M) NaCl solution and the corresponding E_{corr} value shifted toward more active potential region. However, MoO_4^{2-} ions found to suppress both cathodic and anodic currents. A significant cathodic inhibition of current density was observed for sample polarized in metavanadate-containing solution (Hollingsworth 1987). Chloride ions are known to be specifically adsorbed on the oxide film on the top of Al surface and thus react with the metallic phase under the oxide layer.

$$Al^{3+} + 3Cl^- \rightarrow AlCl_3$$

TABLE 17.2
Tafel Parameters of Different Samples in the Presence of VO_3^- and MoO_4^{2-} Inhibitors in Neutral Chloride Solutions

Solution	0.05 (M) NaCl	0.05(M) NaCl + 0.0.1(M) Na_2MoO_4	0.05(M) NaCl + 0.01(M) NH_4VO_3	H_2O	0.01(M) Na_2MoO_4	0.01(M) NH_4VO_3
Sample Al-6061+20% SiC_p						
I_{corr} (μA/cm²)	20.1	7.70	5.30	4.60	2.40	1.80
E_{corr} (V)	−0.618	−0.627	−0.671	−0.444	−0.361	−0.453
C.R. (mm/year)	0.296	0.084	0.068	0.060	0.026	0.025
Sample Al-6061						
I_{corr} (μA/cm²)	2.05	0.73	0.50	0.33	0.27	0.16
E_{corr} (V)	−0.605		−0.796	−0.499	−0.63	−0.577
C.R. (mm/year)	0.030	0.007	0.008	0.005	0.003	0.002
Al-6061+10% Al_2O_3						
I_{corr} (μA/cm²)	9.15	7.54	4.06	3.48	3.10	1.61
E_{corr}(V)	−0.593	−0.577	−0.745	−0.508	−0.513	−0.548
C.R. (mm/year)	0.099	0.0847	0.051	0.038	0.033	0.017
Al-6061+20% Al_2O_3						
I_{corr} (μA/cm²)	20.79	8.34	6.326	4.73	3.177	2.35
E_{corr}(V)	−0.55	−0.523	−0.691	−0.528	−0.475	−0.611
C.R. (mm/year)	0.226	0.091	0.067	0.061	0.034	0.025

Mg_2Si, the intermetallic part of the matrix alloy (6061 AA) being highly reactive toward chloride ion leading to the formation of MgO and SiO_2 of Al-Si-Mg alloy. Molybdate forming a resistive metastable compound of $Al_2(MoO_4)_3$ by the incorporation molybdate ions in the surface film and thus exhibit extensive passive region in the potentiodynamic polarization plot (Rosalbino et al. 2006).

$$Al_2O_3 + 3H_2O + 3MoO_2^- \rightarrow 2Al_2(MoO_4)_3 + 6OH^-; 2Al(OH)_3 + 3MoO_4^{2-} \rightarrow Al_2(MoO_4)_3 + 6OH^-$$

The anodic current density increases only after polarizing beyond this potential due to the breakdown of passivity. The extended passive region of potentiodynamic polarization curve is attributable to the formation of compact resistive coating, by incorporation of MoO_4^{2-} ions and metavanadate ions onto the porous oxide/hydroxide film and thereby securing protection to the Al composite material underneath (Das, Bandyopadhyay, and Blairs 1994). However, under a high positive applied field, the materials lose their transpassive character and the film gets disintegrated due to predominant chemical reaction producing complex anions.

The important criteria of Al alloy and its composites for inhibition of cathodic reactions are primarily the decreasing of the oxygen reduction process. Metavanadate adsorbs on the active sites of the surface and is reduced to form a monolayer of a V^{+4} species (Rosliza 2010). This layer is effective at reducing the activity of both cathodic sites and anodic sites in the matrix. The anodic inhibition is related to the initiation stage of localized corrosion and not propagation. Aluminum metavanadate may be formed under the condition of the reaction.

$$Al(OH)_3 + VO_3^- \rightarrow Al(VO_3)_3 + 6OH^-$$

Table 17.2 also represents the variation of corrosion current with respect to different samples. From the table, it is evident that the corrosion current density in the presence of pure water is of the order of 10^{-6} A/cm^2 for all the samples. In the presence of chloride ion, I_{corr} increases to the maximum level (order of 10^{-5} A/cm^2), whereas minimum values are observed in metavanadate-containing solution. This is due to the formation of resistive compounds of Al in the presence of molybdate and metavanadate ions in the solution. Moreover, with an increase in heterogeneity of the matrix, the I_{corr} values gradually increase so (Brett 1994).

17.4 ELECTROCHEMICAL IMPEDANCE SPECTROSCOPY

17.4.1 Nyquist Plot Analysis

The influence of different electrolytic media on the impedance spectra of the alloy, its 20% SiC_p, its 10% and 20% Al_2O_3-reinforced MMC is shown in Nyquist plots, Figure 17.2a–d. Here Z' is the real part and Z'' is the imaginary part of the impedance. In this figure, capacitive semicircles are obtained with different diameters for different electrolytes. It is well known that high-frequency capacitive semicircles are related to the dielectric properties and thickness of the barrier film and the low-frequency inductive loops are indicative of the specific adsorption of anions (Szklarska-Smialowska 1999). Incomplete inductive loops are often visible which suggest occurrence of localized corrosion in certain media. The semicircle radii were dependent on the inhibitor used and its concentration. In the present investigation, well-defined capacitive semicircle in HCl solution suggests that the corrosion process occurs under activation control (Rosalbino et al. 2006). Further, the incomplete inductive loop is likely to reflect specific adsorption of Cl^- and sequential localized corrosion by the attack of aggressive Cl^- ions into the barrier film (McCafferty 2003). The impedance parameters are presented in Table 17.3. The interface reaction resistance of the reinforced material was slightly lower than for the unreinforced matrix and the real part of the low-frequency region contracted, indicating the formation of an adsorptive film. These results appear to be in good accord with those obtained by electrochemical experiments. From the table, it is evident that

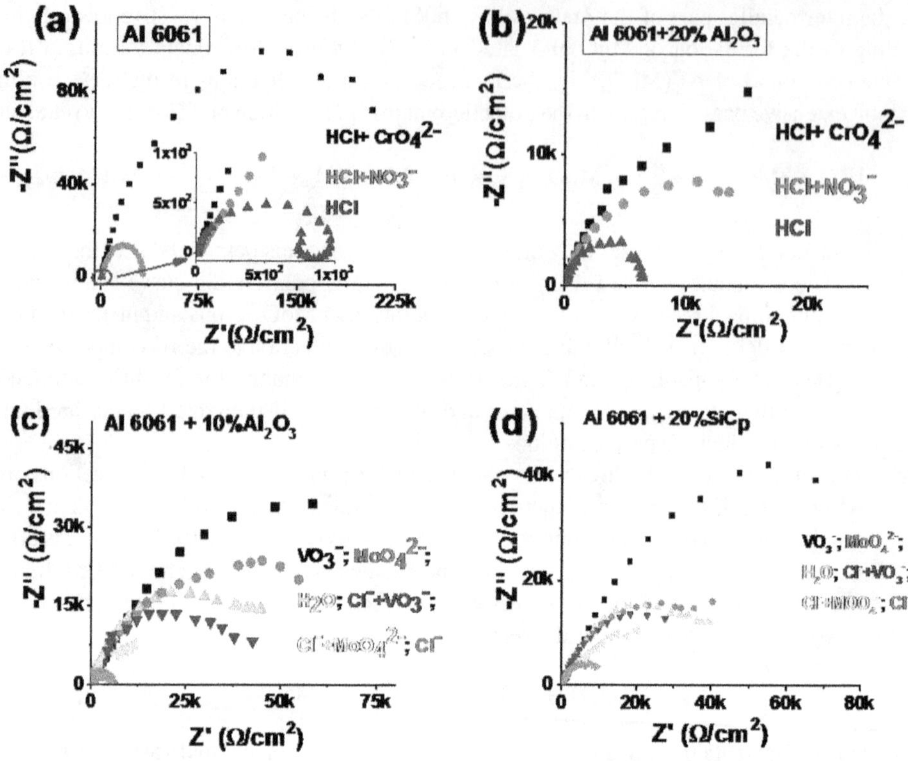

FIGURE 17.2 (a–d) Nyquist plot of different samples in the presence of various electrolytic media

TABLE 17.3
Impedance Parameters as Evaluated in HCl Media in the Presence of Different Inhibitors

Sample		0.01(M)HCl	0.01(M)HCl +0.01(M)$NaNO_3$	0.01(M)HCl +0.01(M)K2CrO_4	H2O
Al-6061 + 20% SiC_p	O.C.P	−0.650	−0.472	−0.554	−0.542
	R_p (ohm/cm²)	1.66	6.37	12.9	4.97
	CPE/F × 10^{-5}	3.99	3.61	2.91	2.6
	n	0.90	0.80	0.77	0.81
Al-6061	O.C.P	−0.646	−0.443	−0.544	−0.620
	R_p (ohm/cm²)	9.38	24.8	36.5	19.2
	CPE/F × 10^{-5}	1.69	1.02	0.81	1.20
	n	0.93	0.80	0.79	0.74
Al-6061 + 10% Al_2O_3	O.C.P	−0.626	−0.452	−0.565	−0.610
	R_p (ohm/cm²)	6.93	19.7	267	9.2
	CPE/F × 10^{-5}	3.08	1.19	1.03	2.23
	n	0.78	0.83	0.68	0.75
Al-6061 + 20% Al_2O_3	O.C.P	−0.652	−0.475	−0.563	−0.633
	R_p (ohm/cm²)	5.3	18.3	29.3	8.2
	CPE/F × 10^{-5}	3.97	1.43	0.94	1.99
	n	0.94	0.84	0.82	0.77

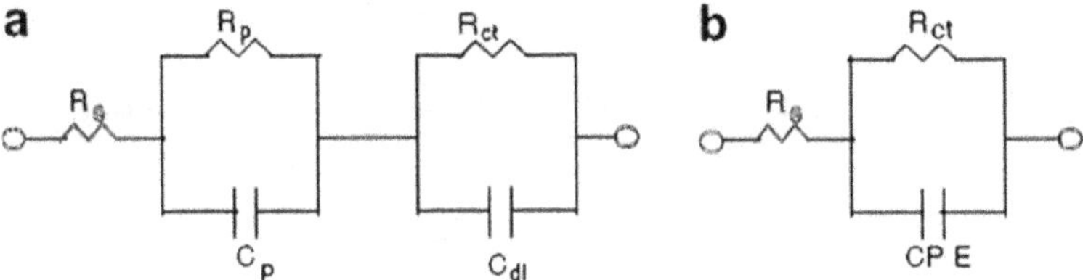

FIGURE 17.3 Equivalent circuit diagram – (a) EC diagram for the system in pure water; (b) EC diagram for the system in electrolytic media

$R_p(HCl+CrO_4^{2-})>R_p(HCl+NO_3^-)>R_p(H_2O)>R_p(Cl^-)$. Impedance spectra probably show the differences between the two inhibitor ions most clearly. The radius of the semicircle is much larger in the case of nitrate. By 0.6 V slightly above the pitting potential, the semicircle for the nitrate-containing solution is now smaller than before and begins to show evidence of film relaxation effects at low frequency. The resistance of the parallel RC combination is much smaller in the absence of nitrate ions and with just chloride ions present. At this applied potential chromate-containing solutions begin to show indications of a capacitance in series with parallel RC combination, which becomes more accentuated at more positive potentials occurs, the current fluctuates widely and it is not possible to record reproducible impedance spectra. The above impedance spectra have been analyzed based on a simple Randles equivalent circuit (EC) model as represented in Figure 17.3a and b. The circuit includes a solution, a solution resistance (R_s), a charge transfer or polarization resistance (R_p) and a constant phase element (CPE), which substitutes the double-layer capacitance for the film. CPE takes into account the phenomena related to the heterogeneous nature of surface and diffusion processes (Kahanda and Tomkiewicz 1990).

Table 17.4 summarizes the electrochemical impedance parameters for all of the materials in neutral solution containing different electrolytic environments. It is evident that for the 20% SiC_p-reinforced MMC, the R_p value for the Cl^--containing solution is 4.67×10^3 which is lower than that for the material in chloride- and molybdate-containing solutions (1.05×10^4) or chloride- and vanadate-containing solutions (2.40×10^4) or that for the pure water (4.26×10^4). The value of the R_p in the presence of pure molybdate-containing solution is found to be of the order of 5.38×10^4 which is again lower than that for pure vanadate-containing solution (1.17×10^5). A similar trend in the electrochemical impedance parameters has been observed for all of the materials as presented in Table 17.4. It can be concluded that in comparison with molybdate ion, vanadate acts as an effective inhibitor for the aluminum samples in neutral environments.

17.4.2 Analysis of the Bode Plots

Distinct changes are also observed in Bode plots in Figure 17.4a–d for different electrolytes. The maximum phase shift, max, for the ionic solutions is in the order of $HCl+CrO_4^{2-}>HCl+NO_3^->HCl$ which indicates departure from capacitive behavior toward approximately resistive behavior for the circuit in the same order. The inductive behavior in chloride solution caused by the localized process is apparent as the impedance reduces at low frequencies. Moreover, for Cl^--containing media, the extensions of phase angle values in the low-frequency region toward the negative Y-axis clearly support the specific adsorption of Cl^- ionic species on barrier oxide film with the consequent dissolution of the Al_2O_3 film. The difference between the high-frequency limit and low-frequency limit in the Bode plot is polarization resistance (R_p).

From the Bode plots, it has been observed that the maximum phase shift (max) in different media is in the order of $Cl^-<Cl^-+MoO_4^{2-}<Cl^-+VO_3^-<H_2O<MoO_4^{2-}<VO_3^-$. The inductive behavior in chloride solution caused by the localized process is apparent as the impedance decreases at low frequencies. The difference between the high-frequency limit and low-frequency limit in the Bode

TABLE 17.4
Impedance Parameters as Evaluated in Neutral Media in the Presence of Different Inhibitors

Solution	0.05 (M) NaCl	0.05(M)NaCl+0.01(M) Na_2MoO_4	0.05(M)NaCl+0.01(M) NH_4VO_3	H_2O	0.01(M) Na_2MoO_4	0.01(M) NH_4VO_3
Sample Al-6061+20% SiC_p						
R_p (ohm/cm²)	4.67	10.57	23.94	42.64	53.86	117.32
CPE/F×10^4	2.98	2.45	0.87	0.12	0.20	0.22
N	0.70	0.58	0.75	0.79	0.68	0.76
Sample Al-6061						
R_p (ohm/cm²)	1.28	4.43	5.51	10.65	22.67	31.63
CPE/F×10^5	2.13	2.11	2.44	1.18	1.08	0.66
n	0.82	0.68	0.5	0.75	0.79	0.77
Al-6061+10% Al_2O_3						
R_p (ohm/cm²)	8.55	24.96	43.64	52.03	71.97	196.77
CPE/F×10^5	3.19	2.86	1.58	1.76	2.21	1.70
n	0.81	0.64	0.73	0.76	0.67	0.53
Al-6061+20% Al_2O_3						
R_p (ohm/cm²)	3.68	8.70	14.53	22.22	29.34	66.38
CPE/F×10^4	1.18	0.39	0.73	0.62	0.43	0.73
n	0.64	0.74	0.67	0.62	0.62	0.65

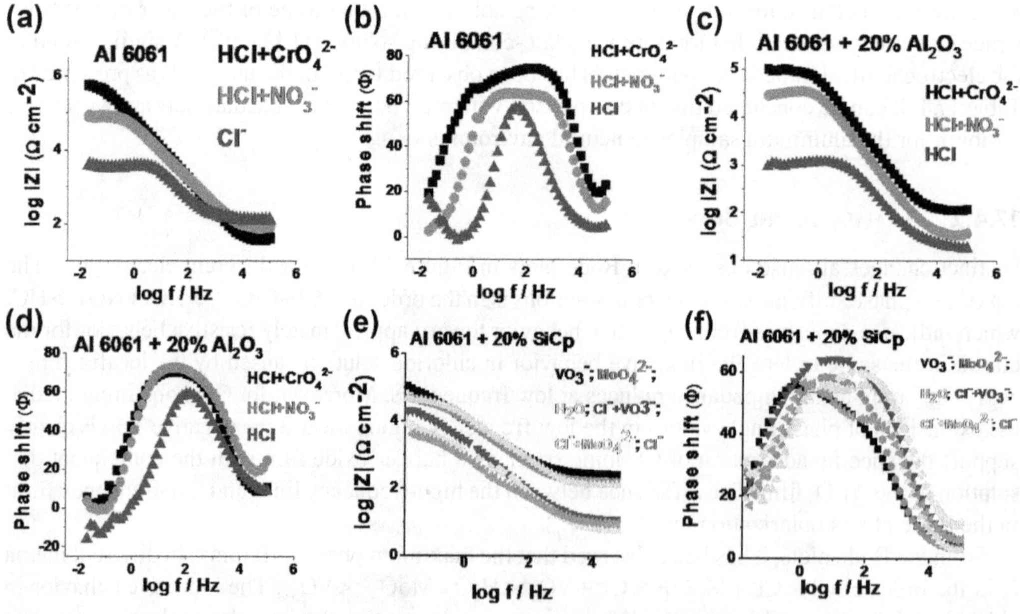

FIGURE 17.4 (a–f) Bode plots of different samples in various electrolytic media

plot is equal to R_p, the polarization resistance, which is associated with the dissolution and repassivation processes occurring at the interface as well as with the electronic conductivity of the film. R_p is significantly lowered in the case of Cl^- ion due to its penetration power through the barrier oxide film and direct contact with the matrix. However, when the VO_3^- and MoO_4^{2-} ions approach the surface, they seemingly get adsorbed in the voids of the oxide film through p-bonding with the metal oxide. The above impedance data have been analyzed based on the simple Randles EC model as represented in Figure 17.3a. The circuit includes a solution resistance (R_s), a charge transfer or polarization resistance (R_p) and a CPE, which substitutes the double-layer capacitance for the film. CPE takes into account the phenomena related to the heterogeneous nature of surface and diffusion processes. The R_p values in different electrolytes for the alloy materials are found to be in the order of $VO_3^- > MoO_4^{2-} > H_2O > Cl^- + VO_3^- > Cl^- + MoO_4^{2-} > Cl^-$.

17.5 THE MICROSCOPIC ANALYSIS OF THE CORRODED SURFACES THROUGH SCANNING ELECTRON MICROSCOPY (SEM) AND ATOMIC FORCE MICROSCOPY (AFM) ANALYSIS

The corroded surface microstructures were analyzed through scanning electron microscopy (SEM) and atomic force microscopy (AFM). The SEM images of the Al-6061 alloy and its reinforced MMC samples before and after corrosion are represented in Figure 17.5a–h and the corresponding AFM images for the alloy matrix are presented in Figure 17.6a–d. The surfaces of the as-received samples are free from any major cracks and pits and only a few polishing lines were seen on the surface of the samples. Extensive damage on the surface along with growth of large pits was visible on the surface of both the samples, when exposed to either HCl- or pure NaCl-containing solution. The extensive damage of the specimen indicates susceptibility of the alloy and the MMC material toward pitting corrosion, particularly in the presence of chloride medium. Chloride ions are known to be specifically adsorbed on the oxide films over the Al surface and thus can easily penetrate the barrier film and reacts with the metallic phase under the oxide layer (Frankel 1998). Moreover, as discussed earlier, Mg_2Si, the intermetallic part of the matrix alloy (Al-6061: Al-Si-Mg alloy) is highly reactive toward chloride ions in aqueous medium leading to the formation of MgO and SiO_2, thereby further enhances the aggressiveness of the environment toward the specimen surface. Addition of the inhibitors to the Cl^--containing environment imparts significant passivity, as shown in the potentiodynamic polarization, further demonstrated through the microscopic analysis. The surface of the materials does not undergo any significant pitting or cracks and remains almost intact with growth of only a small amount of products, as observed in the corresponding SEM images of the samples in CrO_4^{2-}-ion-containing environment.

17.6 CONCLUSION

This comprehensive analysis of the stability behavior of Al-6061 alloy and its 20% SiCp, 10% and 20% Al_2O_3-reinforced MMCs indicates that all of the materials suffer significant pitting corrosion in Cl^--containing environment. Addition of either nitrate or chromate to that solution restricts the damage of the samples. Nitrate ion protects the surface through formation of resistive metastable aluminum nitrates, whereas chromate significantly inhibits the cathodic reactions ($H^+ \rightarrow H_2$) occurring on the surface. Among these two ions, inhibition efficacy of chromate is higher in acidic environment. Addition of Al_2O_3 microspheres in the 20% Al_2O_3-reinforced MMC leads to the presence of more IMPs compared to other samples which makes the sample highly susceptible to localized form of corrosion. In neutral chloride environment, both molybdate and metavanadate reduce the corrosion rate through adsorption and protective film formation, respectively, on the oxide layer of the Al surface. The metavanadate ion can be considered as the effective inhibitor in chloride-containing solution and as an alternative to chromate.

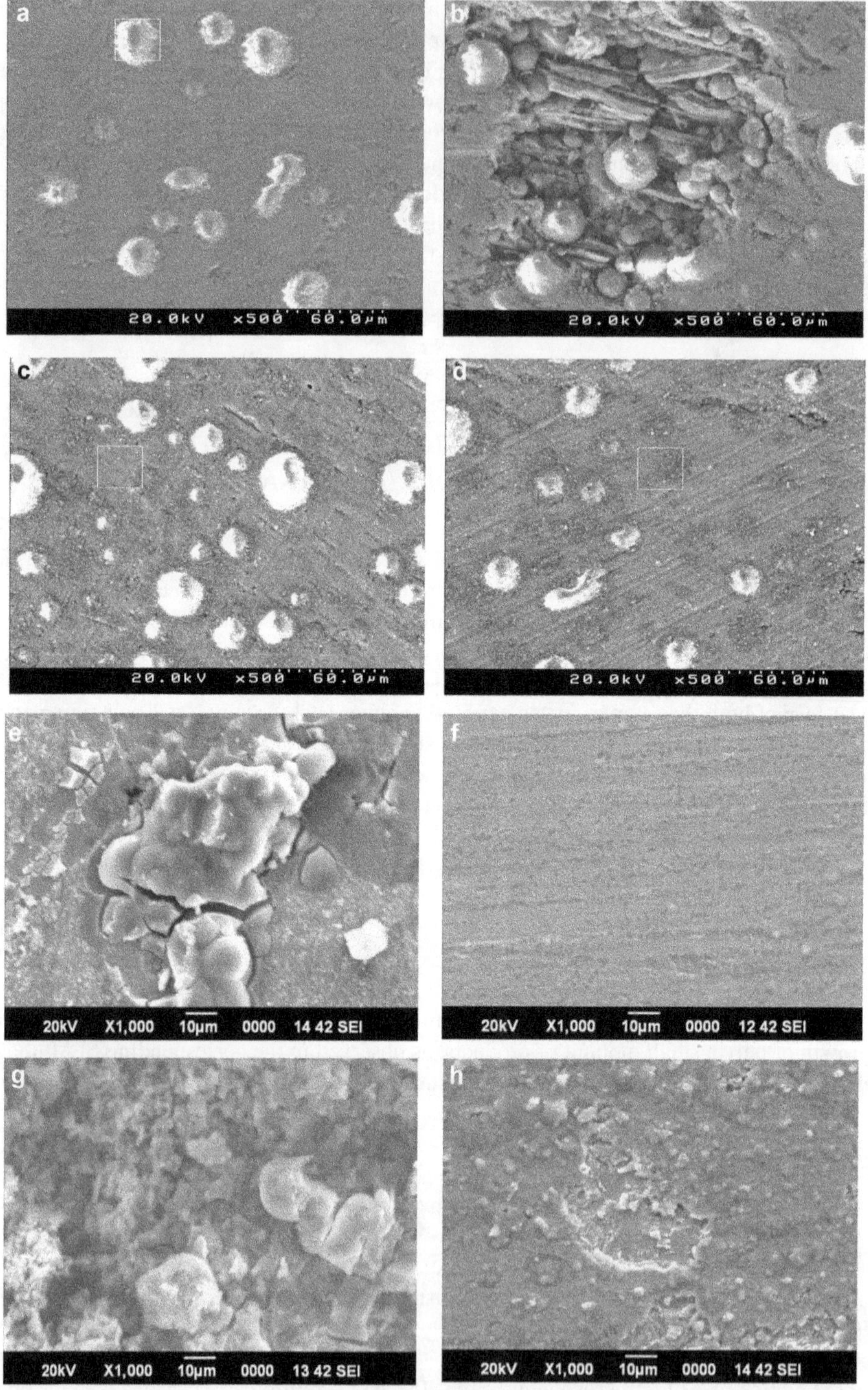

FIGURE 17.5 Scanning electron micrograph of the different samples under different environmental conditions: Al-6061-20% Al_2O_3-reinforced MMC – (a) noncorroded; polarized in (b) 0.1(M) HCl; (c) 0.01(M) HCl + 0.01(M) $NaNO_3$ and (d) 0.01(M) HCl + 0.01(M) K_2CrO_4; Al-6061 – polarized in (e) Cl^-; (f) $Cl^- + VO_3^-$; and Al-6061 + 20% SiCp – polarized in (g) Cl^-; (h) $Cl^- + MoO_4^-$ solution.

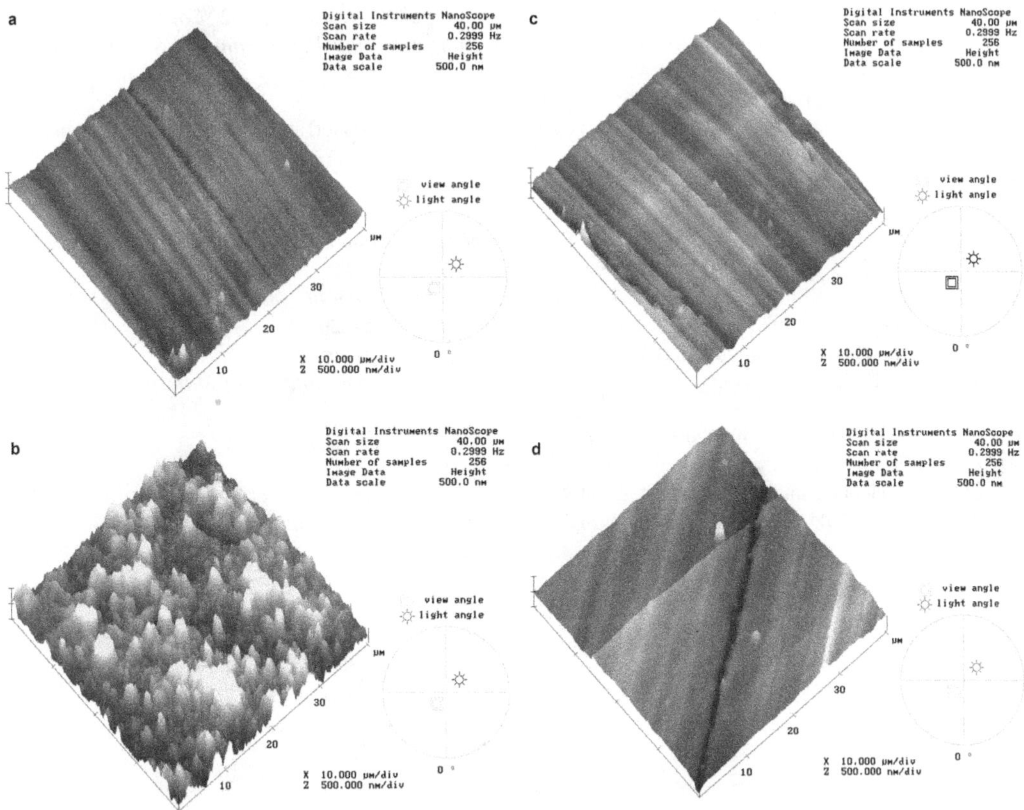

FIGURE 17.6 (a–d) Atomic force micrograph (3-D image) of the Al-6061 alloy sample – (a) noncorroded; polarized in (b) 0.1(M) HCl; (c) 0.01(M) HCl+0.01(M) NO_3^- and (d) 0.01(M) HCl+0.01(M) CrO_4^{2-}

REFERENCES

Ambat, R., A.J. Davenport, G.M. Scamans, and A. Afseth. 2006. *Corrosion Science* 48: 3455.

Baorong, H., Jinglei, Z., Yanxu, L., and Fangying, Y. 2001. *Study on Effect of Seawater Saliniyu on Electrochemical Performance of Al Anodes.* Department of Marine Corrosion, Institute of Oceanology, Chinese Academy of Sciences, China.

Bishop, J.E. and V.K. Kinar. 1995. *Mcfall-Metallurgical Transaction A* 26: 2773–82.

Brett, C.M.A., I.A.R. Gomes, and J.P.S. Martins. 1994. *Corrosion Science* 36: 915–23.

Cai, M., and G.J. Cheng. 2007. *The Journal of The Minerals, Metals & Materials Society* 59: 58–72.

Costa, M., and C.B. Klein. 2006. *Critical Reviews in Toxicology* 36: 155–63.

Da Costa, C.E., F. Velasco, and J.M. Toralba. 2000. *Review Metalurgia Madrid* 36: 179–92.

Danilidis, I., A.J. Davenport, and J.M. Sykes. 2007. *Corrosion Science* 49: 1981–91.

Das, T., S. Bandyopadhyay, and S. Blairs. 1994. *Journal of Materials Science* 29: 5680–88.

Despic, A., V. Parkhutik, J.O.M. Bockris, R.E. White, and B.E. Conway (Eds.). 1989. *Modern Aspects of Electrochemistry.* Plenum, New York, 20.

Frankel, G.S. 1998. *Journal of the Electrochemical Society* 145: 2186–98.

Hollingsworth, E.H., and H.Y. Hunsicker. 1987. Corrosion of aluminum and aluminum alloys. *ASM Metals Handbook* 13: 582–609.

Iannuzzi, M. 2010. Doctoral Thesis, Fontana Corrosion Center, The Ohio State University, Columbus, OH.

Kahanda GL, Tomkiewicz M. 1990. Fractality and impedance of electrochemically grown silver deposits. *Journal of the Electrochemical Society* 137(11): 3423.

Kitowski, C.J. 1993. *An Investigation of the Effect of Chloride on Reinforcing Steel Exposed to Simulated Concrete Solutions.* M.S. Thesis, Department of Mechanical Engineering, University of Texas, Austin, TX.

Koch, G.H., M.P.H. Brongers, N.G. Thompson, Y.P. Virmani, and J.H. Payer. 2002. *Corrosion Costs and Preventive Strategies in the United States. Supplement to Materials Performance*, U.S. Department of Transportation, Federal Highway Administration.

McCafferty E. 2003. *Corrosion Science* 45: 1421-1438.

Miller, W.S., L. Zhuang, J. Bottema, A.J. Witterbrood, P. DeSmet, A. Haszler, and A. Vieregg. 2000. *Materials Science and Engineering A* 280, 37–49.

Pardo, A., M.C. Merino, S. Merino, F. Viejo, M. Carboneras, and R. Arrabal. 2005. *Corrosion Science* 47: 1750–64.

Pardo, A., M.C. Merino, S. Merino, M.D. Lopez, F. Viejo, and M. Carboneras. 2003. *Materials and Corrossion* 54: 311–317.

Rosalbino, F., E. Angelini, S. De Negri, A. Saccone, and S. Delfino. 2006. *Intermetallics* 14: 1487–1492.

Rosliza, R., and W.B. Wan Nik. 2010. *Current Applied Physics* 10: 221–229.

Seah, K.H.W., S.C. Sharma, and M. Krishna. 2006. *Journal of ASTM International* 3: 12394.

Staab, T.E.M., R. Krause-Rehberg, U. Hornauer, and E. Zschech. 2001. *Journal of Materials Science* 41: 1059–66.

Szklarska-Smialowska, Z. 1999. *Corrosion Science* 41: 1743–67.

Uzuo, Y., H. Wang, J. Zhao, and J. Xing. 2002. *Corrosion Science* 44: 13–24.

Vargel, C., M. Jacques, and M.P. Schmidt. 2004. *Corrosion of Aluminium*, Elsevier Ltd.

Winston Revie, R. 2011. *Uhlig's Corrosion Handbook*, Chapter 40, 2nd ed. ISBN: 0-471.

18 Toughening of Polymers

M. Sreejith, G. Santhosh, and R. S. Rajeev
Vikram Sarabhai Space Centre

CONTENTS

18.1 INTRODUCTION

Brittleness is always a problem needed to be addressed in polymer technology. Ebonite, which is produced by adding a large amount of sulphur in natural rubber (polyisoprene), is a rigid material whose toughness increases when the amount of sulphur is reduced. The decrease in crosslink density made the polymer tougher resulting in better impact strength. The first synthetic polymer made tough is cellulose nitrate, where Hyatt and Hyatt (1870) improved the toughness by adding camphor as a plasticizer. Nature also plays a role in making a polymer tough. Casein formaldehyde, a plastic formed by the condensation reaction between casein and formaldehyde, is brittle but will become tough under high humidity (Seymour 1989). Polyester resins such as Glyptal were made tough by incorporating unsaturated vegetable oils (Kienle and Hovey 1929).

In composites, rubber-toughened polymers are used as the matrix resin for fabricating structural components, which are thereby tough to absorb impact energy. Many technologies associated with commercial products, automobiles, and aerospace require ductile components for impact and crack resistance, where toughening is the method by which the targeted properties are achieved. The research on toughening of polymers is a multidisciplinary program involving polymer science and technology, rheology, physics, material science, and engineering. Modelling to predict the properties of the toughened polymers is adopted by many researchers.

Acrylonitrile butadiene styrene (ABS) plastic is one of the earliest rubber-toughened engineering polymers known to have a combination of strength, rigidity, and toughness. However, their heat deflection temperature and solvent resistance are not attractive for many industrial applications. This shows that toughness alone will not qualify a material for many commercial applications. The need for toughening the polymers also arises when the polymer is formulated for other desirable properties, which made them highly brittle, for example, crosslinking. Epoxy resin, which is brittle due to the nature of the crosslinks, can be made tough by adding rubber or blending with other polymers. The classic example of the commercialization of a toughened polymer is high-impact polystyrene (HIPS). One of the early reports of toughening polystyrene is by Iwan (1927) in a patent publication. All polymer composites, which are reinforced by short fibre, continuous fibre,

DOI: 10.1201/9780429330575-18

particulate filler, or nanofiller, require toughening for many practical applications, including those in automotive, construction, and aerospace. Though the improvement in mechanical properties of composites by introducing suitable fillers is studied in detail, nanomodification of the composite matrix has gained much attention in the last decade. Nanomodification also leads to improvement in toughness of the matrix.

18.2 TOUGHENING OF POLYMERS

Thermoplastics and thermosetting polymers are two types of plastic materials that go through different manufacturing processes and have varying qualities based on the constituent elements and manufacturing methods. A thermoplastic is a polymer that remains solid at room temperature and softens upon heating. The melting of the thermoplastic is due to the crystalline re-arrangement of the chains. Thermoplastics are processed into any shape as per the geometry of the mould. The processing techniques include injection moulding, extrusion moulding, blow moulding, and rotational moulding. Once the molten plastic solidifies upon cooling into the desired shape, the shaped article is taken out of the mould. Thermoplastics are known for their recyclability and their ability to undergo reprocessing by reheating. Thus, one can process these materials multiple times. Examples of some commonly used thermoplastics in our daily life are polyethylene, polypropylene, polyethylene terephthalate, polycarbonate (PC), and polyvinyl chloride (PVC). The major disadvantage of thermoplastics is their inability to withstand extremely high temperatures. Thermoplastics generally possess high strength, are flexible and resistant to shrinking, though this varies depending on the type of polymer utilized.

On the other hand, a thermosetting resin is a liquid material that hardens irreversibly when heated (Nicholson and Nolan 1983). When a thermoset is poured into a mould and heated, it solidifies into the desired shape and the process also causes the formation of crosslinks, which hold the molecules in place and modify the material in such a way as to prevent melting. As a result, unlike thermoplastics, thermosets cannot be reprocessed. Thermosets disintegrate without entering a fluid phase when they are overheated. Compression moulding, resin transfer moulding, pultrusion, hand lay-up, and filament winding are the standard processing techniques for thermosetting polymers. Epoxy, cyanate ester, polyurethane, phenolics, and some polyimides are the common thermosetting polymers necessary for composite fabrication. Thermosets generally yield higher chemical and heat resistance and a more robust structure that does not deform easily. Depending on the desired output, curing agents, inhibitors, hardeners, or plasticizers are added to the resin along with reinforcement or fillers.

Toughness, or resistance to crack propagation, is one of the most essential requirements for materials used for load-bearing applications. Because of their high stiffness and strength, thermoset polymers find application as the matrix in structural composites, but their low toughness, especially in the presence of sharp notches, limits their application. Rubber particles have been used to increase the toughness of thermoset resins (Robinette, Ziaee, and Palmese 2004). Rubber-toughened thermosets can have a significant increase in toughness, but not all thermosets are toughened by rubber. Rubber toughening is more critical for thermosets that are lightly crosslinked compared to those having high crosslink density. The introduction of any toughness mechanism compromises the strength and stiffness requirement of the structural composites. Furthermore, the elastomeric phase raises the mixture's melt viscosity, thereby lowering processability. As a result, other toughness methods need to be explored for toughening thermoset polymers.

The total energy necessary to cause a failure, i.e., the total area under the stress–strain curve, is referred to as a specimen's toughness. Toughening the resin by adding appropriate toughening agents or chemical modification is one way to increase its energy absorption capability. This improves the impact resistance, elongation, and crack propagation resistance in addition to toughness. Because engineering polymers require high strength and modulus, the toughening procedure must not cause these qualities to deteriorate significantly.

18.3 TOUGHENING OF THERMOSETS

Toughening of thermosets has been the subject of extensive investigation. Imparting higher glass transition temperature by incorporating high-performance thermoplastics into thermosets has proven to be the most effective strategy for toughening high-performance thermosets to date. This process boosts fracture toughness with slight reduction in other desired properties.

The challenge of toughening thermoset resins is an expanding polymer science area. Mainly two types of strategies are adopted here. The first strategy involves decreasing the crosslink density of the resin matrix because the higher the crosslink density, the more brittle the material is (Van der Sanden and Meijer 1993). However, because the glass transition temperature and modulus tend to drop, this is not always followed. The second and possibly most extensively used technique involves altering the matrix chemistry or composition by introducing a second phase that precipitates during cure, resulting in morphologies capable of triggering a range of toughening mechanisms. The usage of reactive rubber additives has dominated this multiphase toughening of thermoset resins over the years. Although the rubbers are excellent at toughening these systems, they also cause a loss of critical mechanical properties like modulus and glass transition temperature (T_g). Thermoplastic modifiers have recently been used to overcome these issues, especially in thermoset materials with high crosslink densities (Hodgkin, Simon, and Varley 1998). Thermoplastics were initially used because of their high ductility, high modulus, and glass transition temperatures and thus did not compromise the desirable properties of thermosets, particularly the epoxy resin. The investigations on toughening of a majority of epoxies with thermoplastic toughening investigations have focused on the impact of different factors on material toughness, such as molecular weight, functionality, epoxy ductility, the chemistry of thermoplastic backbone and morphology.

While many of these additives efficiently toughen the epoxy resins, they also cause a reduction in other desirable mechanical properties, including modulus and tensile strength. Toughening with engineering thermoplastics has attracted a lot of interest in recent years because it does not seem to affect other mechanical properties as significantly. There have been several types of research utilizing various thermoplastic types and several reviews on the subject in the last decade following the initial work on thermoplastic toughening of epoxy resins.

The addition of even modest amounts of high molecular weight thermoplastic results in a significant increase in the prepolymer viscosity, which is yet to be overcome when this strategy is followed. Processing operations with the blended resin becomes difficult, and subsequently, the associated properties will not be able to be achieved. The addition of low molecular weight thermoplastics is a good choice in such cases. The molecular weight of these low molecular weight polymers plays a major role in improving the toughness of thermoset polymers. There was little or no increase in fracture toughness when oligomers with a molecular weight of 3,000–5,000 g/mol were incorporated. Previously, it was discovered that a low molecular weight linear imide thermoplastic (1,000 g/mol) could be integrated into bismaleimides (BMIs) and epoxy to produce extremely low-viscosity prepolymers that could be cured to produce durable epoxy and BMI thermosets (Gopala, Wu, and Heiden 1998). The hardness of the resulting thermoset improved significantly due to toughening to the extent of 75% for BMIs and up to 220% for epoxy polymers.

Several researchers have looked into the modification of epoxy resins with polyether sulphones (PESs) (Mimura, Ito, and Fujioka 2000). A mixture of epoxy and PES gives a low critical solution temperature. Incorporating higher PES content or curing at more elevated temperatures enhances the phase separation between PESs and epoxy matrix. This will continue till the T_g exceeds the curing temperature where the morphology is frozen. The final morphology is decided by the competition between the increase of molecular weight of epoxy leading to phase separation and the suppression of the phase separation due to crosslinking. Thus, moulding temperature plays a vital role in controlling the phase morphology of PES-modified epoxy resin. Below 140°C, a homogeneous morphology is obtained, whereas a phase-separated morphology is observed above 160°C. The morphologies of the epoxy resins, cured at 140°C and 180°C, are shown in Figure 18.1.

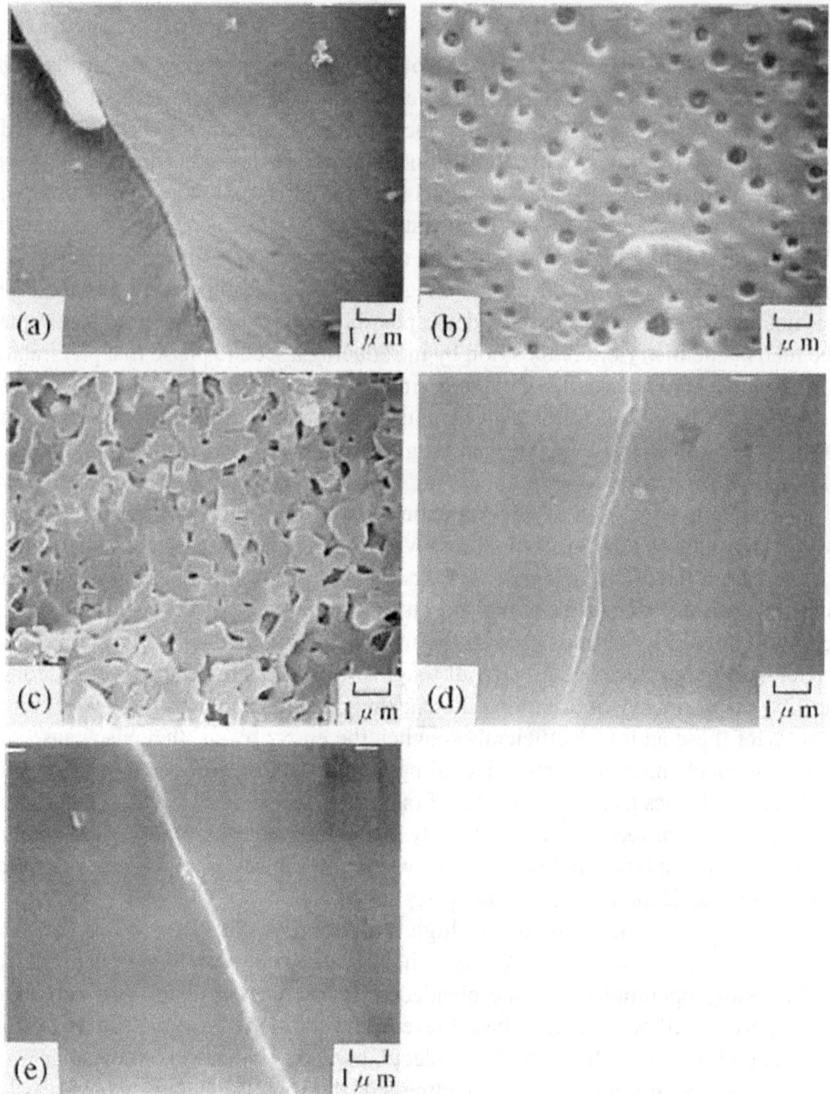

FIGURE 18.1 Fracture surface morphologies of epoxy resin modified with PES, cured at two different temperatures (Reproduced with permission from Mimura, Ito and Fujioka (2000) © 2000 Elsevier Science Ltd.)

The phase separation morphologies are evident in epoxy resin cured at 180°C (Figure 18.1b and c) compared to those cured at 140°C (Figure 18.1d and e). As the PES content increases, the morphology changes from a well-defined dispersed phase (Figure 18.1b) to a co-continuous phase (Figure 18.1c). This change in morphology for the same moulding temperature is attributed to the increase in viscosity with an increase in the concentration of PES. It is generally agreed that such materials show only minor improvements in impact properties compared to unmodified epoxy resins, as long as phase separation and good interfacial adhesion (Goossens and Groeninckx 2007) are maintained.

The toughness of thermoset resins is altered in several ways, mainly under four categories: (i) elastomer modification, (ii) particle modification, (iii) thermoplastic modification, and (iv) other approaches. Incorporating a second elastomeric phase into the glassy epoxy matrix via in situ phase separation has proven to be the most effective toughening approach.

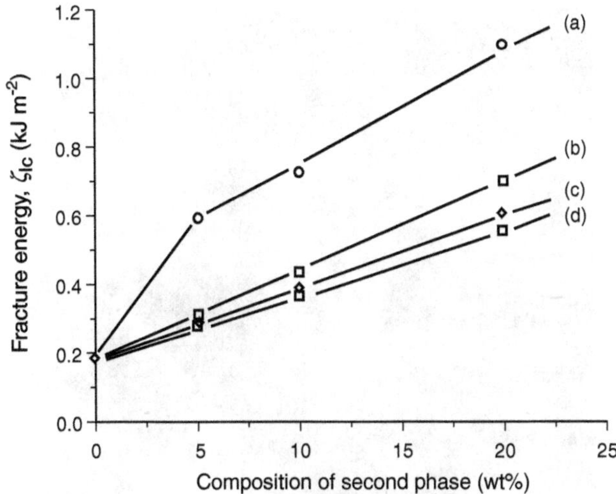

FIGURE 18.2 Fracture energy for initiation for (a) PBT, (b) nylon 6, (c) CTBN, and (d) PVDF (Reproduced with permission from Kim and Robertson (1992) © 1992 Chapman and Hall)

The toughening process using three crystalline polymers of a strongly crosslinked epoxy by particle inclusions, which cannot be toughened otherwise appreciably by the addition of rubber, has been reported in the literature (Kim and Robertson 1992). Poly(butylene terephthalate) (PBT), nylon 6, and poly(vinylidene fluoride) (PVDF) were the three crystalline polymers investigated. With the addition of nylon 6 or PBT to the epoxy, there was no loss of Young's modulus or yield strength, while with the addition of PVDF, there was a minor loss of these properties than with the addition of rubber as a toughening system. Nylon 6 and PVDF toughened the epoxy to a similar extent as in carboxyl-terminated polybutadiene (CTBN) rubber, as seen in Figure 18.2. On the other hand, PBT was found to toughen epoxy nearly twice as compared to other polymers. Toughness improvement is due to a combination of processes and is presumed to be governed by the mechanism termed as "phase transformation toughening."

The fracture surface morphology of one of the typical PBT-epoxy blends is given in Figure 18.3. The blend is processed by mixing 15-μm-sized PBT particles with epoxy resin at room temperature without adding any curing agent. The mixture is then heated to 220°C at a heating rate of 10°C/minute and suddenly cooled to room temperature. The curing agent is then added, and the mixture is cured followed by post-curing. This study was undertaken to understand the phase transformations in PBT-epoxy blend under different process conditions. With the process condition described above, the PBT phase is spherical with a diameter of 8–10 μm. Irrespective of the concentration or the initial particle size (before mixing), the particle size of PBT remains the same in the blend if the above process method is adopted.

18.4 RUBBER TOUGHENING OF THERMOSETS

Poly(siloxane)s, fluoro elastomers, acrylated elastomers, and reactive butadiene–acrylonitrile solid and liquid rubbers are among the elastomeric materials that have been investigated for rubber modification of epoxy resins. Toughening can be accomplished by incorporating reactive liquid polymers or incorporating a small amount of elastomer as a discrete phase. A certain ratio of epoxy resin, curing agent, and curing accelerator is combined at room temperature. To this liquid rubber is added, which is gently stirred until the mixture is uniform and clear. The curing time and temperature differ depending on the system, and that can even be improvised. Heating is usually done in two steps to achieve complete crosslinking, i.e., pre-cure heating followed by post-cure heating. This

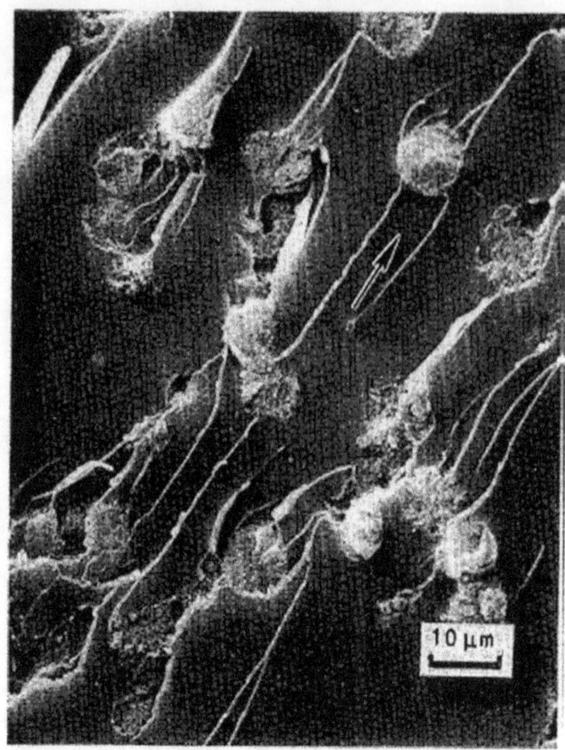

FIGURE 18.3 Scanning electron micrograph of the fracture surface of the PBT-epoxy blend. Arrow shows the direction of crack propagation (Reproduced with permission from Kim and Robertson (1992) © 1992 Chapman and Hall)

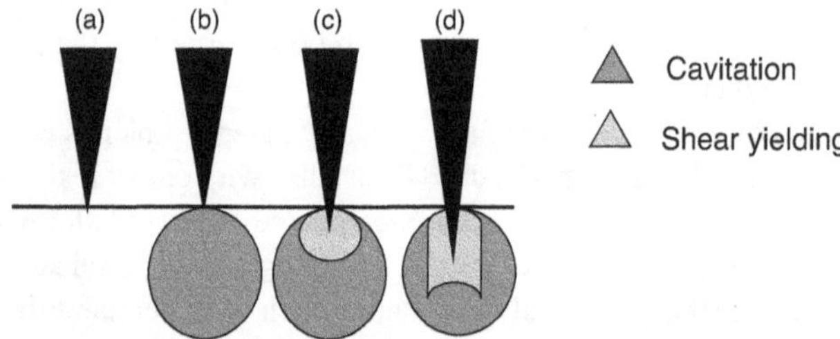

FIGURE 18.4 Schematic of toughening mechanism of core-shell-modified epoxy resin system. From (a)–(d), initiation of crack to shear yielding through the formation of cavitation and initial yielding.

method produces an epoxy–rubber adduct that behaves like epoxy-terminated rubber, which is more compatible with the epoxy matrix. Solid rubber can be blended with resin and hardener to make a solution (Kim and Datta 2013). The mixture can be held under vacuum to allow for simultaneous solvent evaporation and curing further.

The toughening mechanism of a core-shell rubber-modified epoxy resin system is schematically shown in Figure 18.4.

The major toughening mechanism shown in Figure 18.4 is cavitation of rubber particles followed by shear yielding of the matrix. The rubber particle at the crack tip elongates to the same extent as that of the matrix, as is evidenced by microscopy. A thick specimen and a sharp crack cause an increase in the strain constraint, which is relieved in the presence of rubber particles. If the toughening agent can alter the crack-tip stress state from the one that favours brittle fracture to the one that favours shear yielding, it is possible to toughen the highly crosslinked thermosets.

Most engineering thermoset resins have a toughening agent formed by modifying the polymer chain with reactive terminal groups. An example of a rubber having a low molecular weight backbone and reactive groups at the terminal locations is acrylonitrile butadiene rubber (nitrile rubber). Rubbers having reactive groups suspended from the main polymer chain have been thoroughly investigated for toughening properties. Studies on toughening with reactive functional groups and butadiene polymers have been published in open literature. The rubber part of the system is first mixed with the thermoset resin and a suitable hardener in the rubber modification procedure. The rubber creates a copolymer with the resin during the curing procedure, and then, the phase separation occurs. As a result, a distributed rubbery phase exists in the cured thermoset. In terms of terminal functionality, the type of rubber modifier chosen will be determined by the chemical properties of the matrix polymer. Epoxy, phenyl, vinyl, hydroxyl, mercaptan, amine, and carboxyl are a few of the functional groups that have been explored.

However, elastomers having carboxyl activity, for example, CTBN, have been shown to provide the best results with respect to toughness. During cure, the rubbery domains precipitate in situ, resulting in toughened epoxy composites. The use of elastomers to improve the toughness of epoxy matrix necessitates a reaction between the elastomer and the resin, which results in an appropriate binding between elastomeric and epoxy phases. Attempts to use hydroxyl-terminated elastomers to improve the toughness have run into problems because the conditions required to promote the required hydroxyl–epoxy reaction often result in self-polymerization of the epoxy, with the latter occurring at the expense of the former, limiting the extent of the elastomer–epoxy reaction (Bussi and Ishida 1994). A contemporary solution to this problem is to use a curing agent like toluene diisocyanate, which can react with both epoxide and hydroxyl functionalities, forming urethane and oxazolidone groups between elastomeric and epoxide components. The epoxy–rubber contact can also be improved with silane coupling agents (Kaynak, Celikbilek, and Akovali 2003). The fracture surface morphology of silane coupling agent-treated rubber particle-incorporated epoxy resin is shown in Figure 18.5 where the rubber particles are from recycled automobile tires, which are ground and incorporated into the resin matrix after surface treatment.

In case of rubber-toughened epoxy resins, the particle size and particle size distribution play a significant role in imparting effective toughness, especially in some rubber-toughened commercial epoxy formulations that make a conscious effort to create bimodal particle size morphology. Levita et al. (1996) demonstrated the toughening effect of bisphenol A in rubber-modified epoxy systems.

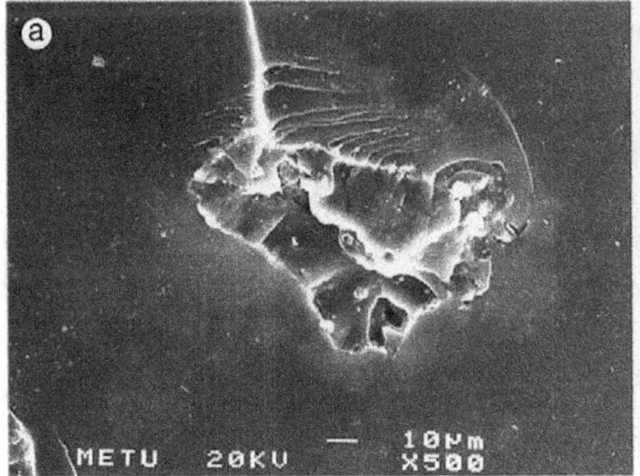

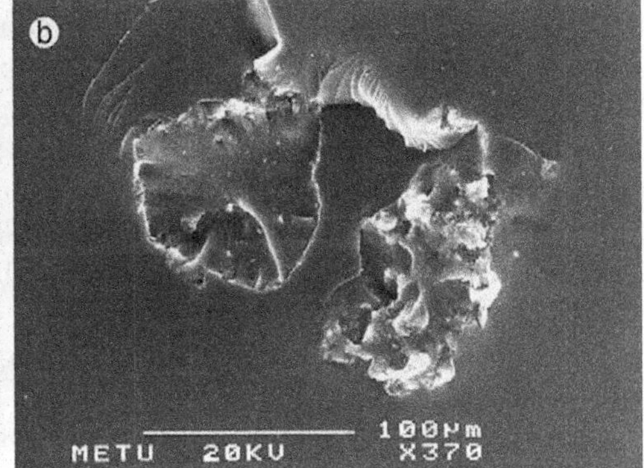

FIGURE 18.5 Fracture surface morphology of surface-treated rubber particle-incorporated epoxy resin matrix. (a) Triaminopropyl triethoxysilane-treated and (b) trimethyl silylpropyl methacrylate-treated rubber particle (Reproduced with permission from Kaynak, Celikbilek, and Akovali (2003) © 2003 Elsevier Science Ltd.)

The CTBN–diglycidyl ether of bisphenol A (DGEBA) liquid epoxy resin–piperidine exhibited a bimodal dispersion of rubber particles (Akbari, Beheshty, and Shervin 2013). The addition of bisphenol A to rubber-modified epoxy resin resulted in a significant increase in toughness, which was linked to a bimodal particle size distribution. Two main toughening mechanisms induced by rubber particles in epoxy matrix are the debonding or internal cavitation with subsequent plastic growth of voids and the shear bands running between rubber particles due to localized shear deformation (Pearson and Yee 1991).

18.5 OTHER TOUGHENING METHODS OF THERMOSETS

The addition of particle fillers like silica and alumina trihydrate to crosslinked thermosets can improve the toughness. Particulates contribute to a considerable increase in modulus compared to elastomeric modification, which results in a decrease in modulus. The volume fraction and the particle size and shape of the filler were found to influence the degree of toughness enhancement (Dittanet and Pearson 2013). In the case of filled polymers, it has been shown that fracture energy peaks at a given volume fraction of the added particles, thus lowering the elongation at break and impact resistance. Particulate toughening works in a somewhat different way than elastomeric modification. Although particle reinforcement has improved toughness slightly, it has not attained the same level of improvement as elastomeric modification. Particulate modification has a significant modulus increase that goes hand in hand with toughness improvement (Chikhi, Fellahi, and Bakar 2002). Several researchers have looked into the possibility of merging particle and rubber modification of epoxies.

Thermoplastic modification of thermosets has mostly been attempted with PES and polyetherimide. Bucknall and Partridge (1983) have investigated the link between the structure and mechanical properties of PES-modified epoxies using trifunctional epoxy and tetraglycidyl diamino diphenyl methane. They noticed that the morphology was influenced by the resin type and concentration as well as the curing agent used. Unfortunately, adding PES to epoxy only provided minimal benefits in fracture energy; the addition of up to 40 phr of resin to epoxy has resulted in a fracture energy gain of less than 100%. The level of toughness improvement achieved by thermoplastic modification is often inferior to that produced by rubber modification, particularly with low-T_g systems based on difunctional resins. Polyphenylene oxide was employed by Pearson and Yee (1990) as an epoxy resin modifier. Epoxy resin has also been toughened with polyether esters. The fracture toughness of high-performance epoxy treated with poly(etherimide) was shown to increase (Shin and Jang 1997). A detailed analysis of epoxy that facilitates decomposition by blending with thermoplastics has been carried out. A significant improvement is seen in the thermal and mechanical performance of a toughened DGEBA– diaminodiphenyl sulfone (DDS) system incorporating a tetrafunctional epoxy and thermoplastic (Unnikrishnan and Thachil 2006).

Epoxidized soybean oil has been used to toughen epoxy resins that have been cured at room temperature. Hu et al. (2021) used epoxidized soybean oil, which was modified using fatty acids to improve the toughness of epoxy resins. Here, saturated fatty acids from renewable resources were first grafted onto epoxidized soybean oil. This biorubber is then blended with DGEBA and cured using an amine hardener, and it was found that the fracture toughness was significantly improved for the modified epoxy resin (Figure 18.6).

The reinforcement of DGEBF by the liquid crystalline diglycidyl ether of 4,4-dihydroxybiphenol has been widely researched with relevance to the strengthening of epoxy resins utilizing liquid crystalline epoxies. To increase the toughness of DGEBA and diamino diphenyl methane epoxies, polyesters produced by direct polycondensation from bisphenol A and aliphatic dicarboxylic acids (adipic acid, sebacic acid, and dodecanoic acid) have been utilized. Cresol novolacs catalyzed by triphenyl phosphine have been used to modify epoxies. Tensile, flexural, thermal, and hardness properties have been demonstrated for blends of methylene dianiline (MDA)–cured DGEBA epoxy oligomer and ethylene–vinyl acetate copolymer rubber. The impact and shear strengths of

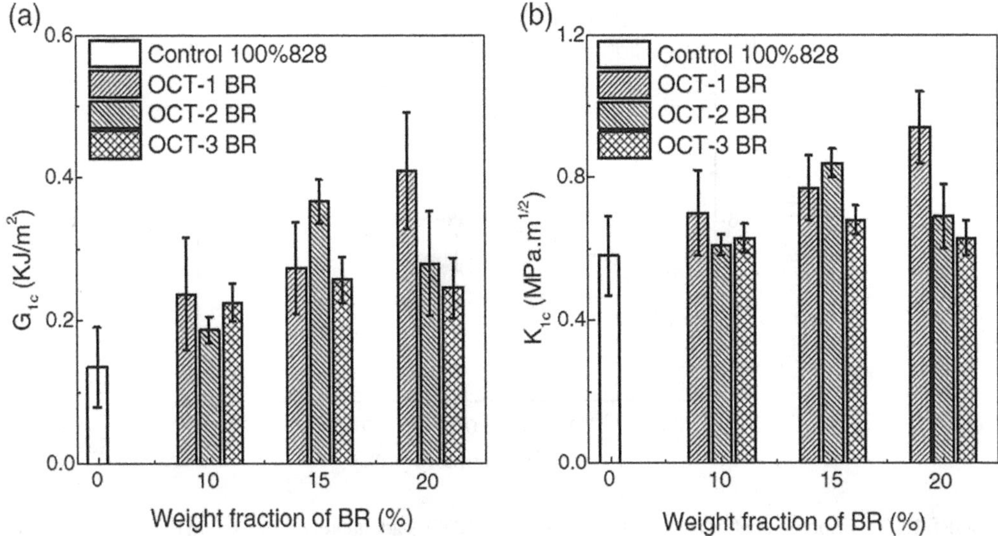

FIGURE 18.6 Fracture toughness improvement demonstrated in epoxy resin modified using fatty acid-grafted epoxidized soybean oil (Reproduced with permission from Hu et al. (2021) © 2021 Wiley Periodicals LLC)

the modified epoxy resin E 44 were apparently increased after a succession of poly(butyl acrylate)–poly(methyl methacrylate) core-shell elastomer particles were utilized as toughening modifiers (Hu et al. 2021). MDA-cured epoxy resin toughened with acrylate-based liquid rubbers has been examined in terms of impact characteristics and morphology. Park et al. (2003) established the role of electron beam radiation in the toughening process of CTBN-modified epoxy. The interpenetrating network (IPN) has been widely explored for blending two thermosets. Using meta-(xylene diamine) and benzoyl peroxide curing agents, toughened IPN materials based on unsaturated polyester and DGEBA-type resin have been created. It has also been claimed that combining thermotropic hydroxy ethyl cellulose with epoxy resin can toughen the thermoset polymer systems.

The IPN grafting of a polyurethane prepolymer based on hydroxyl-terminated polyesters to improve the toughness of epoxies resulted in enhanced mechanical capabilities (Harani, Fellahi, and Bakar 1998). Due to an effective molecular weight build-up by a chain-extension process, aliphatic hydroxyl polyester increased fracture toughness more than amine- or acid-terminated polyester. Hydroxyl-terminated polyesters are regarded as a subset of polyols that have been proven to improve the impact strength and fracture toughness of epoxy. PC polyurethane has also been used as an epoxy modifier. There has been much research towards toughening epoxies with polyurethane as a second phase to make IPNs.

Aromatic polyesters and hydroxyl-terminated polyesters have been employed as epoxy resin modifiers. Investigations on chain-extended ureas as both curing agents and toughness modifiers for epoxies have been carried out with reasonable success. Epoxy resins have been modified with plastisols based on PVC and diethyl hexyl phthalate (Lopez et al. 1998). Polybutadiene with an isocyanate terminal has also been used as a toughness modifier in epoxy resins.

Recent studies have been reported on the use of nanoparticle- and nanoplatelet-based toughening of thermosets. Tang et al. (2013) dispersed reduced graphene oxide (RGO) in epoxy matrix by first dispersing RGO in ethanol solution by sonication followed by subjecting the RGO/ethanol mix in the epoxy resin by sonication. Planetary ball milling is carried out afterwards, where balls of different sizes were used to ensure that the RGO sheets are exfoliated to the maximum in epoxy resin. They could achieve dispersion by strictly controlling the process parameters such as rpm of ball mill. The ethanol was then removed under a prolonged vacuum, and the resin/RGO mix was cured

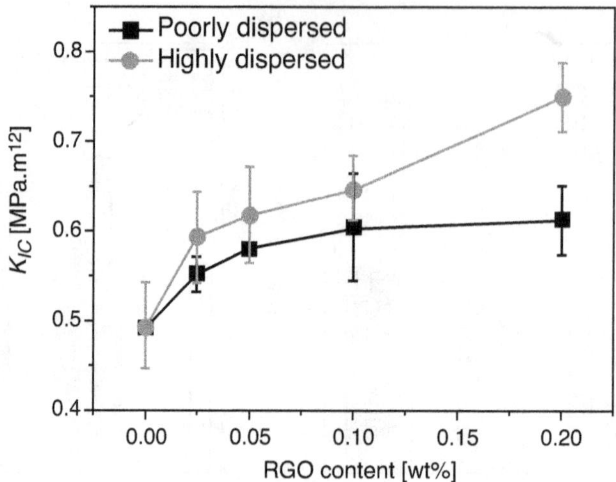

FIGURE 18.7 Improvement of fracture toughness when RGO is adequately dispersed in the epoxy matrix by controlling the process parameters (Reproduced with permission from Tang et al. (2013) © 2013 Elsevier Ltd.)

using a suitable hardener. To prove that proper dispersion of RGO in epoxy is the key for improving the toughness of the latter, they prepared controlled coupons where no ball milling was used for dispersing RGO in epoxy. From Figure 18.7, it is evident that proper dispersion of RGO in epoxy improves the fracture toughness of cured epoxy. The analysis of the fractured surface revealed a rougher surface for composite containing properly dispersed RGO compared to that having poorly dispersed RGO.

18.6 MECHANISM OF TOUGHENING

The toughness increase in rubber-modified epoxy resins have been attributed to a number of causes. These methods explain why a thermosetting polymer with a multiphase microstructure of scattered rubber particles may have better fracture energy or fracture toughness. It is essential to consider each of these mechanisms and evaluate how well they work. One of the most effective mechanisms proposed to understand the toughening mechanism for rubber-modified thermoset polymers is rubber particle tear/deformation. The system's toughness is determined by the energy required to rupture the particles and shatter in the glassy matrix.

Small rubber particles (particle size <0.1 μm) do not toughen epoxies, whereas large particles (particle size 1–20 μm) increase the fracture toughness by at least an order of magnitude. Although evidence of stretched rubber particles in the vicinity of cracks has been identified in several microscopic investigations of rubber-modified epoxies, this mechanism was previously regarded as irrelevant to toughened thermoplastics. The quantity of elastic energy retained in the rubber particles during two-phase system loading indicated the toughness enhancement is provided by rubber particle inclusion. The existence of the second phase, which enhances toughness, aids the primary deformation mechanism in the matrix. Rubber particles contained in the epoxy matrix stretch and tear during failure, resulting in significant energy absorption (Kim and Robertson 1990). Microcavitation at the crack tip is one of the toughening mechanisms in these cases.

Another technique for toughening thermoset polymers is multiple crazing. A crack area will have a refractive index different from its surroundings. It contains an interconnected void network of polymer fibrils drawn across, normal to the crazed surface. Bucknall and Partridge (1983)'s multiple crazing theory suggested that toughness improvement is related to the development and efficient termination of crazes by rubber particles. Optical microscopic investigations on thermoplastics, such as

HIPS, have demonstrated this phenomenon. Crazing is considered as the primary toughening mechanism in rubber-modified epoxies (Kinloch et al. 1983). Bucknall and Smith (1965) hypothesized that crazing could be a toughening mechanism in rubber-modified epoxies in addition to shear deformation. During tensile creep testing on rubber-modified epoxies, they saw increases in specimen volume and longitudinal extension. This increase in bulk could be linked to the specimen's extreme crazing. However, when the length of the polymer chain diminishes, tests on thermoplastics have revealed an apparent change from crazing to a shear yielding process. Crazing would be reduced in thermosets with a high crosslink density and, as a result, a short-chain length between crosslinks. The creation of crazes was attributed to the commonly observed stress-whitening phenomena.

Newman and Strella (1965) proposed a theory on toughness mechanism based on the work on acrylonitrile–butadiene–styrene thermoplastics. This theory proposes that shear deformation takes place either as shear bands or as a diffuse form of shear yielding where the yielding is initiated at the site of stress concentration due to the presence of rubber particles. This serves as the primary source of energy absorption and hence toughness improvement. The embedded soft rubber particles cause increased stress concentration, which enhances the plastic shear yielding in the resin matrix. The rubber particles are intended to produce the required triaxial tension in the resin matrix so that the local free volume is increased, and hence, shear yielding and drawing initiated. Rigid particulate fillers like carbon black, silica, and glass beads may also cause an increase in shear yielding of the matrix. However, toughness increase is not as high as that offered by rubber particles.

The shear yielding theory is not taking into consideration the stress-whitening phenomenon since the former is a constant volume deformation. In fact, shear yielding occurs simultaneously with crazing in many polymers during deformation, where crazing is accounted for the stress-whitening effect. The most applicable theory on toughening mechanism in rubber-modified epoxies is a dual-mode mechanism which is based on the shear yielding of the matrix and rubber particle cavitation.

18.7 RECENT ADVANCES IN TOUGHENING OF POLYMERS

Though rubber toughening still continues to be the most common method of toughening polymers, recent studies focus on the toughening of polymers using nanomaterials. Nanomaterials include carbon nanotube, carbon nanofibres (CNFs), rubber particles in the nanometre size, and graphene. Ravindran et al. (2018) found that multiscale fillers such as CNFs and short carbon fibres (SCFs) improve the mode-1 fracture toughness of epoxy composites. They used vapour-grown CNFs of diameter and length in the range of 70–300 nm and 30–200 μm, respectively, along with SCFs of diameter 7 μm, which were obtained by cutting continuous T300 carbon. The composites were made by mixing liquid epoxy with either CNF or SCF or both CNF and SCF to get the multi-scale modification. Both hand mixing and mixing of the constituents in a three-roll mill have been adopted for proper dispersion of the fibre in the matrix. These fibres were aligned in the matrix using externally applied alternating current. The responses of the fibres to the AC electric field, while suspended in the liquid epoxy resin, were observed using an optical microscope. Figure 18.8 shows the optical microscopy images of the both 0.10 wt.% CNF and 0.10 wt.% SCF-incorporated liquid epoxy resin before and after exposure (photograph taken 300 seconds after exposure) to an external AC electric field.

It is observed that the mode-1 fracture toughness of epoxy resin increases with the incorporation of both CNF and SCF up to 1.5 wt.%, beyond which agglomeration causes a decrease in the toughness values. Alignment of the CNFs or SCFs under the influence of the external electric field resulted in a higher toughness compared to the composite with random fibre orientation (without exposure to electric field). Here, the improvement in toughness is attributed to the many toughening mechanisms induced by the CNFs and SCFs. These mechanisms include intrinsic toughening mechanisms such as interfacial debonding and plastic void growth as well as extrinsic toughness mechanisms. Interfacial debonding dissipates the stored strain energy followed by yielding the matrix in the vicinity of the fillers. The extrinsic mechanisms include fibre snubbing (for the SCFs),

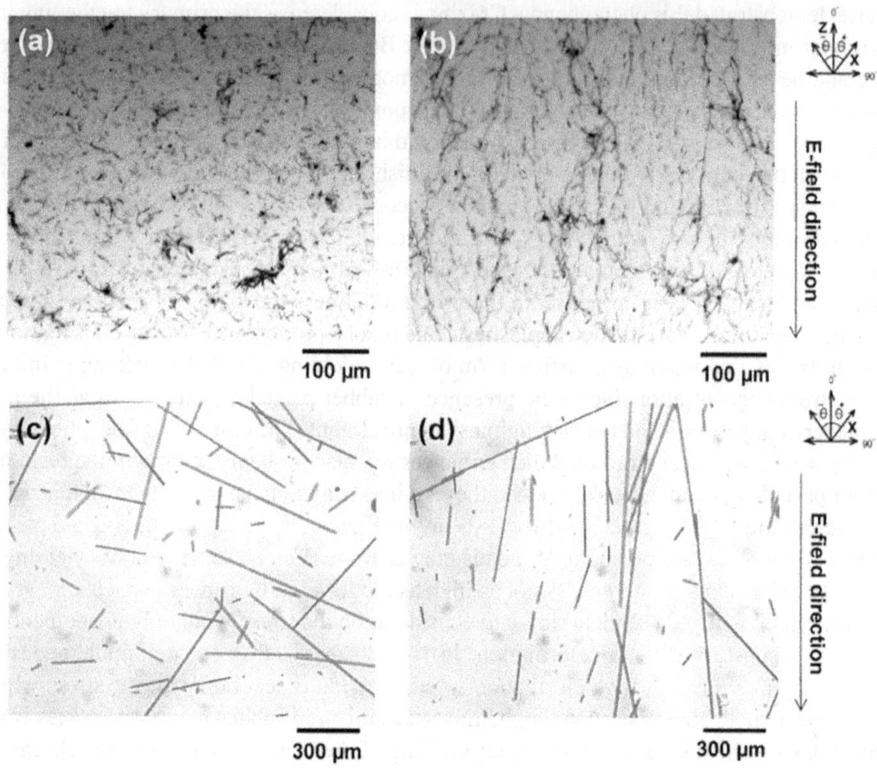

FIGURE 18.8 Optical microscopy images of CNF and SCF before alignment (a and c, respectively) and after alignment (b and d, respectively) when subjected to external electric field (Reproduced with permission from Ravindran et al. (2018) © 2018 Elsevier Ltd.)

fibre bridging, fibre pull-out, and fibre rupture. Some of these extrinsic mechanisms are influenced by the external electric field as well. Wang et al. (2019) summarized a host of polymers toughened using graphene and graphene derivatives, whereas Dong et al. (2021) toughened polylactide (PLA) using epoxy-functionalized core-shell starch nanoparticles where 54 times improvement in toughness with reference to neat PLA observed.

18.8 CONCLUSION

Toughening of polymers and polymer composites is a vast research area that started with early polymers such as cellulose nitrate. The research on polymer toughening is continuing where the focus now is more on achieving multifunctional properties by incorporating organic and inorganic nanoparticles. Among the many toughening methods, rubber toughening of epoxies has achieved broad interest because of the overall application prospects of epoxy polymers. Recent studies on rubber toughening of epoxies include modification using rubber nanoparticles and graphene nanosheets. Apart from improving toughness, properties such as conductivity are also imparted in the later. Such multifunctional toughened polymers find extensive application in aerospace, industrial, and commercial areas.

ACKNOWLEDGEMENT

The authors would like to thank the Director, Vikram Sarabhai Space Centre, Thiruvananthapuram, for granting permission to publish this work and the Deputy Director, VSSC (PCM), for all the support.

REFERENCES

Akbari, R, M H Beheshty, and M Shervin. 2013. Toughening of dicyandiamide-cured DGEBA-based epoxy resins by CTBN liquid rubber. *Iranian Polymer Journal* 22 (5): 313–24.

Bucknall, C B, and I K Partridge. 1983. Phase separation in epoxy resins containing polyethersulphone. *Polymer* 24 (5): 639–44.

Bucknall, C B, and R R Smith. 1965. Stress-whitening in high-impact polystyrenes. *Polymer* 6 (8): 437–46.

Bussi, P, and H Ishida. 1994. Partially miscible blends of epoxy resin and epoxidized rubber: Structural characterization of the epoxidized rubber and mechanical properties of the blends. *Journal of Applied Polymer Science* 53 (4): 441–54.

Chikhi, N, S Fellahi, and M Bakar. 2002. Modification of epoxy resin using reactive liquid (ATBN) rubber. *European Polymer Journal* 38 (2): 251–64.

Dittanet, P, and R A Pearson. 2013. Effect of bimodal particle size distributions on the toughening mechanisms in silica nanoparticle filled epoxy resin. *Polymer* 54 (7): 1832–45.

Dong, X, Z Wu, Y Wang, T Li, X Zhang, H Yuan, B Xia, P Ma, M Chen, and W Dong. 2021. Toughening polylactide using epoxy-functionalized core-shell starch nanoparticles. *Polymer Testing* 93: 106926.

Goossens, S, and G Groeninckx. 2007. High melting thermoplastic/epoxy resin blends: Influence of the curing reaction on the crystallization and melting behavior. *Journal of Polymer Science Part B: Polymer Physics* 45 (17): 2456–69.

Gopala, A, H Wu, and P Heiden. 1998. Investigation of readily processable thermoplastic-toughened thermosets. III. toughening bmis and epoxy with a comb-shaped imide oligomer. *Journal of Applied Polymer Science* 70 (5): 943–51.

Harani, H, S Fellahi, and M Bakar. 1998. Toughening of epoxy resin using synthesized polyurethane prepolymer based on hydroxyl-terminated polyesters. *Journal of Applied Polymer Science* 70 (13): 2603–18.

Hodgkin, J H, G P Simon, and R J Varley. 1998. Thermoplastic toughening of epoxy resins: A critical review. *Polymers for Advanced Technologies* 9 (1): 3–10.

Hu, F, S K Yadav, J J La Scala, J Throckmorton, and G R Palmese. 2021. Epoxidized soybean oil modified using fatty acids as tougheners for thermosetting epoxy resins: Part 1. *Journal of Applied Polymer Science* 138 (24): 50570.

Hyatt, J W, and I S Hyatt. 1870. *Improvement in Treating and Molding Pyroxyline.* U.S. Patent 105338, July.

Iwan, O. 1927. *Process for Manufacturing Plastic Compositions and Products Obtained Thereby.*

Kaynak, C, C Celikbilek, and G Akovali. 2003. Use of silane coupling agents to improve epoxy–rubber interface. *European Polymer Journal* 39 (6): 1125–32.

Kienle, R H, and A G Hovey. 1929. The polyhydric alcohol-polybasic acid reaction. I. Glycerol-phthalic anhydride. *Journal of the American Chemical Society* 51 (2): 509–19.

Kim, J K, and S Datta. 2013. Rubber-Thermoset blends: Micro and nano structured. In *Advances in Elastomers I*, 229–62. Springer.

Kim, J, and R E Robertson. 1992. Possible phase transformation toughening of thermoset polymers by poly (butylene terephthalate). *Journal of Materials Science* 27 (11): 3000–09.

Kim, J, and R Robertson. 1990. *Polymer Materials Science & Engineering* 63 (301).

Kinloch, A J, S J Shaw, D A Tod, and D L Hunston. 1983. Deformation and fracture behaviour of a rubber-toughened epoxy: 1. Microstructure and fracture studies. *Polymer* 24 (10): 1341–54.

Levita, G, A Livi, P A Rolla, and G Gallone. 1996. Time evolution of dielectric parameters during the cross-link of epoxy resins. *Polymers for Advanced Technologies* 7 (12): 873–78.

Lopez, J, S Gisbert, S Ferrandiz, J Vilaplana, and A Jimenez. 1998. Modification of epoxy resins by the addition of PVC plastisols. *Journal of Applied Polymer Science* 67 (10): 1769–77.

Mimura, K, H Ito, and H Fujioka. 2000. Improvement of thermal and mechanical properties by control of morphologies in PES-modified epoxy resins. *Polymer* 41 (12): 4451–59.

Newman, S, and S Strella. 1965. Stress—Strain behavior of rubber-reinforced glassy polymers. *Journal of Applied Polymer Science* 9 (6): 2297–2310.

Nicholson, J W, and P F Nolan. 1983. The behaviour of thermoset polymers under fire conditions. *Fire and Materials* 7 (2): 89–95.

Park, J S, P H Kang, Y C Nho, and D H Suh. 2003. Characterization in the toughening process of CTBN modified epoxy resins induced by electron beam radiation. *Journal of Macromolecular Science, Part A* 40 (6): 641–53.

Pearson, R A, and A F Yee. 1990. *Polymer Materials Science & Engineering* 63 (311).

Pearson, R A, and A F Yee. 1991. Influence of particle size and particle size distribution on toughening mechanisms in rubber-modified epoxies. *Journal of Materials Science* 26 (14): 3828–44.

Ravindran, A R, R B Ladani, S Wu, A J Kinloch, C H Wang, and A P Mouritz. 2018. Multi-scale toughening of epoxy composites via electric field alignment of carbon nanofibres and short carbon fibres. *Composites Science and Technology* 167: 115–25.

Robinette, E J, S Ziaee, and G R Palmese. 2004. Toughening of vinyl ester resin using butadiene-acrylonitrile rubber modifiers. *Polymer* 45 (18): 6143–54.

Seymour, R B. 1989. *Origin and Early Development of Rubber-Toughened Plastics*. ACS Publications.

Shin, S, and J Jang. 1997. Toughness improvement of high-performance epoxy resin using aminated polyetherimide. *Journal of Applied Polymer Science* 65 (11): 2237–46.

Tang, L-C, Y-J Wan, D Yan, Y-B Pei, L Zhao, Y-B Li, L-B Wu, J-X Jiang, and G-Q Lai. 2013. The effect of graphene dispersion on the mechanical properties of graphene/epoxy composites. *Carbon* 60:16–27.

Unnikrishnan, K P, and E T Thachil. 2006. Toughening of epoxy resins. *Designed Monomers and Polymers* 9 (2): 129–52.

Van der Sanden, M C M, and H E H Meijer. 1993. Deformation and toughness of polymeric systems: 3. Influence of crosslink density. *Polymer* 34 (24): 5063–72.

Wang, J, X Jin, C Li, W Wang, H Wu, and S Guo. 2019. Graphene and graphene derivatives toughening polymers: Toward high toughness and strength. *Chemical Engineering Journal* 370: 831–54.

19 Free Vibration of Laminated Composite Stiffened Shallow Shells

Amar N. Nayak
Veer Surendra Sai University of Technology

Janen N. Bandyopadhyay
Indian Institute of Technology Kharagpur

CONTENTS

19.1 INTRODUCTION

Stiffened shells are widely used in civil, marine, aerospace and other engineering applications. The cylindrical shell, which has singly curved surface, and the conoidal, elliptic paraboloid, hypar and hyperbolic paraboloid shells, which have doubly curved surfaces, are generally used as roofing units. Figure 19.1 shows the various forms of shell structures. In order to maintain the objective of the design with minimum weight along with the economy in material, these shells are added with stiffeners, so that greater strength can be accomplished with comparatively lower quantity of material. Moreover, in the recent years, stiffened shells are fabricated from advanced composite materials like laminated fibre-reinforced polymer (FRP) composites due to their high specific strength and stiffness, high corrosion and fatigue resistance and easy in fabrication.

DOI: 10.1201/9780429330575-19

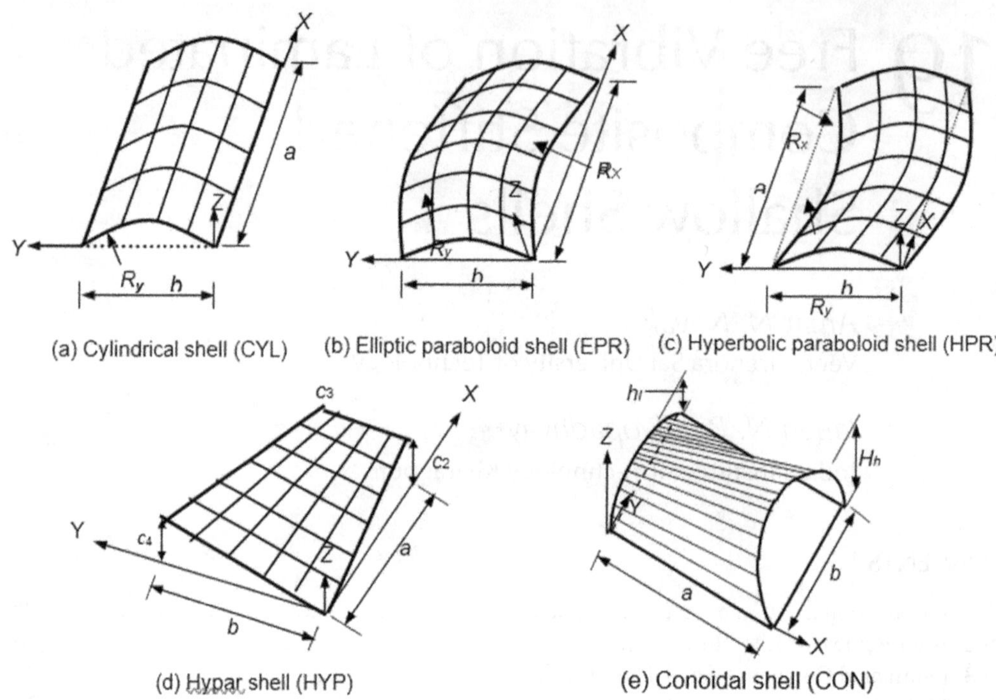

FIGURE 19.1 Schematic diagrams of five shells showing their forms with curvatures, ruled surfaces and other dimensional parameters: (a) cylindrical shell with one positive curvature and one ruled surface; (b) elliptic paraboloid shell with two positive curvatures; (c) hyperbolic paraboloid shell with one negative and one positive curvatures; (d) hypar shell with two ruled surfaces and twist curvature; (e) conoidal shell with one ruled surface, one varying curvature and twist curvature.(Nayak and Bandyopadhyay 2002a,b).

The laminated composite stiffened shell structures are often subjected to various dynamic loading during their working period. Hence, the adequate knowledge of its dynamic behaviour is very much essential from the analysis and design point of view. Especially, the dynamic properties, such as natural frequencies and their corresponding mode shapes, are also very much impotent to predict in-service performance and resonance frequency. Moreover, cut-outs are inevitable in structures for various purposes such as ventilation, inspection, providing accessibility to other part of structure and other engineering utilities. But, cut-outs in structural members may result in a change in dynamic characteristics. Undesirable vibrations may cause sudden failures due to resonance in the presence of cut-outs. Hence, it is very much essential to know the natural frequencies of these members appropriately with cut-outs also. In order to exploit the full potential of materials, a proper and systematic free vibration analysis of these laminated stiffened shells is very much required.

In order to fulfil the above requirements, a systematic free vibration study on the laminated composite stiffened shells along with a brief idea on shell theories and methods of analysis is presented in this chapter. A review of the past research works is presented very briefly in Section 19.2 in order to give a state of art on this subject. The shell theories and methods of analysis adopted in these shells are discussed in Sections 19.3 and 19.4, respectively. An appropriate finite element method (FEM) formulation is described in Section 19.5 to analyse the laminated composite stiffened shells. Section 19.6 contains some numerical results and discussion. Finally, concluding remarks are furnished in Section 19.7.

19.2 BRIEF REVIEW OF LITERATURE

Several researchers conducted extensive research works on the free vibration study of isotropic stiffened shells with/without cut-out by using numerical and analytical methods since the beginning of second half of 20th century. Due to development of advanced laminated FRP composite at a later stage, the researchers diverted their attentions from isotropic stiffened shells to composite shells. FEM was considered by Nayak and Bandyopadhyay (2005) for the study of vibration behaviour of laminated shallow stiffened shells. Goswami and Mukhopadhyay (1995) considered FEM for the estimation of the vibration phenomena of stiffened laminated cylindrical and spherical shells. The vibration aspects of orthogonally and ring stiffened cylindrical composite shells with cut-outs were analysed by Lee and Kim (1998). Prusty and Ray (2004) considered FEM for the analysis of the vibration responses of stiffened shells.

Qing et al. (2006) provided semi-analytical solution to analyse the vibration responses of stiffened cylindrical shells. Sahoo and Chaktravorty (2006) also employed FEM to conduct the free vibration study of laminated stiffened hypar shells. Sahoo (2013) also conducted the free vibration analysis of laminated composite stiffened conoidal shallow shells with cut-out by using FEM. Wang and Lin (2006) developed analytical solution to estimate the vibration responses of ring stiffened laminated cylindrical shells. Torkamani et al. (2009) analysed the vibration responses of the cross-stiffened cylindrical shells by the development of scaling laws using the similitude theory. Kayran and Yavuzbalkan (2010) analysed the vibration phenomena of shells with ring stiffeners considering multisegment numerical integration technique. Hemmatnezhad et al. (2014) conducted the free vibration study of composite cylindrical shells with grid-type stiffeners by applying a unified analytical approach. Hemmatnezhad et al. (2015) conducted the research on the vibrational characteristics of composite cylindrical shells with stiffeners experimentally, numerically and analytically. Tuan et al. (2016) analysed the free vibration behaviour of laminated stiffened cylindrical shell using FEM. Ni et al. (2019) examined the vibrational responses of laminated shells with stiffeners considering modified Ritz along with a semi-analytical approach.

19.3 LAMINATED COMPOSITE SHELL THEORIES

In the past, many researchers were engaged themselves in analysing the vibration problem of laminated shells based on the three-dimensional (3D) model. In this model, the free vibration analysis of the composite plates/shells is conducted by solving the 3D elasticity equations. Few notable examples of 3D solutions for layered shells are due to Ye and Soldatos (1994), Bhimaraddi (1991) and Wu et al. (1996). However, these models become very complex with the most general case of geometry, boundary, laminate layout and loading conditions, and hence, the solution for these problems is difficult. Moreover, the computational cost of these analyses is very high due to which these theories are rarely applied to the practical problems.

In order to overcome the above difficulties, many researchers considered various assumptions to simplify these problems depending upon their implementation and to convert the problems of 3D–2D accurately. The 2D theories developed for laminated composite shells are divided into two categories: equivalent single-layer models and layer-wise models. The notable review articles and monographs on these theories have been published by Noor and Burton (1990), Toorani and Lakis (2000), Reddy (2003) and Qatu (2004). A brief discussion on these theories is also reported by Tran and Trinh (2017).

The equivalent single-layer shell theories are further divided into three subheads, that is classical shell theory (CST), first-order shear deformation theory (FSDT) and higher-order shear deformation theory (HSDT). The CST is developed on the basis of the Kirchhoff–Love assumptions. In this theory, the transverse normal and shear deformations are considered to be zero. Various thin-shell theories are developed using this theory, out of which most common shell theories are Love's, Donnell's, Reissner's, Novozhilov's, Sanders', Flugge's and Vlasov's shell theories. These theories were described in detail in the monograph by Leissa (1993).

When the thickness of the shell changes from thin to moderately thick, the above theory is not suitable. The FSDTs are developed considering the effects of transverse deformations. However, these theories could not account properly for the parabolic shear stress distribution across the thickness of the shell, and therefore, a shear correction factor is applied for accounting this shear distribution. The HSDTs are developed not to consider the shear correction factor in the analysis. Some of the notable works considering shear deformable shell theories for doubly curved laminated composite shells are due to Reddy (1984) and Khdeir and Reddy (1997).

19.4 METHODS OF ANALYSIS

The exact solution is very rarely considered to estimate the stresses and deflections of the plates/shells with stiffeners, because it is very tedious and calculation procedure is also difficult. Accordingly, the early two types of models of stiffened plates/shells have been intended to idealize the structure to simpler ones for which solution methods are known. In the first type of modelling, there is replacement of the stiffened plate/shell is made by an equivalent orthotropic plate/shell whose thickness remains same. The orthotropic theory of plates/shells insisted that the stiffeners should be equally spaced, very close to each other and orthogonally placed. Also, the stiffeners are to be of identical profiles. Moreover, it is very much difficult to evaluate the stress and strain of the stiffeners along with plates/shells due to the conversion of these to an "equivalent plate/shell" structure. Due to these problems, the functionality of this technique is restricted. Using this technique, Szilard (2004) and many other investigators have analysed stiffened plates and shells.

The second type of modelling is more practical and precise. Here the stiffened plates/shells are replaced by the discrete plate/shell and beam idealization, where they are modelled separately maintaining the compatibility at the interface of the two. The plate/shell-beam idealization can make the analysis more complex, but it can give better structural behaviour. However, the analysis becomes more easy and appropriate with development of the sophisticated approaches of the structural analysis and the arrival of digital computers. The stiffeners along with the plate/shell are treated as separate units in many of the numerical methods. Few notable methods of them are Ritz method (Ni et al. 2019; Shi et al. 2015), Rayleigh–Ritz method (Boyd and Brugh 1977), differential quadrature method (Zeng and Bert 2001), collocation method (Mecitoglu and Dokmeci 1992), Galerkin method (Jam et al. 2011) and FEM (Nayak and Bandyopadhyay 2002a, b, 2005, 2006; Goswami and Mukhopadhyay 1995; Lee and Kim 1998; Sahoo and Chakravorty 2006; Sahoo 2013).

Among all these methods, the most adaptable method is the FEM because of its applicability for any kind of geometry, incorporating complicating effects, ease of formulation and including wide range of elements. The FEM for a stiffened structure has various approaches, such as finite strip method, discrete stiffener approach, lumped model approach and arbitrary-oriented stiffener approach. Out of these approaches, the arbitrary-oriented stiffener approach is considered to be the most flexible model as at any place of the plate element, the stiffeners can be easily located and nodal connectivity is not necessary. Hence, the mesh division is not dependent on the position of the stiffeners and can be made according to the requirements of the stress output and resolutions sought. In this chapter, this method is considered to model the stiffened shells, which is explained later in the section.

19.5 FINITE ELEMENT FORMULATION

A stiffened shell is divided into a number of stiffened shell elements for the present finite element analysis. Each element of the stiffened shell is obtained by combining appropriately the eight-/nine-nodded isoparametric thin shallow laminated shell element, which is doubly curved, and three-nodded isoparametric curved laminated beam element(s) if the shell element consists of any stiffeners. The stiffeners, which are considered as beam elements, can be conveniently placed at any suitable position with reference to the overall shell geometry and nodal configurations. For the stiffened shell element, the overall element matrices are evaluated by combining the element matrices of the shell element appropriately and the contribution of those matrices of the beam elements at the shell nodal points.

19.5.1 Shell Element

A shallow laminated shell element considered for analysis is doubly curved with uniform thickness h. R_x and R_y represent the radius of curvature of the shell along x- and y-directions, respectively, and R_{xy} represent the twist radius of curvature. The shell is projected along the xy plane in the rectangular form with dimensions represented by a and b. Five degrees of freedom (DOFs) including three translations u, v and w along x, y and z directions, respectively, and two rotations α and β along x and y axes, respectively, are considered at each node of the shell element.

The relationship between strain and displacement can be expressed as per the modified Sanders' first-approximation theory of thin shells, in the following manner:

$$\{\varepsilon\}=[\varepsilon_x^0\,\varepsilon_y^0\,k_x\,k_y\,k_{xy}\,\gamma_{xz}^0\,\gamma_{yz}^0]^T$$

$$=\left[\left(\frac{\partial u}{\partial x}+\frac{w}{R_x}\right)\left(\frac{\partial v}{\partial y}+\frac{w}{R_y}\right)\left(\frac{\partial u}{\partial y}+\frac{\partial v}{\partial x}+\frac{2w}{R_{xy}}\right)\left(\frac{\partial \alpha}{\partial x}\right)\left(\frac{\partial \beta}{\partial y}\right)\left(\frac{\partial \alpha}{\partial y}+\frac{\partial \beta}{\partial x}\right)\left(\alpha+\frac{\partial w}{\partial x}\right)\left(\beta+\frac{\partial \beta}{\partial y}\right)\right]^T \tag{19.1}$$

The force strain relation of the shell element can be written as follows:

$$\{F\}=[D]\{\varepsilon\}=[D][B]\{\delta_i\} \tag{19.2}$$

where

$$\{F\}=\left[N_x\;\;N_y\;\;N_{xy}\;\;M_x\;\;M_y\;\;M_{xy}\;\;Q_{xz}\;\;Q_{yz}\right]^T \tag{19.3}$$

$[B]$ is obtained by differentiating the shape functions and termed strain–displacement matrix, $\{\delta_i\}$ constitutes the nodal displacements and $[D]$ is the elasticity matrix of the shell element, which can be derived from the macro-mechanical analysis of the laminated shell, which is furnished in Chattopadhyay et al. (1992). N_x, N_y and N_{xy} are the in-plane stress resultants; M_x, M_y and M_{xy} represent the moments; and Q_{xz} and Q_{yz} are the transverse shear resultants.

On the basis of minimum energy principle, the element stiffness $[K_{she}]$ and mass $[M_{she}]$ matrix are derived in terms of local natural coordinates ξ and η as follows:

$$[K_{she}]=\int_{-1}^{+1}\int_{-1}^{+1}[B]^T[D][B]|J|d\xi\,d\eta \tag{19.4}$$

$$[M_{she}]=\int_{-1}^{+1}\int_{-1}^{+1}[N]^T[m][N]|J|d\xi\,d\eta \tag{19.5}$$

Here, $[N]$ is the shape function matrix, $|J|$ is the Jacobian matrix and $[m]$ represents the inertia matrix of the shell element.

The element matrices are obtained by representing the integrals in the form of local natural coordinates ξ, η, and the reduced integration of 2×2 Gaussian quadrature is performed to avoid the shear locking (Zienkiewicz et al. 1971).

19.5.2 Stiffener Element

The stiffener element comprised of a curved beam element of rectangular section. R_x and R_y represent the radius of curvature of the x-direction and y-direction stiffener, respectively. Each node of the beam element has four DOFs, that is u_{sx}, w_{sx}, α_{sx} and β_{sx} for x-directional stiffener and v_{sy}, w_{sy}, α_{sy} and β_{sy} for y-directional stiffener, respectively.

The strain–displacement equations of both the stiffeners mentioned above can be derived from the generalized displacements of the corresponding stiffeners.

The generalized force–strain relationship for the x-direction stiffener can be represented as follows:

$$\{F_{sx}\}=[D_{sx}]\{\varepsilon_{sx}\}=[D_{sx}][B_{sx}]\{\delta_{sxi}\} \tag{19.6}$$

where

$$\{F_{sx}\}=\begin{bmatrix}N_x^{sx} & M_x^{sx} & T_x^{sx} & Q_{xz}^{sx}\end{bmatrix}^T \tag{19.7}$$

$$\{\varepsilon_{sx}\}=\begin{bmatrix}u_{sx,x} & \alpha_{sx,x} & \beta_{sx,x} & (\alpha_{sx}+w_{sx,x})\end{bmatrix}^T \tag{19.8}$$

$[D_{sx}]$ is the constitutive matrix of the x-direction stiffener and derived from the macro-mechanical analysis of laminated beam, which is furnished in the literature (Chattopadhyay et al. 1992). The local derivatives of shape functions can be expressed by $[B_{sx}]$ for the x-stiffener element. N_x^{sx} is the axial force resultant, M_x^{sx} represents moment resultant, T_x^{sx} is the torsional resultant and Q_{xz}^{sx} denotes for the shear resultant for the x-stiffener element. Further, for the y-direction stiffeners, the similar relationship between force and strain can also be obtained.

The principle of minimum potential energy has been employed for the derivation of the element stiffness $[K_{xe}]$ and mass $[M_{xe}]$ matrices of the x-stiffener element, which are expressed as follows:

$$[K_{xe}]=\int_{-1}^{1}[T_{sx}]^T[B_{sx}]^T[D_{sx}][B_{sx}][T_{sx}]|J|\,d\xi \tag{19.9}$$

$$[M_{xe}]=\int_{-1}^{1}[T_{sx}]^T[N_{sx}]^T[m_{sx}][N_{sx}][T_{sx}]|J|\,d\xi \tag{19.10}$$

where $[T]$ is the transformation matrix given as follows:

$$[T_{sx}]=[T_{csx}][T_{sh}] \tag{19.11}$$

in which, the curvature effect of the x-stiffener can be considered by the transformation matrix $[T_{csx}]$ as per Sinha and Mukhopadhyay (1994), and the transformation matrix that considers the random location of the x-direction stiffener within the shell element is represented by $[T_{sh}]$ as considered in Palani et al. (1992). $[N_{sx}]$ and $[m_{sx}]$ represent the shape function and inertia matrices of the x-direction stiffener, respectively.

The element matrices of the y-stiffener can also be obtained in the similar procedure as that of the x-direction stiffener.

The element stiffness and mass matrix, that is $[K_e]$ and $[M_e]$, respectively, for the stiffened shell can be demonstrated as follows:

$$[K_e]=[K_{she}]+[K_{xe}]+[K_{ye}] \tag{19.12}$$

and

$$[M_e]=[M_{she}]+[M_{xe}]+[M_{ye}] \tag{19.13}$$

The global stiffness matrix $[K]$ and mass matrix $[M]$ can be obtained by assembling the corresponding element matrices. The eigen value form of the undamped free vibration equation can be represented as follows:

$$\left([K]-\omega^2[M]\right)\{d\}=\{0\} \tag{19.14}$$

where ω is the natural frequency in *r/s* which can be obtained by solving Eq. (19.14) using subspace iteration technique, and the mode shape is represented by $\{d\}$.

The automatic mesh generation programme has been developed to employ for the free vibration analysis of laminated stiffened shell with square/rectangular cut-out. The progressive refinement of the sizes of the mesh near the cut-out is done in the programme.

19.6 NUMERICAL RESULTS AND DISCUSSION

Free vibration analysis of laminated stiffened shells with/without cut-out is furnished in this section. The convergence of the finite element formulation for both the shell elements (N8 and N9) for laminated composite cross-stiffened shells is discussed in Section 19.6.1. The validation of the developed computer codes has been performed by the comparison of the present results with those of the published papers in Section 19.6.2. The effects of various parameters on the vibration phenomena of stiffened laminated shells with/without cut-out are considered in Section 19.6.3.

19.6.1 Convergence Study

The convergence of the FEM formulation with both the shell elements (N8 and N9) is investigated with respect to mesh size for the five different laminated shell forms, that is cylindrical (CYL), elliptic paraboloid (EPR), hyperbolic paraboloid (HPR), hypar (HYP) and conoidal (CON) shells, stiffened with single centrally placed stiffener along each x- and y- direction with simply supported (SS) and clamped (CC) boundary conditions. For the sake of brevity, the convergence results are not presented here. But, it is seen from these results that the fundamental frequencies of both the elements converge to nearly the same value. But, the nine-node shell element (N9) converges faster, that is with coarser mesh size, in comparison to the eight-node shell element (N8). The frequencies of the stiffened shells of all forms with the eight-node shell element (N8) converge with the mesh sizes of 8×8 and 10×10 for SS and CC boundary conditions, respectively. The corresponding converged mesh sizes for the nine-node shell element (N9) are 6×6 and 8×8.

19.6.2 Validation

The comparison of the free vibration results obtained by the author with those of others taking several problems of earlier published papers has been performed for validation purpose. For the comparison purpose, the problems of Reddy (1984), Tuan et al. (2016), Prusty (2008) and Chakravorty et al. (1998) are considered to validate the present computer code.

The ϖ of SS composite cylindrical and spherical shells obtained by both formulations (N8 and N9) of the present investigation along with those available in the existing literature (Reddy 1984) are presented in Tables 19.1 and 19.2, respectively, which compares well. Similarly, the present frequencies of the CC laminated composite cross-and angle-ply doubly curved shells (Table 19.3) are found to be compared well with those of Tuan et al. (2016).

TABLE 19.1
ϖ of SS Laminated Composite Square Cylindrical Shells

R_x/a	Lamination: 0⁰/90⁰			Lamination: 0⁰/90⁰/0⁰		
	(Reddy 1984)	Present authors		(Reddy 1984)	Present authors	
		N8 (8×8)	N9 (8×8)		N8 (8×8)	N9 (8×8)
3	24.516	24.595	24.595	27.116	27.225	27.225
4	19.509	19.559	19.559	22.709	22.722	22.722
5	16.668	16.704	16.704	20.332	20.344	20.344
10	11.831	11.845	11.845	16.625	16.634	16.634
10^{30}	9.6893	9.6893	9.6892	15.183	15.192	15.192

$R_y/R_x=0$, $a/h=100$, $E_{11}=25\,E_{22}$, $G_{12}=G_{13}=0.5\,E_{22}$, $G_{23}=0.2\,E_{22}$, $\nu_{12}=0.25$ and $\varpi=\omega\,a^2\,\{\rho/E_{22}h^2\}^{½}$.

TABLE 19.2
ϖ of SS Laminated Composite Square Spherical Shells

R_x/a	Lamination: 0⁰/90⁰			Lamination: 0⁰/90⁰/0⁰		
	(Reddy 1984)	Present authors		(Reddy 1984)	Present authors	
		N8 (8×8)	N9 (8×8)		N8 (8×8)	N9 (8×8)
3	46.002	46.025	46.025	47.265	47.316	47.316
4	35.228	35.245	35.245	36.971	37.008	37.008
5	28.825	28.839	28.839	30.993	31.022	31.022
10	16.706	16.712	16.712	20.347	20.358	20.358
10^{30}	9.687	9.688	9.688	15.183	15.184	15.184

$R_x/R_y=1$, other parameters are similar to those used in Table 19.1.

TABLE 19.3
ϖ of Laminated Composite Doubly Curved Shell with Orthogonal Stiffeners

	R/a=5			R/a=10			R/a=100		
Stacking Sequence	Present FEM N9	Tuan et al. (2016)	Prusty (2008)	Present FEM	Tuan et al. (2016)	Prusty (2008)	Present FEM N9	Tuan et al. (2016)	Prusty (2008)
0/90/0/90	3.544	3.509	3.584	2.694	2.752	2.660	2.341	2.306	2.408
45/-45/45/-45	3.348	3.436	3.440	2.558	2.637	2.687	2.225	2.300	2.376
75/-75/75/-75	3.473	3.503	3.158	2.628	2.667	2.296	2.271	2.318	1.918
15/-15/15/-15	3.473	3.503		2.628	2.667		2.271	2.318	

$a=b=0.5\,\text{m}$, $a/h=50$, $b_{st}=0.01\,\text{m}$, $d_{st}=0.015\,\text{m}$, $E_{22}=10\,\text{GPa}$, $E_{11}=25\,E_{22}$, $G_{12}=0.5\,E_{22}$, $G_{23}=0.2E_{22}$, $\nu=0.25$ *and* $\varpi=\omega\,\{\rho/E_{22}\,h\}^{½}$.

In addition, the author's results of the ϖ of SS cross-ply antisymmetric composite (CYL), (EPR), (HPR), (HYP) and (CON) shells with/without concentric cut-out (Table 19.4) are found to compare well with those of Chakravorty et al. (1998) except those of the conoidal shell (CON). The deviations of the present results from those of Chakravorty et al. (1998) may be due to the difference of curvatures considered for the elements of the conoidal shell (CON). Chakravorty et al. (1998) assumed constant curvatures for any particular shell element, which are the average values of four

TABLE 19.4
ϖ for SS Cross-ply $[0^0/90^0]_4$ Laminated Square Shells with Concentric Cut-Out

Shell Form	Cut-out Size (a'/a)	Chakravorty et al. (42)	Present author	
			N8	N9
CYL	0.0	26.990	27.038	27.037
	0.2	27.291	27.251	27.249
	0.4	28.711	28.604	28.600
EPR	0.0	47.109	47.384	47.384
	0.2	48.823	49.031	49.030
	0.4	53.789	53.977	53.977
HPR	0.0	14.721	14.743	14.743
	0.2	13.544	13.506	13.503
	0.4	12.560	12.595	12.592
HYP	0.0	50.829	52.003	52.002
	0.2	50.434	51.413	51.412
	0.4	47.244	47.844	47.840
CON	0.0	75.450	81.099	81.097
	0.2	73.668	80.640	80.639
	0.4	61.824	65.757	65.754

Note: $a/h=100$, $a'/b'=1$; for CYL, $h/R_x=0$ and $h/R_y=1/300$, for EPR, $h/R_x=h/R_y=1/300$; for HPR, $h/R_x=-h/R_y=1/300$; for HYP, $\Delta f/a=0.2$, for CON, $a/H_h=2.5$, $h_l/H_h=0.25$; $E_{11}=25\ E_{22}$, $G_{12}=G_{13}=0.5\ E_{22}$, $G_{23}=0.2E_{22}$, $\nu=0.25$ and $\varpi=\omega\ a^2\ (\rho/E_{22}h^2)^{1/2}$.

curvatures at each of 2×2 Gauss points, while the author considered the actual curvatures in all the nodal points of any shell element. The validation of the authors' FEM code for laminated composite shells of different curvatures with/without cut-out is established as most of the present free vibration results for the above shells are well compared with those of the existing literature.

19.6.3 Parametric Study

In order to do a thorough investigation on the free vibration phenomena of stiffened composite shell, a parametric study is conducted considering different boundary conditions and curvatures of five forms including conoidal (CON), cylindrical (CYL), hyperbolic paraboloid (HPR), elliptic paraboloid (EPR) and hypar (HYP) shells, different fibre orientations of both shells and stiffeners.

The above examples have two constant/varying nondimensional parameters a/b and d_{st}/h (depth-of-stiffener-to-thickness-of-shell ratio), in addition to the following constant nondimensional parameters:

$b/h=100$, $b_{st}/h=2$ and $\nu=0.15$ for all the above shells; $R_y/R_x=0$ and $R_y/b=3$ for CYL; $R_y/R_x=1$ and $R_y/b=3$ for EPR; $R_y/R_x=-1$ and $R_y/b=3$ for HPR; $\Delta f/b=0.2$ for HYP; and $H_h/b=0.20$ and $hl/H_h=0.25$ for CON shells.

where a, b and h are span, width of chord and thickness of shell, respectively, R_x and R_y are radii of curvature of shell along x- and y-directions, respectively, ν is Poisson's ratio, b_{st} and d_{st} are width and depth of stiffener and the symbols Δf, h_l and H_h are defined in Figure 19.1.

The following nondimensional material properties of the graphite/epoxy lamina are used for the shell and stiffeners: $E_{11}=25\ E_{22}$, $G_{12}=G_{13}=0.5\ E_{22}$, $G_{23}=0.2\ E_{22}$ and $\nu_{12}=\nu_{13}=\nu_{23}=0.25$;

where E_{11} and E_{22} are Young's modulus of laminated shell and stiffener along the directions parallel and perpendicular to the direction of fibre orientation, respectively, G_{12}, G_{13} and G_{23} are shear modulus of laminated shell and stiffener, and ν_{12}, ν_{13} and ν_{23} are Poisson's ratios of laminated shell and stiffener in the plane of fibres.

TABLE 19.5
ϖ of SS Square Antisymmetric Laminated Cross-stiffened Shells Having Eccentricity at Top

Stacking Sequence	Shell Forms	$n=1$	$n=2$	$n=3$	$n=4$	$n=6$	$n=10$
$[0^0/90^0]_n$	CYL	38.891	42.212	42.736	42.919	43.052	43.123
	EPR	52.548	54.861	55.237	55.366	55.457	55.504
	HPR	32.404	36.078	36.683	36.898	37.056	37.144
	HYP	39.452	42.402	42.878	43.043	43.162	43.225
	CON	68.350	77.076	77.843	78.113	78.339	78.521
$[45^0/-45^0]_n$	CYL	56.347	66.746	67.525	67.767	67.930	68.010
	EPR	77.841	85.714	90.112	92.281	94.725	97.313
	HPR	32.806	34.672	35.193	35.337	35.428	35.469
	HYP	55.250	71.757	73.944	74.513	74.910	75.123
	CON	84.608	99.776	102.11	102.88	103.45	103.76

Note: $d_{st}/h=4$, $n_x=n_y=1$ and $\varpi=\omega\, b^2\, (\rho/E_{22}h^2)^{½}$. n_x and n_y represent the number of stiffeners along x and y-directions, respectively.

19.6.3.1 Effects of Lamina Stacking Sequence

Non-dimensional fundamental frequencies $[\varpi=\omega b^2(\rho/E_{22}h^2)^{½}]$ of (SS) antisymmetric cross-ply $[(0^0/90^0)_n]$ and angle-ply $[(45^0/-45^0)_n]$ shells are presented in Table 19.5 for five forms with "eccentric at top" cross-stiffeners along both x- and y- directions with respect to different stacking sequences of shell and stiffeners. It is found out that ϖ increases with the increase of number of layers maintaining the constant thickness of shell and the depth of stiffener for both the cases of laminations and also for all the five forms. However, more than 8 layers, that is repeating the antisymmetric $(0^0/90^0)/(45^0/-45^0)$ unit more than four times in both the shell and stiffeners, marginally improve the value of ϖ. This is due to the reason that the coupling effects of an antisymmetric laminate reduce with the increase in the layer number and become insignificant when it exceeds 8.

19.6.3.2 Effects of Curvatures of Shell

Table 19.6 presents the values of ϖ for SS and clamped antisymmetric cross-ply $[(0^0/90^0)_4]$ and angle-ply $[(45^0/-45^0)_4]$ laminated cross-stiffened shells having eccentricity at the top with different curvature parameters (R_y/b for CYL, EPR and HPR, $b/\Delta f$ for HYP and b/H_h for CON).

It is observed from the study of above results that the introduction of curvatures (reducing the curvature parameters) is beneficial for all the stiffened shells in increasing the fundamental frequencies from their respective basic values for SS and CC stiffened plates (shells with zero curvature), except for the SS HPR. The gradual reduction of the fundamental frequencies of SS HPR from their corresponding basic values of cross- and angle-ply plates with the introduction of curvatures may be justified due to the presence of opposite curvatures (anticlastic) of this particular form when simply supported. The clamped boundaries of HPR somehow overcome the influence of the opposite curvatures in reducing the ϖ, and hence, it is to be kept in mind for the effective utilization of this form.

19.6.3.3 Influence of Numbers, Types and Orientations of Stiffeners

In this section, the antisymmetric cross-ply $(0^0/90^0)_4$ or angle-ply $(45^0/-45^0)_4$ equi-spaced stiffeners are oriented along any of x-, y- or orthogonal directions for all five shell forms (CYL, EPR, HPR, HYP and CON). The concentric, eccentric at top and eccentric at bottom type of stiffeners along the three above-mentioned orientations thus give rise to nine different cases for each of the five forms taken up here. The stiffener numbers are also varied up to ten along both x/- and y/- directions to notice its influence on the dynamic stiffness. Figures 19.2a–e present the values of ϖ versus

TABLE 19.6
ϖ of SS and CC Antisymmetric Laminated Square Cross-stiffened Shells Having Eccentricity at Top with Different Curvature Parameters

Shell	Curvature	$[0^0/90^0]_4$		$[45^0/-45^0]_4$	
Form	Parameter	SS	CC	SS	CC
CYL	$R_y/b=10^{30}$ (Plate)	37.548	67.484	36.379	65.409
	$R_y/b=10$	38.209	73.532	41.465	68.586
	$R_y/b=3$	42.919	105.15	67.767	94.870
EPR	$R_y/b=10^{30}$ (Plate)	37.548	67.484	36.379	65.409
	$R_y/b=10$	39.803	79.253	51.257	75.829
	$R_y/b=3$	55.336	118.06	92.281	131.72
HPR	$R_y/b=10^{30}$ (Plate)	37.548	67.484	36.379	65.409
	$R_y/b=10$	37.474	78.982	36.280	67.486
	$R_y/b=3$	36.898	110.02	35.337	85.489
HYP	$b/\Delta f=10^{30}$ (Plate)	37.548	67.484	36.379	65.409
	$b/\Delta f=20$	37.920	67.746	41.620	68.016
	$b/\Delta f=5$	43.043	71.434	74.513	95.887
CON	$b/H_h=10^{30}$ (Plate)	37.548	67.484	36.379	65.409
	$b/H_h=20$	46.328	95.469	51.738	84.457
	$b/H_h=5$	78.113	129.36	102.88	156.43

Note: $d_{st}/h=4$, $n_x=n_y=1$ and $\varpi=\omega b^2(\rho/E_{22}h^2)^{1/2}$.

number of stiffeners of square clamped cross-ply $(0^0/90^0)_4$ stiffened shells. Identical values of ϖ have reduced the number of plots to five for HPR and HYP and seven for EPR from a total number of nine as for CYL and CON. The respective inset legends of Figure 19.2b–d show the grouping of the stiffeners exhibiting the identical values of ϖ by adopting the same symbols. The identical values of ϖ in all these cases are largely due to the symmetric nature of structures.

Figures 19.2a–e clearly reveal that for all the shell forms, the values of ϖ are increasing with the increase in stiffener numbers, as normally expected, for all the types and orientations of stiffeners taken up here. However, in the initial stage, the rate of increase of ϖ is though appreciable, it gradually reduces with the increase in the stiffener number from 2 onwards, for all the five forms.

A close scrutiny of the above figures further reveals the excellent performance in improving ϖ for each of the five shell forms with the introduction of these stiffeners more than 2. Figures 19.2a–e indicate the best performance of CON showing the values of ϖ around 202 (Figure 19.4e) followed by EPR, HPR, CYL and HYP for a maximum number of ten "eccentric at top" stiffeners. Moreover, CON with the x-directional stiffeners should preferably be avoided, as there is marginal increase of ϖ. On the other hand, HYP should always be stiffened to enhance its fundamental frequency to a great extent, though such increased ϖ of HYP is nearly in the same order as those of EPR, HPR, CON and CYL without stiffeners (Figure 19.2a–c and e).

19.6.3.4 Influence of d_{st}/h Ratio

Out of the five different forms of the shell, that is CYL, EPR, HPR, HYP and CON, the excellent cases with respect to the orientations and types of stiffener which give the higher ϖ for same value

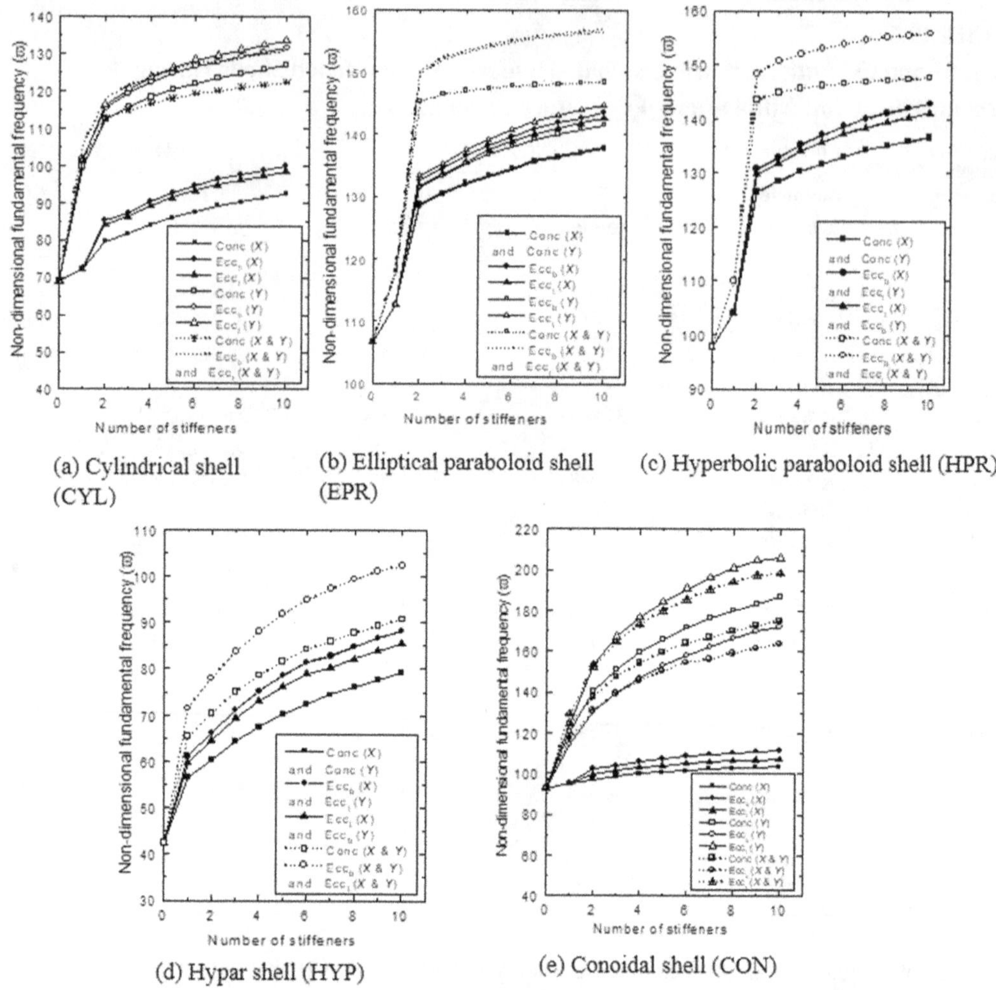

(a) Cylindrical shell (CYL) (b) Elliptical paraboloid shell (EPR) (c) Hyperbolic paraboloid shell (HPR)

(d) Hypar shell (HYP) (e) Conoidal shell (CON)

FIGURE 19.2 Variation of ϖ of CC stiffened square laminated shells with stiffener numbers [Conc: concentric, Ecc_b: eccentric at bottom, Ecc_t: eccentric at top, $d_{st}/h=4$ and $\varpi=\omega\, b^2\,(\rho/E_{22}h^2)^{1/2}$]. Five figures showing the influence of number, types and orientation of stiffeners on ϖ for clamped CYL, EPR, HPR, HYP and CON laminated shells.

of $d_{st}/h=4$ have considered for the determination of the respective values of ϖ when d_{st}/h value varied up to ten. These values are presented in Figure 19.3a–e to investigate the influence of the d_{st}/h on ϖ.

Figure 19.3 represents that as the value of d_{st}/h increases, ϖ also increases. Furthermore, two distinct types of behaviour are observed in the gain in ϖ value with reference to the increase in stiffener numbers. Figures 19.3b and c show that as the number of cross-stiffeners increases from 1 to 2, there is notable enhancement in the ϖ value though the increase in rate of ϖ is reduced when the number of stiffeners increased further. Therefore, EPR and HPR shells show improved performance with the addition of two numbers of cross-stiffeners. Moreover, from Figure 19.3a, d and e, it is observed that there is monotonic increase in the value of ϖ and also with the increase in the stiffener number; the increase rate is gradually reduced. Figures 19.3a–e will thus be very much helpful for the designers in the determination of the stiffener depth for a particular value of ϖ after the selection of the stiffener numbers and depth of shell.

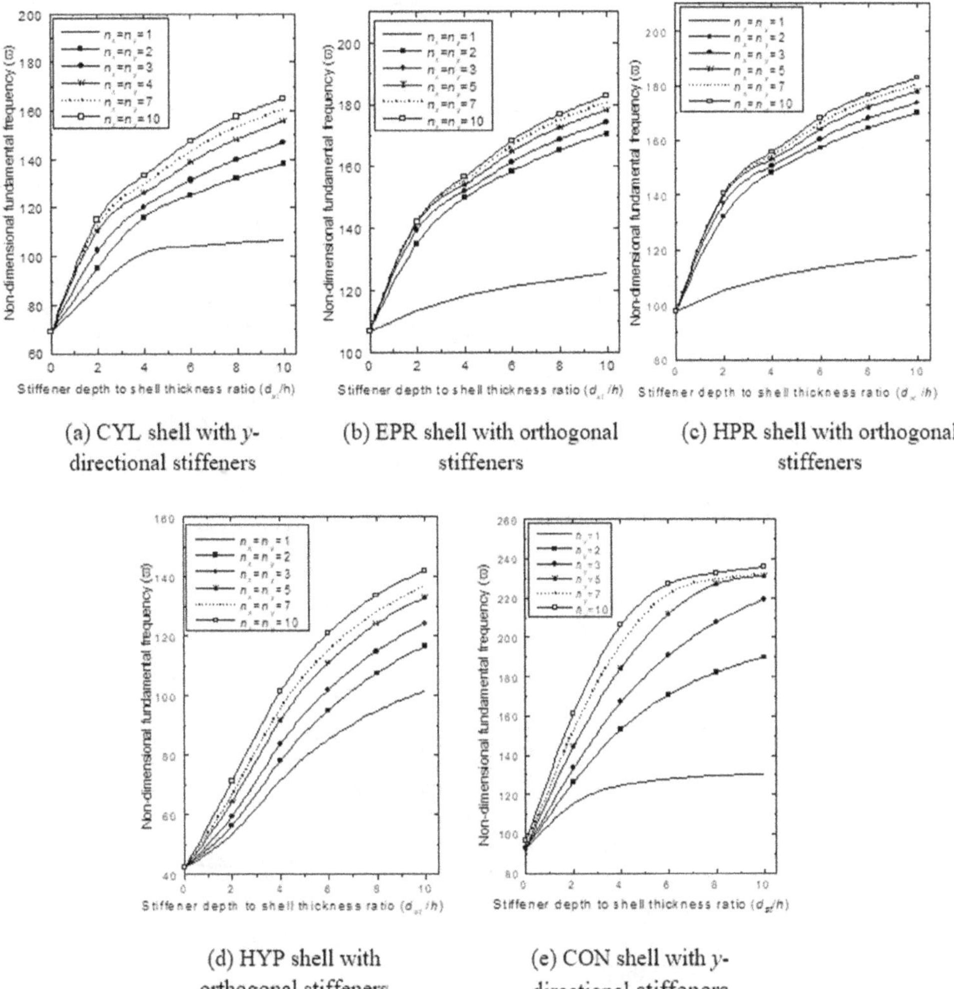

(a) CYL shell with y-directional stiffeners

(b) EPR shell with orthogonal stiffeners

(c) HPR shell with orthogonal stiffeners

(d) HYP shell with orthogonal stiffeners

(e) CON shell with y-directional stiffeners

FIGURE 19.3 Variation of ϖ with d_{st}/h ratio of CC stiffened square laminated shells having "eccentric at top" stiffeners [Stacking sequence: $[0^0/90^0]_4$, and $\varpi=\omega\, b^2\, (\rho/E_{22}h^2)^{½}$]. Five figures showing the increasing trend of fundamental frequency with increase in the dst/h ratio for clamped CYL, EPR, HPR, HYP and CON laminated stiffened shells.

19.6.3.5 Effects of Cut-Out

The free vibration study of laminated composite stiffened shells with concentric cut-out is carried out for all five shell forms considered here in this section. The fundamental frequencies are obtained for these shells with concentric cut-out and presented below.

Figures 19.4a–e present the variation of ϖ with reference to the size of the concentric cut-out of clamped square cross-ply shells with/without the "eccentric at top" orthogonal cross-ply stiffeners (unstiffened, 1×1, 4×4 and 9×9) for the shell forms (CYL, EPR, HPR, HYP and CON). Each of the shells has been analysed for discrete cut-out sizes, that is with the variation of a'/a ratio from 0.0 to 0.5. A thorough investigation of the variations of ϖ with respect to cut-out sizes from these figures reveals the following:

- The ϖ of all unstiffened shell forms ($n_x=n_y=0$) with cut-out are lower than those of the corresponding unstiffened shells without cut-out except the hyperbolic paraboloid (HPR) and hypar (HYP) shells for any cut-out size ranging from $a'/a=0.0$ to $a'/a=0.5$. The value of ϖ

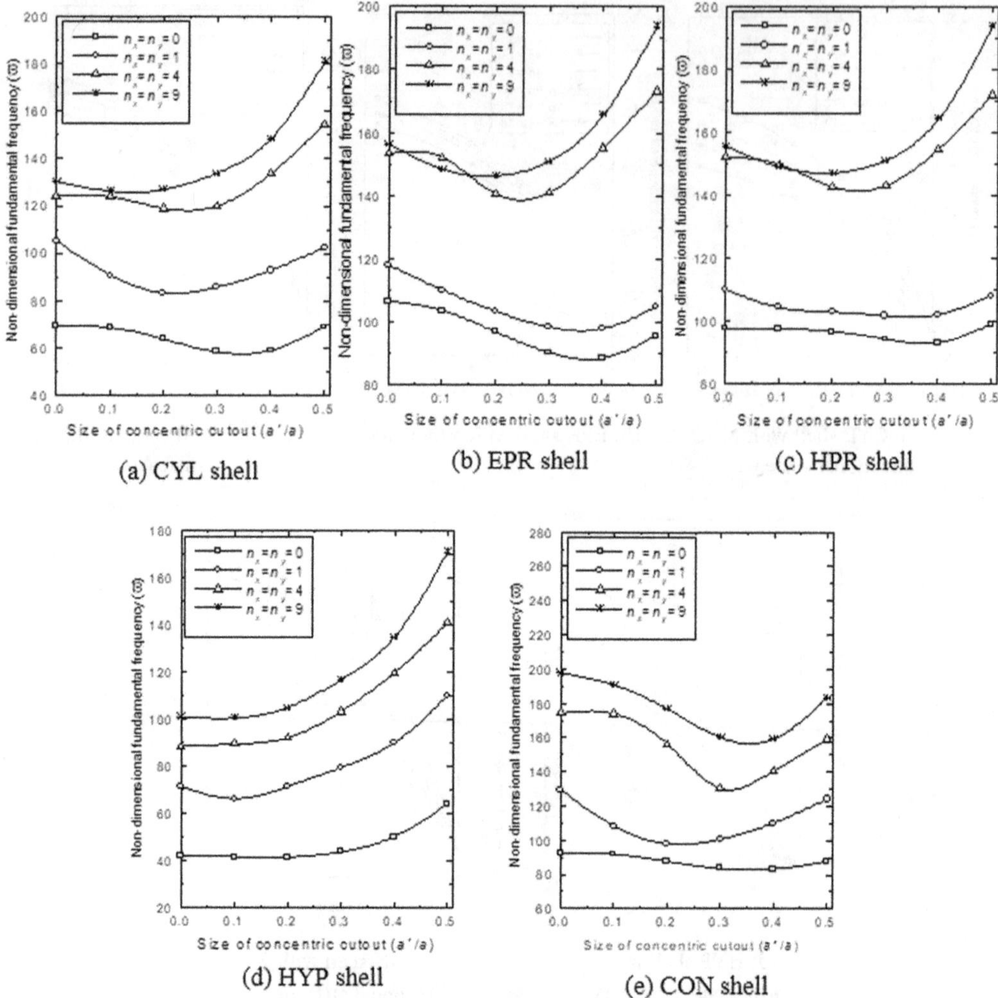

FIGURE 19.4 Variation of ϖ versus cut-out size of CC square laminated shells with/without "eccentric at top" orthogonal stiffeners [ply orientation of shell and stiffener: $[0^0/90^0]_4$, $d_{st}/h=4$ and $\varpi=\omega b^2\ (\rho/E_{22}h^2)^{1/2}$]. Five figures showing the effects of the a'/a value (cut-out size) on ϖ with the increase in the number of orthogonal stiffeners for clamped CYL, EPR, HPR, HYP and CON laminated stiffened shells.

initially decreases with the increase of cut-out size up to the value of *a'/a* equals to 0.2 and 0.3 for HYP (Figure 19.4d) and CYL (Figure 19.4a), respectively, and 0.4 for EPR (Figure 19.4b), HPR (Figure 19.64c) and CON (Figure 19.4e) shells. Thereafter, the increasing trend of ϖ is again observed up to $a'/a=0.5$.

- Similar trends of initial reduction and then increase of ϖ up to $a'/a=0.5$ are seen in the case of stiffened shells having 1 number of stiffener in each orthogonal direction ($n_x=n_y=1$) with cut-out. The corresponding *a'/a* are 0.1 for HYP, 0.2 for CYL and CON, 0.3 for HPR and 0.4 for EPR up to which the reduction trends are observed.
- For four numbers of stiffeners in each orthogonal direction ($n_x=n_y=4$) having cut-out in these shells, the reduction of ϖ is observed up to *a'/a* equal to 0.3 for CYL, EPR, HPR and HYP and 0.4 for CON, and then the increasing trend of ϖ is seen up to $a'/a=0.5$. The values of ϖ for CYL, EPR, HPR and HYP for the cut-out size $a'/a=0.5$ exceed their corresponding values without cut-out.

- The value of ϖ for these shells having nine stiffeners in each orthogonal direction decreases as the (a'/a) increases, that is a'/a equal to 0.1 for HYP, 0.2 for CYL, EPR and HPR, 0.4 for CON shells and then increases up to cut-out size $a'/a=0.5$, showing higher values of ϖ for all shell forms except CON shell with respect to those of the corresponding shells without cut-out.
- The above discussion reveals that the conoidal (CON) shell form has a deterioration of ϖ for any cut-out size ranging from $a'/a=0.0$ to $a'/a=0.5$ with/without stiffener and hence is a matter of concern. On the other hand, the hypar (HYP) shell form with/without stiffener shows encouraging results in improving ϖ with the increase in the a'/a value.

The introduction of cut-out to the shells with/without stiffeners may decrease or increase the values of ϖ depending on the predominant "stiffness action" or "mass action" because of reduction of stiffness and mass of the structure as they are the most important parameters in the determination of fundamental frequency as given in Eq. (19.14).

19.7 CONCLUDING REMARKS

In this chapter, an efficient method of analysis, that is a FEM-based code, is developed for the laminated stiffened shallow shells and also validated with the published works. It is found that this efficient method can suitably be employed for obtaining the vibration responses of stiffened laminated shells with/without cut-out. Moreover, the detailed discussion for the improvement of fundamental frequency with respect to number, orientations and types of stiffeners, and stacking sequence of lamina will be very much helpful for the designers for the selection of the above parameters suitably for achieving optimum design.

REFERENCES

Bhimaraddi, A. 1991. Free vibration analysis of doubly curved shallow shells on rectangular planform using three-dimensional elasticity theory. *International Journal of Solids and Structures* 27(7): 897–913.

Boyd, D.E. and R.L. Brugh 1977. Vibrations of stiffened cylinders with cutouts. *Journal of Sound and Vibration* 52(1): 65–78.

Chakravorty, D., P.K. Sinha, and J.N. Bandyopadhyay 1998. Applications of FEM on free and forced vibration of laminated shells. *Journal of Engineering Mechanics* 124(1): 1–8.

Chattopadhyay, B., P.K. Sinha, and M. Mukhopadhyay 1992. Finite element free vibration analysis of eccentrically stiffened composite plates. *Journal of Reinforced Plastics and Composites* 11(9): 1003–1034.

Goswami, S., and M. Mukhopadhyay 1995. Finite element free vibration analysis of laminated composite stiffened shell. *Journal of Composite Materials* 29(18): 2388–2422.

Hemmatnezhad, M., G.H. Rahimi, and R. Ansari 2014. On the free vibrations of grid-stiffened composite cylindrical shells. *Acta Mechanica* 225(2): 609–623.

Hemmatnezhad, M., G.H. Rahimi, M. Tajik, and F. Pellicano 2015. Experimental, numerical and analytical investigation of free vibrational behavior of GFRP-stiffened composite cylindrical shells. *Composite Structures* 120: 509–518.

Jam, J.E., M. Yusef Zadeh, and B. Eftari 2011. Vibration analysis of grid-stiffened circular cylindrical shells with full free edges. *Polish Maritime Research*: 23–27.

Kayran, A., and E. Yavuzbalkan 2010. Free-vibration analysis of ring-stiffened branched composite shells of revolution. *AIAA Journal* 48(4): 749–762.

Khdeir, A.A., and J.N. Reddy 1997. Free and forced vibration of cross-ply laminated composite shallow arches. *International Journal of Solids and Structures* 34(10): 1217–1234.

Lee, Y.-S., and Y.-W. Kim 1998. Vibration analysis of rotating composite cylindrical shells with orthogonal stiffeners. *Computers & Structures* 69(2): 271–281.

Leissa, A. 1993. *Vibration of Shells Acoustical Society of America. New York.*

Mecitoglu, Z., and M. Cengiz Dokmeci 1992. Free vibrations of a thin, stiffened, cylindrical shallow shell. *AIAA Journal* 30(3): 848–850.

Nayak, A.N., and J.N. Bandyopadhyay 2002a. Free vibration analysis and design aids of stiffened conoidal shells. *Journal of Engineering Mechanics* 128(4): 419–427.

Nayak, A.N., and J.N. Bandyopadhyay 2002b. On the free vibration of stiffened shallow shells. *Journal of Sound and Vibration* 255(2): 357–382.

Nayak, A.N., and J.N. Bandyopadhyay 2005. Free vibration analysis of laminated stiffened shells. *Journal of Engineering Mechanics* 131(1): 100–105.

Nayak, A.N., and J.N. Bandyopadhyay 2006. Dynamic response analysis of stiffened conoidal shells. *Journal of Sound Vibration* 291(3–5): 1288–1297.

Ni, Z., K. Zhou, X. Huang, and H. Hua 2019. Free vibration of stiffened laminated shells of revolution with a free-form meridian and general boundary conditions. *International Journal of Mechanical Sciences* 157: 561–573.

Noor, A.K., and W. Scott Burton 1990. Assessment of computational models for multilayered composite shells. *Applied Mechanics Reviews* 43(4): 67–97.

Palani, G.S., N.R. Iyer, and T.V.S.R. Appa Rao 1992. An efficient finite element model for static and vibration analysis of eccentrically stiffened plates/shells. *Computers & Structures* 43(4): 651–661.

Prusty, B.G. 2008. Free vibration and buckling response of hat-stiffened composite panels under general loading. *International Journal of Mechanical Sciences* 50(8): 1326–1333.

Prusty, B.G., and C. Ray 2004. Free vibration analysis of composite hat-stiffened panels by method of finite elements. *Journal of Reinforced Plastics and Composites* 23(5): 533–547.

Qatu, M.S. 2004. *Vibration of Laminated Shells and Plates*. Elsevier.

Qing, G., Z. Feng, Y. Liu, and J. Qiu, 2006. A semianalytical solution for free vibration analysis of stiffened cylindrical shells. *Journal of Mechanics of Materials and Structures* 1(1): 129–145.

Reddy, J.N. 1984. Exact solutions of moderately thick laminated shells. *Journal of Engineering Mechanics* 110(5): 794–809.

Reddy, J.N. 2003. *Mechanics of Laminated Composite Plates and Shells: Theory and Analysis*. CRC Press.

Sahoo, S. 2013. Dynamic characters of stiffened composite conoidal shell roofs with cutouts: design aids and selection guidelines. *Journal of Engineering*.

Sahoo, S., and D. Chakravorty 2006. Stiffened composite hypar shell roofs under free vibration: Behaviour and optimization aids. *Journal of Sound and Vibration* 295(1–2): 362–377.

Shi, P., R.K. Kapania, and C.Y. Dong 2015. Free vibration of curvilinearly stiffened shallow shells. *Journal of Vibration and Acoustics* 137(3): 031006.

Sinha, G., and M. Mukhopadhyay 1994. Finite element free vibration analysis of stiffened shells. *Journal of Sound and Vibration* 171(4): 529–548.

Szilard, R. 2004. Theories and applications of plate analysis: classical, numerical and engineering methods. *Applied Mechanics Reviews* 57(6): B32–B33.

Toorani, M.H., and A.A. Lakis 2000. General equations of anisotropic plates and shells including transverse shear deformations, rotary inertia and initial curvature effects. *Journal of Sound and Vibration* 237(4): 561–615.

Torkamani, S., H.M. Navazi, A.A. Jafari, and M. Bagheri 2009. Structural similitude in free vibration of orthogonally stiffened cylindrical shells. *Thin-Walled Structures* 47(11): 1316–1330.

Tran, M.T., and A.T. Trinh 2017. Static and vibration analysis of cross-ply laminated composite doubly curved shallow shell panels with stiffeners resting on Winkler–Pasternak elastic foundations. *International Journal of Advanced Structural Engineering* 9(2): 153–164.

Tuan, T.A., T.H. Quoc, and T.M. Tu 2016. Free vibration analysis of laminated stiffened cylindrical panels using finite element method. *Vietnam Journal of Science and Technology* 54(6): 771–771.

Wang, R.-T., and Z.-X. Lin 2006. Vibration analysis of ring-stiffened cross-ply laminated cylindrical shells. *Journal of Sound and Vibration* 295(3–5): 964–987.

Wu, C.-P., J.-Q. Tarn, and S.-M. Chi 1996. Three-dimensional analysis of doubly curved laminated shells. *Journal of Engineering Mechanics* 122(5): 391–401.

Ye, J., and K.P. Soldatos 1994. Three-dimensional vibration of laminated cylinders and cylindrical panels with symmetric or antisymmetric cross-ply lay-up. *Composites Engineering* 4(4): 429–444.

Zeng, H., and C.W. Bert 2001. A differential quadrature analysis of vibration for rectangular stiffened plates. *Journal of Sound and Vibration* 241(2): 247–252.

Zienkiewicz, O.C., R.L. Taylor, and J.M. Too 1971. Reduced integration technique in general analysis of plates and shells. *International Journal for Numerical Methods in Engineering* 3(2): 275–290.

20 Toughening Effect of Aluminium Particles in Conductive Polyester Composites

Zahir Bashir, Arfat Anis, and Saeed M. Al-Zahrani
King Saud University

CONTENTS

20.1 INTRODUCTION TO FILLER COMPOSITES

Composites of high-modulus, high-strength materials in the continuous-fibre form give the ultimate in high mechanical properties, combined with low density. However, only a few high-modulus, high-strength materials such as glass, carbon, polyaramid and basalt are available in the continuous-fibre form. The manufacture of composites with complex shapes from continuous fibres involves slow and capital-intensive processes. Most high-modulus, high-strength materials are available only as irregular particles, spherical particles, platelets and short fibres or whiskers. These are still used as they are convenient to make shaped articles by injection moulding. The lowest use is a cheap filler to counter the cost of the thermoplastic polymer. However, there are also targeted property enhancements sought from fillers, such as increased modulus, strength, hardness, toughness, scratch resistance, ultraviolet resistance, chemical resistance, anti-microbial activity, and electrical and thermal conductivity. While gaining other functionalities, an eye has to be kept on mechanical properties; if their enhancement is not possible, too much degradation has to be avoided.

Ahmed and Jones (1990) reviewed the mechanical properties of particle-filled plastics. The following trend is generally observed in mechanical properties when rigid fillers are incorporated in high T_g, rigid plastics. The tensile modulus increases modestly, by ~2× that of the matrix polymer. The modulus is a low strain property and is not significantly affected by the quality of the adhesion of the filler to the matrix (Ahmed and Jones 1990). The strength on the other hand has a strong dependence on the adhesion. If the adhesion is poor – as is often the case with fillers in plastics – the tensile strength shows a decrease with volume fraction. The particle in effect is like a void. The elongation-at-break always decreases, and this is attributed to the immobilisation of the matrix. Most often, the decrease in elongation-at-break signifies a decrease in toughness.

DOI: 10.1201/9780429330575-20

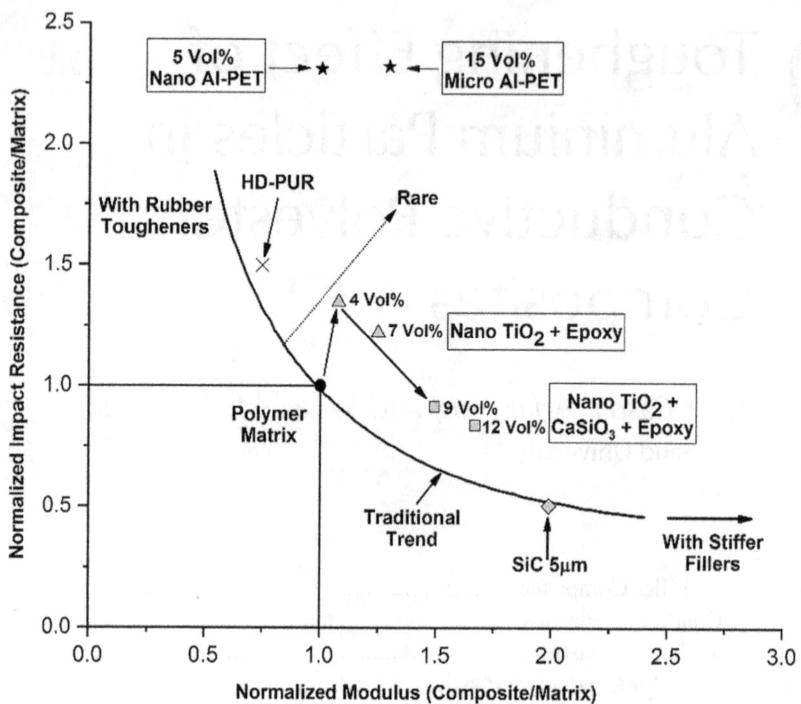

FIGURE 20.1 Plot of normalised impact resistance versus normalised modulus. Usually, in particle-filled composites, increasing composite stiffness with stiffer particles leads to decrease in impact resistance. Rubber tougheners increase toughness, but there is a reduction in the modulus. Rare exceptions which show simultaneous increase in modulus and impact resistance are nano-TiO_2 in epoxy indicated by X (Wetzel et al. 2002) and aluminium-filled amorphous PET indicated by ★ (Anis et al. 2020). (Reproduced from Anis et al. 2020.)

In general, it is well known that increasing modulus and toughness go in opposite directions. This trend is shown graphically in Figure 20.1 (adapted from Wetzel et al. (2002)). Adding very stiff mineral fillers like SiC increases modulus, but the impact resistance comes down. Most unfilled plastics have a notched Izod impact resistance under 100 J/m, typically 20–70 J/m. Exceptions are super-tough plastics like the polycarbonate of bisphenol A, and ultra-high-molecular-weight polyethylene (PE), which can reach a notched Izod impact resistance of over 1,000 J/m. The mechanical property portfolio of materials may be visualised in three classes: (i) very strong but not tough, (ii) strong and tough and (iii) weak and brittle. In most applications, it is most desirable to have category (ii).

Engineering plastics with high T_g-like poly(ethylene terephthalate) (PET) may be tough when unnotched, but are often notch-brittle (Tanrattanakul et al. 1997). The notched Izod impact resistance is well under 50 J/m, and these are toughened by adding rubber. Rubber toughening is a well-known concept, but it always involves a reduction in modulus and strength (Bora, McKinney, Faust 1994). This relies on the capacity for internal cavitation of the rubber particles when deformed. The critical parameters for the efficacy of rubber toughening are fine rubber particle size and critical interparticle distance (Pecorini and Calvert 2000). In rubber-modified nylons and polyesters, grafting improves the adhesion between rubber particles and the matrix (Bartczak 2002). Grafting also creates a tighter particle size distribution and a smaller range of particle sizes, for the same volume fraction and mixing conditions. The toughening mechanism involves cavitation and shear banding (Bartczak 2002).

Another method of increasing toughness is done by rigid polymer toughening concept, which allows the possibility of retention of stiffness (Pearson and Yee 1993). An example is polyphenylene oxide (PPO) particles in polypropylene (PP). Wei et al. (2000) showed that rigid thermoplastic particles of PPO in a PP matrix can play a role similar to rubber particles, allowing toughening via either massive crazing and/or shear banding, without sacrificing stiffness.

Rigid inorganic fillers with nanoparticle size can also allow simultaneous increase in modulus, strength and impact toughness. Wetzel et al. (2002) showed (Figure 20.1) that if nano-TiO_2 instead of micron-sized TiO_2 is used in epoxy, an increase in modulus and impact strength is observed. Likewise, Fu and Wang (1992) reported that nanocalcium carbonate increases the toughness of PE without decreasing the modulus and strength, while micron-sized does not have the same effect. These are of course rare cases where the rigid filler does not decrease the tensile strength and toughness.

Michler and von Schmeling (2013) showed that even nanovoids of ~20 nm size with inter-void distance of 20 nm and 10% content (i.e. a nanofoam) can be tougheners, as the polymer strands between the voids deform in a homogeneous and ductile manner. However, thus far, there is no practical method to produce nanofoams with regular, homogeneously distributed nanovoids of similar size. Foaming processes give a large distribution of voids up into the micron range, and then, the mechanical properties of the voided material are generally dominated negatively by the larger voids (Michler and von Schmeling 2013).

We found when compression moulding PET between aluminium (Al) foils, the adhesion was so strong that the foils could not be peeled off the PET sheet. This is unlike the case of PP or PE moulded between Al foils, where the latter can be peeled off easily. Hence, we thought that if a plastic with the capacity to adhere to a ductile metal is selected, a fine dispersion of such metal particles may lead to a toughening method. Indeed, we found that Al particles in amorphous PET acted like a rubber toughener without the drop in modulus (Anis et al. 2020). Figure 20.1 shows this in the graph of normalised impact resistance versus normalised modulus. With Al, the effect was observed with nano- and micron-sized particles. Figure 20.1 shows if the values of the normalised impact resistance and normalised modulus of the PET matrix are (1, 1), with 15 volume % of micro-nodular Al particles in amorphous PET, the modulus increased by 1.29× the value of the matrix, while the impact increased by 2.32× the value of the matrix. The strength either increased or held level with the matrix. These data will be discussed further.

However, we realised it is unlikely that Al particles will be used purely as a toughening agent in polyesters in place of rubber. While Al has the lowest densities among metals, there will be an increase in density of the composite compared with using a rubber. However, the use of Al in plastics can be coupled with its electrical and thermal conductivity. With Al in polyesters, it is then possible to design electrically and thermally conductive plastics with enhanced mechanical properties of higher toughness combined with good modulus and strength. To provide a context for Al–polyester composites, below we say a few words on conductive plastics and how the combination of electrically and thermally conductive plastics with enhanced mechanical properties would be useful.

20.2 ELECTRICALLY CONDUCTIVE PLASTICS

These are polymers filled with a fully conducting material such as metal powders or metal wires (Bhattacharya and Chaklader 1982), or carbon materials (carbon black, graphite, carbon fibres, carbon nanotubes (CNTs) and graphene (Huang 2002; Ren et al. 2014; Sandler et al. 2003; Nakamura et al. 1997; Bagotia, Choudhary and Sharma 2018; Al-Saleh 2015; Tony et al. 2005; Irina et al. 2010; Wei and Bai 2015; Wang et al. 2020; Li et al. 2006; MayPat et al. 2012; Shabafrooz et al. 2018; Aoyama et al. 2020). The loading at which percolation (electrical connectivity) is attained has to be minimised to lower cost, and damage to mechanical properties. Electrically conductive plastics are used for anti-static, electrostatic charge dissipation and electromagnetic interference (EMI) shielding (Bhattacharya and Chaklader 1982).

TABLE 20.1
Three Types of Electrically Conductive Plastics

Conductive Plastic	Surface Conductivity (S/m)
Anti-static	10^{-10} to 10^{-4}
Electrostatic dissipation (ESD) for packaging of electronics and for avoiding explosions	10^{-4} to 10^{-2}
Electromagnetic interference (EMI) shielding of enclosures with electronics.	$<10^{-2}$

Static accumulation leads to the clinging of fabrics and films. It can also lead to discharge which is familiar as sparks jumping from fingers to a metal door handle, for example. Accumulated static charge can halt mechanical processes like sieving by clogging the flow of powders. Electrostatic discharge can damage or destroy electronic components, erase or alter magnetic media, and cause explosions or fires where inflammable vapours are present.

In electronic circuitry which is adjacent or near to other electrical and electronic devices, EMI is a problem as it can distort signals and create noise. It is required that electromagnetic waves from external sources do not penetrate into the boxes housing electronic circuitry as it would cause interference. This can cause corruption of data in information storage systems, inaccuracy in medical diagnostic equipment and malfunctioning of medical devices such as pacemakers. Metal boxes provide the ultimate shielding, but the weight is a penalty, and the transport sector requires weight minimisation for reduction in fuel consumption. In this case, conductive plastics are sought for EMI shielding.

The ability of materials to conduct electricity is measured in terms of either their electrical conductivity or their resistivity. Electrical conductivity is represented by σ, and its units are S/m (siemens/m). Electrical resistivity ρ is the reciprocal quantity of conductivity, and the unit is ohm-m. Metals have conductivity in the range of 1.45×10^{6} S/m (stainless steel)–6.30×10^{7} S/m (silver). Al has an electrical conductivity of 4.10×10^{7} S/m. The electrical conductivity of a plastic-like PET is 10^{-21} S/m. Conductive plastics fall in the range of 10^{-10}–10^{-2} S/m.

Conductive plastics prevent static accumulation by reducing a material's electrical resistance. The static is dissipated slowly and continuously rather than accumulated and discharged rapidly with a spark. There are categories (see Table 20.1) for electrical conductivity in plastics: (i) anti-static, (ii) electrostatic dissipation (ESD) and (iii) EMI shielding ranges.

Stearate additives are typically used to achieve anti-static performance levels. For ESD and EMI shielding, conductive fillers like carbon black, graphite, metal powders, carbon fibres and nickel-coated carbon fibres, and steel fibres are used. CNTs and graphenes may be used, but generally, their cost is too high.

20.3 THERMALLY CONDUCTIVE PLASTICS

For some applications, heat dissipation is important. Examples are housings of light-emitting diodes, where thermally conductive plastics prevent hot spots and extend the lifetime of the housing, and heat sinks where the cooling capacity is limited by convection. In these cases, conductive plastics offer lightweight solutions.

Table 20.2 shows the thermal conductivities of plastics and potential conductive fillers. Polymers typically have a low thermal conductivity of 0.1–0.5 W/m K. The metals like Al, gold, copper and silver fall in the range of 247–429 W/m K. The carbon forms like diamonds reach 2,000 W/m K. CNTs and graphenes exceed diamonds in individual form, but the conductivity is reduced when macroscopic or aggregated forms are made. Commercial, thermally conductive polymers made with some of the cheaper fillers have a conductivity of 2–10 W/m K.

Thermal conductivity can be achieved with the same fillers used for electrical conductivity: carbon black, carbon fibres, metal wires and powders, CNTs and graphenes. However, much higher

TABLE 20.2
Thermal Conductivity of Plastics and Conductive Fillers

Material	Thermal Conductivity (W/m.K)
PET and plastics	0.1–0.5
Conductive plastics	2–10
Aluminium	247
Copper	398
Gold	315
Silver	429
Stainless steel	25

loadings (typically >70 vol. %) are needed to reach percolation for thermal conductivity. When only thermal conductivity is sought without electrical conductivity, Al_2O_3 and boron nitride are used. Again, conductivity has to be balanced with loadings that do not cause a very low strength and brittleness.

20.4 INDUSTRIAL DRIVERS FOR CONDUCTIVE PLASTICS

The industrial demand for conductive (electrical and thermal) plastics is being driven by the industries such as automotive industry and medical equipment. In the emerging electrical vehicle sector, there are applications with interactive property demands of electrical (EMI shielding) and thermal conductivity (heat dissipation from electronics) with lightweighting. In the electric car, which relies on a massive battery stack, heat dissipation would be beneficial as the battery's discharging cycle is longer at lower temperature. Battery trays made of thermally conductive plastics can give a 10°C lower temperature than a non-conductive plastic.

The latest generation of conductive fillers are the CNTs and graphene (Bagotia, Choudhary, and Sharma 2018; Al-Saleh 2015; Tony et al. 2005; Irina et al. 2010; Wei, Pingfu, and Shibing Bai 2015; Wang et al. 2020; Li et al. 2006; MayPat et al. 2012; Shabafrooz et al. 2018; Aoyama et al. 2014, 2020; Gorassi et al. 2014). These materials have low density and intrinsic electrical conductivity and mechanical properties that are much higher than even metals. Yet, despite 25 years of effort, these materials have rarely been cost-effective or industrially practicable. It is observed that they can be incorporated up to about 5–15 vol. %. Electrical percolation can be achieved at even 1%–2% (Bagotia, Choudhary, and Sharma 2018), but the modulus realised in composites with CNTs and graphene is no more than about ~ 3 GPa (short glass fibre composites can reach 10–12 GPa at 40 vol. % loading). The tensile strength may stay constant or decrease, and impact resistance usually decreases. For example, Shabafrooz et al. (2018) reported that the tensile modulus of amorphous PET increased from ~1.8 to ~2.9 GPa with 10 wt. % graphene nanoplatelets. The tensile strength increased from 53 MPa for amorphous PET to 57 MPa with 10 wt. % graphene nanoplatelets. The elongation-to-break decreased from 100% for amorphous PET to 4% with 10 wt. % graphene (Shabafrooz et al. 2018). The toughness of the amorphous PET was ~750 kJ/m^3, and this decreased to 0.4 kJ/m^3, with 10 wt. % graphene (Shabafrooz et al. 2018). That is, the mechanical properties showed the conventional trend found with micron-sized fillers. Further, the gain in modulus and conductivity with CNTs and graphene is not commensurate with the cost. The causes for the reduction in mechanical properties are poor dispersion (i.e. agglomeration occurs at higher loadings) and poor adhesion. Hence, despite individual CNTs showing the highest mechanical properties theoretically possible, these have not been translatable to a composite with outstanding mechanical properties. To solve the adhesion problem with CNTs and graphene, various chemical modifications and dispersion procedures have been tried (Aoyama et al. 2020; Gorassi et al. 2014), but they are cumbersome and impractical for industrial use, and they just add to the cost of already expensive material. Even with these methods to aid dispersion and adhesion, the mechanical properties of the composites with CNTs and graphene have not been astoundingly higher than what can be obtained with conventional fillers, albeit at a higher loading.

We emphasise that for engineering use, specific functionality (in this case electrical and/or thermal conductivity) has to be combined with good mechanical properties and an acceptable cost. Toughness is an important design criterion in engineering use, and often, it is not sufficient to have a specific functionality. In the following, we propose that a ductile metal like Al would be a more cost-effective filler than CNTs or graphene, for attaining conductive plastics with uncompromised mechanical properties, providing the polymer matrix that is chosen is capable of adhesion with Al.

20.5 AL–POLYESTER, A MECHANICALLY ROBUST SYSTEM SUITABLE FOR ELECTRICALLY AND THERMALLY CONDUCTIVE PLASTICS

It is expected that Al-filled plastics will show enhanced electrical conductivity at some critical loading. This has been demonstrated with Al-filled PVC (Dutta, Singh, and Misra, 1992; Mamunya et al. 2002; Bishay, Abd-El-Messieh, and Mansour 2011), Al-filled nylon 6 (Pinto, Gabriel, and Ana Jiménez-Martín 2001), Al-filled poly(methyl methacrylate) (Álvarez, Poblete, and Rojas 2010) and Al-filled PP (Boudenne et al. 2004; Osman and Mariatti 2006). Some works report only the electrical conductivity without the mechanical properties. Osman and Mariatti (2006) compounded Al flakes and spherical Al powder with PP. Microscopy of fractured cross sections showed pull-out of the Al flakes – which is a sign of poor adhesion with PP (Osman and Mariatti 2006). Likewise, Nicodemo and Nicolais (1983) observed poor adhesion between styrene–acrylonitrile (SAN) and Al powder. This was deduced by comparing the tensile strength versus volume fraction with theoretically predicted values for particulate composites. In contrast, we found that the mechanical properties of Al–PET were unusual and interesting (Anis et al. 2020) in that it was possible to increase modulus, strength and impact resistance, despite a large drop in elongation-to-break. We believe that it stems from Al's ductility and its good adhesion to PET.

Al powders are made by gas atomisation. Depending on the atomising gas used to spray the molten Al, the particles can have irregular or slightly elongated (nodular) shape, or be mostly spherical. For the automotive reflective paint industry, flattened flakes are made from the spherical powder by ball milling. Figure 20.2 shows the particle shapes of the Al powders we have compounded with PET. Al fibres (Danes, Garnier, and Dupuis 2003) have been used in poly(butylene terephthalate) (PBT) to increase thermal conductivity. The particle shape and size have an influence on the maximum loading possible and on the percolation threshold. Spherical shapes should give lower melt viscosity, but nanosizes limit extrusion to ~5 vol. %. Incorporation of flakes and fibres leads to higher melt viscosity. Figure 20.3 shows the tensile modulus of amorphous PET as a function of volume fraction of micron-sized, irregular (nodular) Al particles. The tensile modulus showed an increase with volume fraction which is in the range typical of rigid plastics filled with rigid fillers. This in itself is nothing out of the usual.

Figure 20.4 shows the tensile strength of Al–PET. Depending on the powder, there may be a slight decrease or an increase in the strength of the filled composite. When Al flakes are used, we observed an increase in strength. In contrast, in many filler composites, there is a drastic decrease in tensile strength due to the filler. The drop in strength occurs when the adhesion is not good enough, and the filler particle acts like a void. This can be seen with Al in PP (Boudenne et al. 2004), Al in SAN (Nicodemo and Nicolais 1983), and other metal powders in plastics, and indeed in many filled systems with inorganic fillers like mica (Sahai and Pawar 2017) and clay (Ronkay et al. 2019), CNTs (Li et al. 2006; MayPat et al. 2012; Bagotia, Choudhary, and Sharma 2018) and graphene (Shabafrooz et al. 2018). Osman and Mariatti (2006) found that for two types of Al powders (flakes and spherical), a monotonic drop in tensile strength occurred when incorporated in PP. The PP itself

FIGURE 20.2 Types of aluminium powders: (a) micron-sized, nodular aluminium, (b) spherical nanoparticles and (c) flakes. (Reproduced from Anis et al. 2020.)

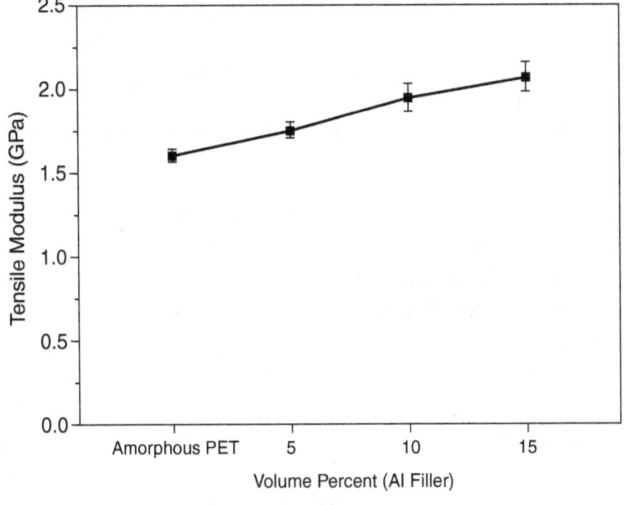

FIGURE 20.3 Comparison of tensile modulus of amorphous PET filled with the nodular Al particles of Figure 20.2a. The modulus shows an increase as found with other plastics filled with rigid fillers.

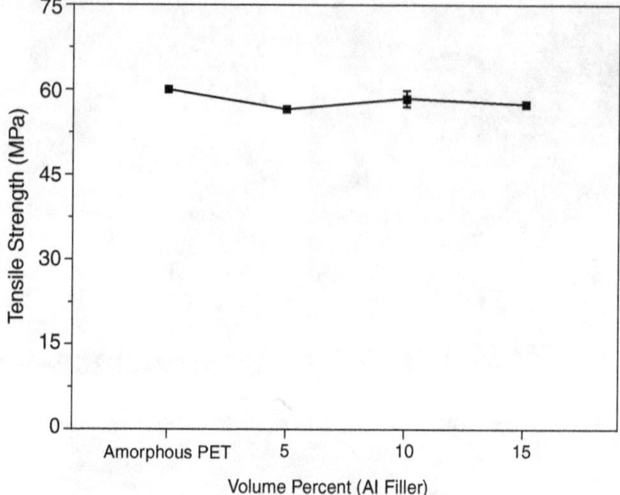

FIGURE 20.4 Tensile strength of amorphous PET filled with nodular Al particles of Figure 20.2a. The strength did not drop drastically as it does in many filled systems.

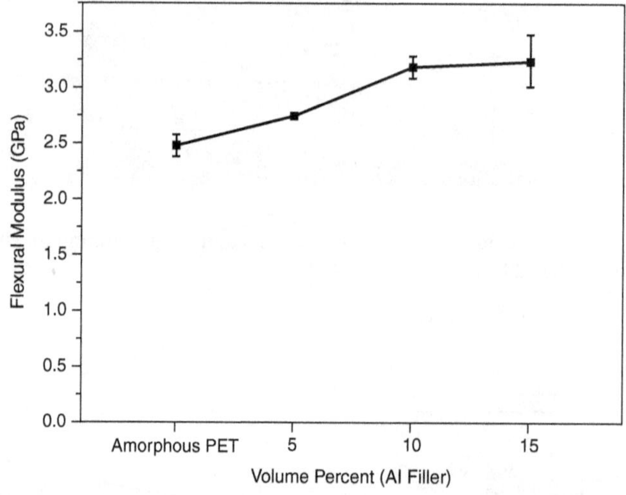

FIGURE 20.5 Flexural modulus of amorphous PET filled with nodular Al particles of Figure 20.2a.

had a low strength of ~32 MPa, and with 30 vol. % of Al powder, it had dropped to ~20 MPa (Osman and Mariatti 2006). This is not the case in Figure 20.4 for Al–PET.

Nicodemo and Nicolais (1983) measured the strength of a SAN copolymer when filled with Al and iron powders. In contrast to Al–PET composites (Anis et al. 2020), both the Al–SAN and Fe–SAN composites showed a drop in tensile strength, with the Al–SAN composite showing a higher fall. With 20 vol. % of Al, the strength was 50% of the value of the unfilled SAN. Nicodemo and Nicolais (1983) attributed the drop in strength to poor adhesion and inferred from the two curves of tensile strength versus volume fraction that the adhesion with Al was worse than with Fe, as the drop was worse for it.

In the Al–PET, a similar trend (as in the tensile modulus and strength) was seen in the flexural modulus and flexural strength – that is, there was an increase in the flexural modulus and the flexural strength was approximately constant (see Figures 20.5 and 20.6). When Al flakes were used

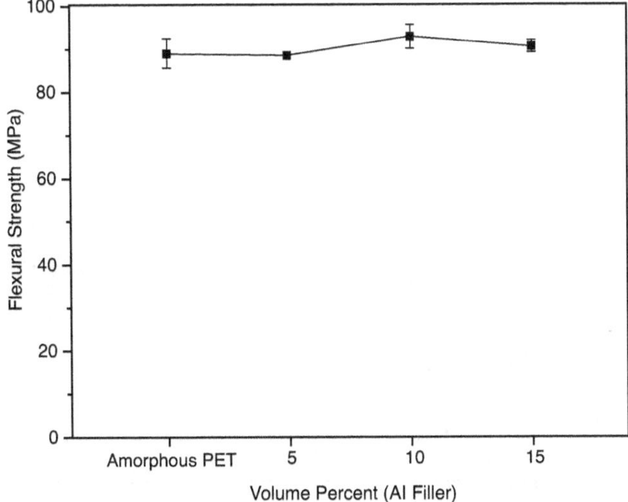

FIGURE 20.6 Flexural strength of amorphous PET filled with nodular Al particles of Figure 20.2a. The flexural strength did not drop drastically.

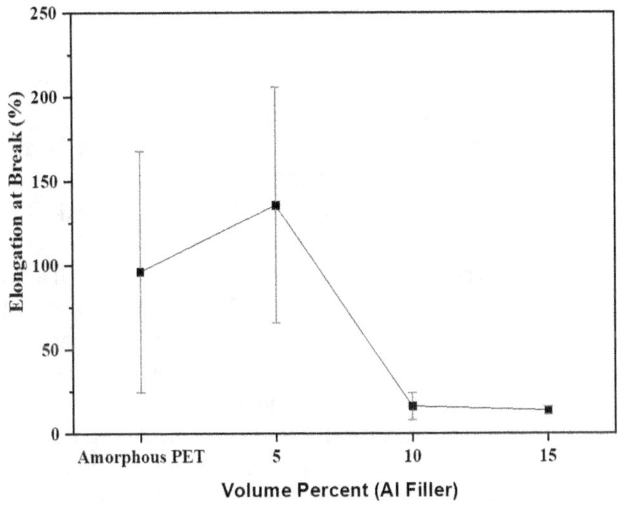

FIGURE 20.7 Elongation-at-break of amorphous PET and amorphous PET filled with nodular Al particles.

instead of nodular particles, an increase in flexural strength along with a very large increase in flexural modulus was observed due to the orientation of the flakes (Anis et al. 2022).

Figure 20.7 shows the tensile elongation-at-break of amorphous PET filled with nodular Al powder. There is a large drop in extensibility when filled. A drastic drop in elongation-at-break is observed when rigid fillers – of whatever type (inorganic minerals, metals, CNTs, clay and graphene) – are incorporated in rigid plastics. This is caused by the immobilisation of the matrix by the filler (Ahmed and Jones 1990). Most high T_g thermoplastics have an elongation-at-break of between 5% and 40%. However, amorphous PET necks and has a much higher elongation of ~100% at a crosshead speed of 50 mm/minute, while at 2 mm/minute, it can stretch up to 550% even at room temperature (i.e. below the T_g). In the Al-filled amorphous PET, at loadings up to 5 vol. %, this high extensibility may be maintained or even can increase somewhat (Anis et al. 2020) depending on the

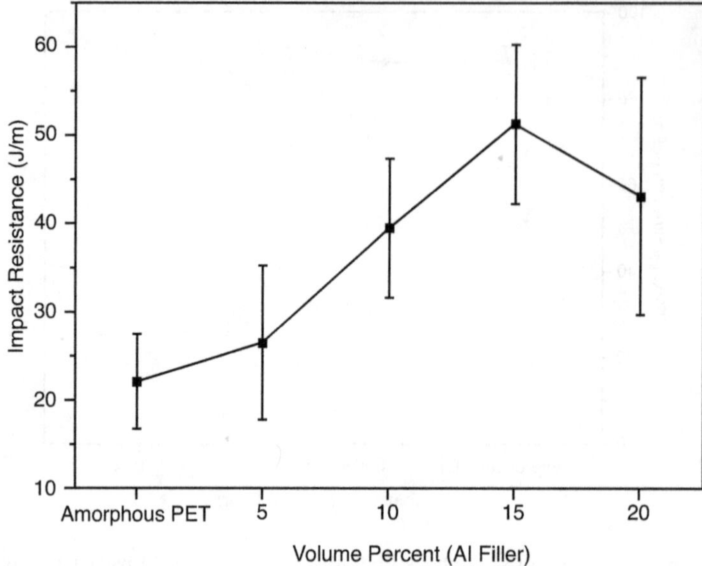

FIGURE 20.8 The notched Izod impact resistance of amorphous PET filled with nodular Al particles. The Al-filled PET shows an increase above the value of the matrix reaching a maximum at 15 vol. %, despite the drop in elongation-to-break shown in Figure 20.7.

particle shape and size. However, beyond 5% of Al filler, whatever the particle shape, the extension-to-break plummets (like in other filled polymers), to under ~5%.

Many polymer matrices have an inherent stretchability of ~40%, and these also show a drop to below ~5% when filled. Since the extension of unfilled amorphous PET drops from upto 550%, to below ~5% when filled, it means that the *relative drop* in elongation-to-break of the Al–PET composites is in fact even higher than the relative drop in elongation-to-break in filler composites with other polymer matrices. Very often, the decrease in elongation-to-break goes with a decrease in toughness and impact resistance and is often taken as synonymous. That is, the elongation-to-break is taken as a simple parameter to gauge the toughness. However, we shall show that is not the case with Al–PET, and despite the larger relative drop in elongation-to-break, its impact resistance does not drop. It is also worth noting that there are other types of impact toughness which may be relevant, for example a projectile impact, or a tensile impact. In a projectile impact, the Al–PET shows denting behaviour at higher volume fractions. We have not tested the tensile impact.

Figure 20.8 shows the notched Izod impact resistance of PET filled with the micron-sized, nodular Al particles of Figure 20.2a. The trend is that the impact resistance increases and reaches a peak at about 15 vol. % (52 from 22 J/m in amorphous PET) after which there was some decrease due to agglomeration; however, it did not decrease below the value of the matrix. A statistical significance test was done, and this showed that all compositions had higher impact values than the unfilled amorphous PET. The 10, 15 and 20 vol. % of Al may be considered statistically indistinguishable, but all have values above the matrix (Anis et al. 2020). The mechanical tests (tensile, flexural and impact) were done strictly following ASTM standards.

Moreover, such a trend (i.e. increase of impact resistance) was observed with Al powders of different shapes and sizes, from different suppliers. Figures 20.9–20.11 show just three of the key mechanical properties for amorphous PET filled with a nodular Al powder of a finer size than shown in Figure 20.2a. The tensile and flexural moduli increase as usual (not shown), and only the trends in tensile strength, flexural strength and notched Izod impact resistance are shown.

Again, the tensile and flexural strengths showed either a small decrease or constancy (Figures 20.9 and 20.10), while the notched Izod impact (Figure 20.11) showed a doubling at

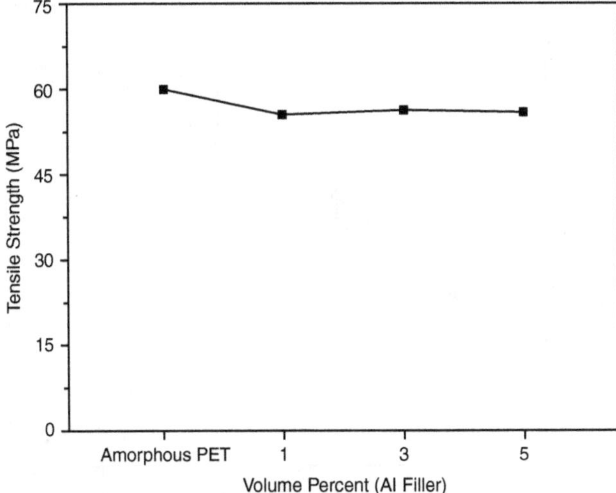

FIGURE 20.9 Tensile strength of amorphous PET filled with a finer nodular Al than in Figure 20.2a. The tensile strength shows a small decrease only.

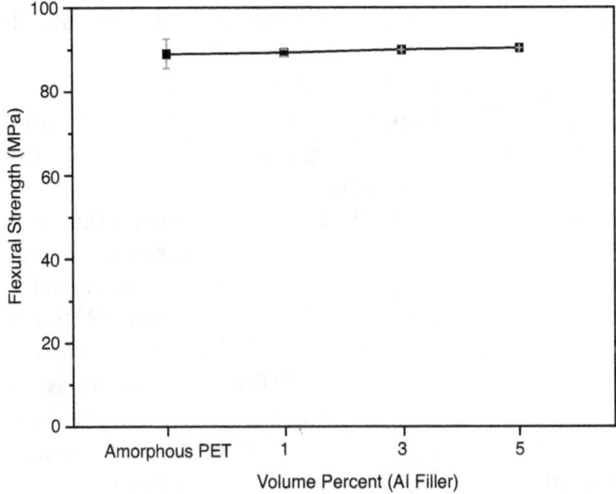

FIGURE 20.10 Comparison of flexural strength of amorphous PET filled with a finer nodular Al than Figure 20.2a. The flexural strength is approximately constant.

5 vol. % of Al. The tensile and flexural modulus showed a small increase, while the elongation-to-break showed the familiar decrease by 5 vol. % (not shown), which is the normal trend. Thus, the mechanical property combination has been confirmed as repeatable with different Al powders in PET, although there may be extra effects due to particle shape and size. For example, Al flakes show flow orientation of particles leading additionally to high flexural modulus of up to 8 GPa and flexural strength higher than the matrix (Anis et al. 2022).

In summary, it appears that the Al acts as a toughener in PET – but without the decrease in modulus and strength. Further work (Alhamidi et al. 2022) has shown that a similar behaviour occurs in Al-filled PET-PBT blends – that is, a simultaneous increase in tensile and flexural modulus, tensile and flexural strengths, and impact resistance is observed despite a drastic drop in elongation-to-break. The PET-PBT allowed a conductive composite where the matrix is crystalline instead of

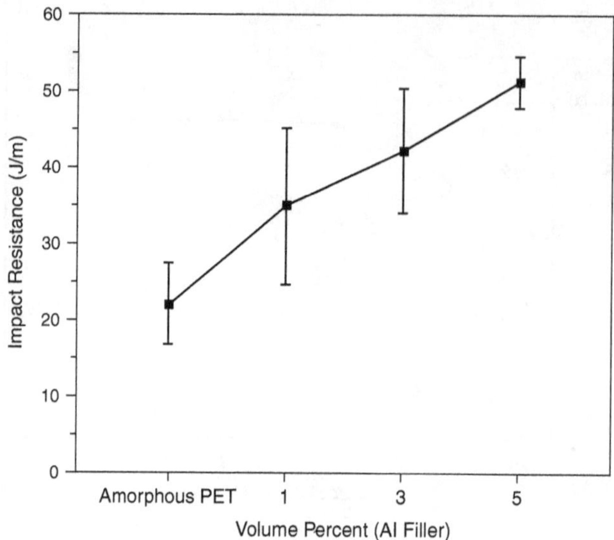

FIGURE 20.11 Notched Izod impact resistance of amorphous PET filled with a finer nodular Al than in Figure 20.2a. Again, the impact increased despite a drop in elongation-to-break.

amorphous, after injection moulding. Electrical conductivity in the static dissipation range was obtained in Al-filled PET-PBT blends (Alhamidi et al. 2022).

The increase or doubling in impact resistance due to Al is not to a super-tough level (50 instead 500 J/m), but it is now conceivable with particle size control (nano instead of micron-sized Al) and with a small quantity of rubber as a co-toughener, super-toughened conductive composites without a drop in modulus and strength may be possible.

We have also used another fabrication method, hot powder compaction, to make Al–PET articles (Al Fayez et al. 2020). While injection moulding allows a maximum Al loading of up to 25 vol. %, hot powder compaction allows loadings up to 80 vol. % of the filler, and with this method, it is easier to make Al–PET articles where the PET phase is crystalline. With more than 30 vol. % of Al, the Al–PET composite takes on the behaviour of Al. Thus, on dropping the composite, it dented like Al, instead of chipping like highly crystalline PET (to be published). Al has another interesting property in that the impact at liquid nitrogen temperature increases. This was seen also in Al–PET composites with high loadings of Al. Further, we 3D-printed Al–PET powder mixtures by selective laser sintering (Gu et al. 2022) and made articles with accurate shape.

20.6 THE ADHESION BETWEEN AL AND POLYESTERS, AND ITS RELATION TO STRENGTH AND IMPACT

The answer to how one gets an increase in modulus, tensile strength, and impact resistance – despite a drop in relative elongation-to-break that was even higher than with other filled plastics – lies in the nature of Al and its capacity to bond somehow with polyesters. Figures 20.12–20.14 show the scanning electron microscope pictures of the fracture surfaces of amorphous PET filled with Al particles of different shapes and sizes (micron-sized nodular; nanospherical and flakes, respectively). In all cases, dropouts and pull-outs were rare, indicating the adhesion was good. Inspection of other pictures showed evidence of some Al particles fracturing leaving the cup-and-cone fragments that are typical of ductile fracture in metals.

Pure Al has an intermediate-high modulus of 70 GPa, a strength of 90 MPa and an elongation-to-break of ~45% (Krishnaiah, Chakkingal, and Kim 2008). The modulus and strength are low for metals in absolute terms, yet Al's density is low (2.7 g/cm^3) among metals, so the specific modulus

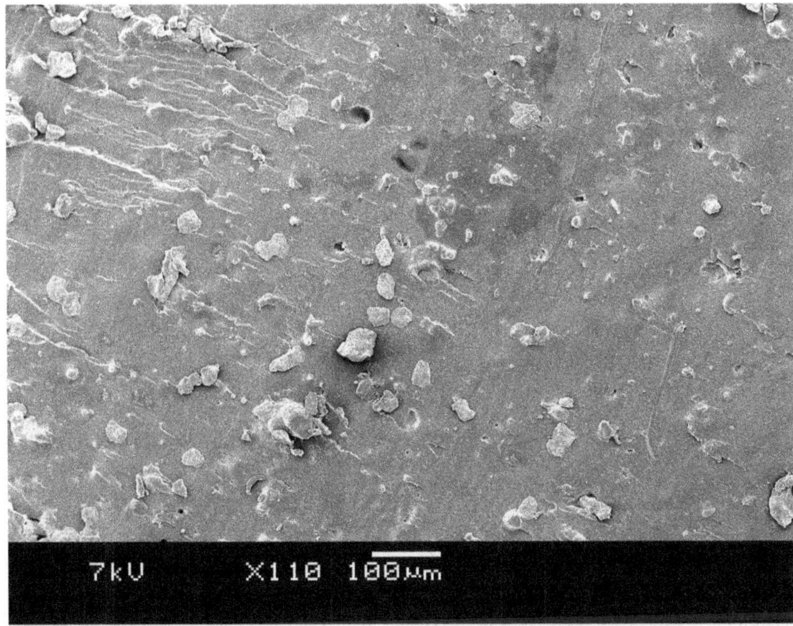

FIGURE 20.12 SEM picture of fracture surface of amorphous PET filled with 15 vol. % of nodular Al of Figure 20.2a. There are very few dropouts (dark void) of the Al particles, indicating that the adhesion is good. (Reproduced from Anis et al. 2020.)

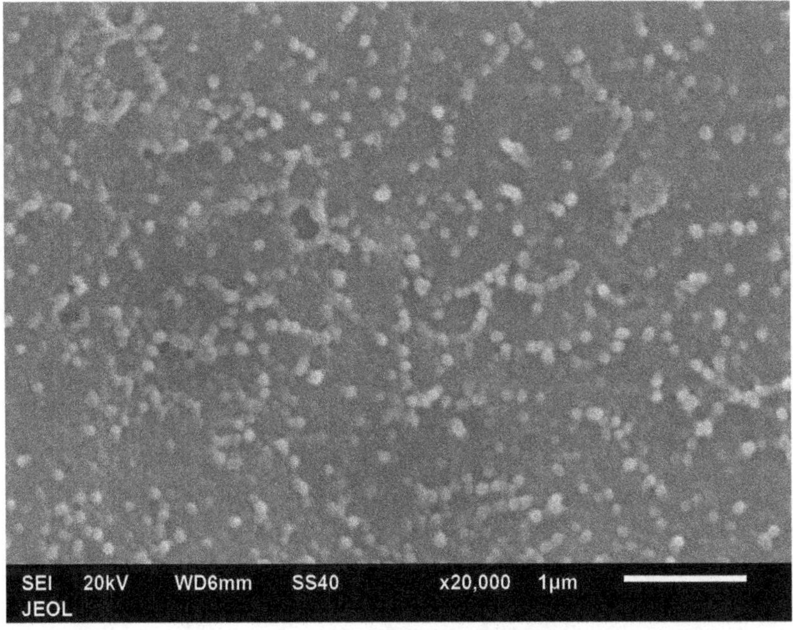

FIGURE 20.13 SEM picture of fracture surface of amorphous PET filled with 3 vol. % of nanospherical Al particles. There are few dropouts of the Al particles, indicating that the adhesion is good.

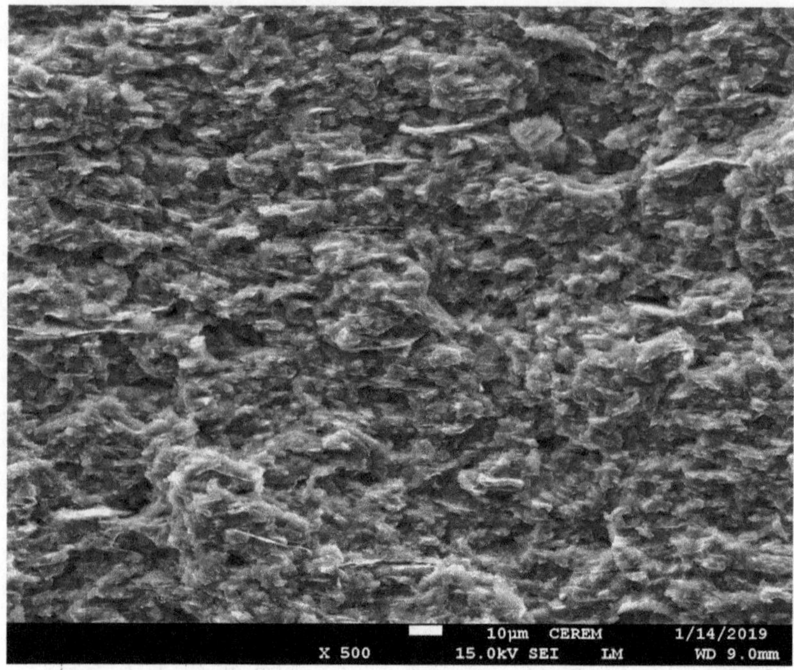

FIGURE 20.14 SEM picture of fracture surface of amorphous PET filled with Al flakes. There are no pull-outs or dropouts of the flakes. The flakes are seen edge-on.

becomes comparable to steels. The strength of pure Al means that it is only comparable to an unreinforced and unoriented plastic, and hence, it is not used for structural purposes. However, Al alloys with Si, Mg, etc., are structural metals as the strength is increased to 300–600 MPa, while the elongation decreases to ~28%.

E-glass has a similar modulus as Al (of 72 GPa), but it has a strength of 3.45–3.79 GPa, an elongation-to-break of 4.8% and a density of 2.6 g/cm^3. That is, pure Al has a similar modulus and density as E-glass fibre; however, its strength is much lower than glass, but it scores in ductility as it has high elongation-to-break compared with glass, which is strong but brittle.

If Al is incorporated into a thermoplastic matrix, and providing the adhesion is good, it could act as a sort of rubber toughener that does not cause a drop in modulus. Even without the microscope pictures in Figures 20.12–20.14, one has to deduce that the adhesion between Al and amorphous PET is good, to maintain the tensile strength and have an increase in impact resistance – despite the elongation-to-break decreasing. If not, there would be a drop in tensile strength and impact resistance as reported with Al and PP (Osman and Mariatti 2006), or Al in SAN (Chen, Das, and Battley 2014).

Although Al (like other metal powders) had been investigated as a conductive filler in several plastics since the 1980s (Osman and Mariatti 2006), it was not tried with PET as a matrix, till we examined it (Anis et al. 2020). This is because PET crystallises slowly and is not considered an injection moulding polyester. PBT is instead the favoured commercial polyester for injection moulding. Hence, earlier, Danes, Garnier, and Dupuis (2003) examined the use of Al fibres in a PBT (instead of PET matrix) to make a thermally conductive composite. PET has been used as an injection moulding matrix only in more recent works with clays, CNTs and graphenes as fillers (Li et al. 2006; MayPat et al. 2012; Shabafrooz et al. 2018).

On the mechanism of the adhesion, it is not directly between PET and the Al (it is unlikely organometallic bonds are formed). Al spontaneously reacts with oxygen and water and forms a coating which is a mixture of Al oxide and hydroxide. This oxide coating, however, prevents further corrosion of Al. The outer surface of the oxide (Al_2O_3) layer is hydroxylated with OH groups. The hydroxide groups are useful for the adhesion of paint and lacquers. van den Brand et al. (2004) used infrared reflection spectroscopy to study the adhesion of molecular ester compounds (dimethyl adipate) with Al. From the shift of vibration bands, they showed that the C=O of the ester group hydrogen-bonds with the –OH groups of the hydroxylated oxide layer on the Al. This is most likely to be also the mechanism by which a polyester like PET bonds strongly with Al.

We would also like to contrast the adhesion mechanism suggested above between the Al and the PET, with that reported between PET and calcium terephthalate (Dominici et al. 2020), and also PET and graphene (Aoyama et al. 2019). These two fillers themselves contain aromatic units, and it is proposed that there is a π–π interaction between the aromatic rings of graphene or calcium terephthalate and PET, and this immobilises the chains. In the Al–PET, if there is hydrogen bonding interaction between the ester C=O and the hydroxylated Al oxide, it is likely to be weaker than the π–π interaction between the aromatic rings of graphene and PET, both in the melt state and in the solid state. The intermediate level of interaction of the PET in the melt state and in the solid state with ductile Al platelets is more advantageous. This intermediate adhesion between Al and PET allows higher loadings for extrusion because the melt viscosity will not rise as with calcium terephthalate and graphene platelets, while in the solid state, the Al's ductility is transferred through bonding via the oxide layer, resulting in maintenance of tensile strength and increase in impact resistance.

20.7 CONCLUSIONS

The study of Al-filled polyesters was motivated by the desire to make conductive plastics, which combine electrical and thermal conductivity with good mechanical properties, together with ease of processing. It is well known that conductive fillers like carbon blacks, carbon fibres, graphite, metal powders, CNTs and graphenes can provide conductivity above a certain critical volume fraction. However, in almost all cases, the filler degrades the tensile strength and impact resistance.

Pure Al has a high electrical and thermal conductivity and a relatively low density. It has a moderately high modulus of 70 GPa combined with high ductility. If the adhesion between a matrix polymer and the Al is good, it should allow the goal of conductivity without drop in tensile strength and toughness.

Al-filled polyesters appear to have good natural adhesion, and this allows a system where there is simultaneous and unusual increase in modulus, strength and impact resistance, despite a large decrease in elongation-to-break. The Al acts as a toughening agent without reducing the modulus like a rubber filler. The mechanism of adhesion of PET appears to be through hydrogen bonding of the ester carbonyl with the OH groups on the oxide layer in Al.

Further work is planned to investigate the possibility of increasing the toughening behaviour of Al in polyesters through optimisation of the particle size and shape (nano-Al with optimum interparticle distance). The inclusion of rubber co-tougheners might allow super-toughness to be attained in a conductive composite.

Although much effort today is focussed on conductive plastics based on CNTs and graphene as fillers, these materials are beset by problems such as high cost, poor adhesion and expensive pre-treatments for dispersion. Hence, despite many years of research, these materials have limited use still. Al in polyester matrices may be a better combination for attaining conductivity without degraded mechanical properties like toughness and strength.

REFERENCES

Ahmed, S., and F.R. Jones, 1990. A review of particulate reinforcement theories for polymer composites. *Journal of Materials Science* 25(12): 4933–4942.

Al Fayez, F., T. Ahmad, Z. Bashir, H. Gu, S. Al-Zahrani, A. El Nour, A. Anis, 2020. *Shaped Object comprising Polyester and Aluminium.* Patent WO2020/234481 A1.

Alhamidi, A., Anis, A., Al-Zahrani, S.M., Bashir, Z., Alrashed, M.M. 2022. Conductive Plastics from Al Platelets in a PBT-PET Polyester Blend Having Co-Continuous Morphology. *Polymers* 14: 1092.

Al-Saleh, M.H. 2015. Electrically conductive carbon nanotube/polypropylene nanocomposite with improved mechanical properties. *Materials and Design* 85: 76–81.

Álvarez, M.P., V.H. Poblete, P.A. Rojas, 2010. Structural, electrical and percolation threshold of Al/polymethylmethacrylate nanocomposites. *Polymer Composites* 31(2): 279–283.

Anis, A., A.Y. Elnour, M.A. Alam, S.M. Al-Zahrani, F. AlFayez, Z. Bashir, 2020. Aluminum-filled amorphous-PET, a composite showing simultaneous increase in modulus and impact resistance. *Polymers* 12: 2038.

Anis, A., Elnour, A.Y., Alhamidi, A., Alam, M.A., Al-Zahrani, S.M., AlFayez, F., Bashir, Z. 2022. Amorphous Poly(ethylene terephthalate) Composites with High-Aspect Ratio Aluminium Nano Platelets. *Polymers* 14: 630.

Aoyama, S., I. Ismail, Y. Tae Park, C.W. Macosko, T. Ougizawa, 2019. Higher-order structure in amorphous poly (ethylene terephthalate)/graphene nanocomposites and its correlation with bulk mechanical properties. *ACS Omega* 4(1): 1228–1237.

Aoyama, S., I. Ismail, Y.T. Park, C.W. Macosko, T. Ougizawa, 2020. PET/graphene compatibilization for different aspect ratio graphenes via trimellitic anhydride functionalization. *ACS Omega* 5: 3228–3239.

Aoyama, S., Y. Tae Park, T. Ougizawa, C.W. Macosko, 2014. Melt crystallization of poly (ethylene terephthalate): Comparing addition of graphene vs. carbon nanotubes. *Polymer* 55(8): 2077–2085.

Bagotia, N., V. Choudhary, D.K. Sharma, 2018. A review on the mechanical, electrical and EMI shielding properties of carbon nanotubes and graphene reinforced polycarbonate nanocomposites. *Polymers for Advanced Technologies* 29(6): 1547–1567.

Bartczak, Z, 2002. Mechanisms of toughness improvement of semi-crystalline polymers. *Journal of Macromolecular Science, Part B* 41(4–6): 1205–1229.

Bhattacharya, S.K., A.C.D. Chaklader, 1982. Review on metal-filled plastics. Part 1. Electrical conductivity. *Polymer-Plastics Technology and Engineering* 19: 21–51.

Bishay, I.K., S.L. Abd-El-Messieh, S.H. Mansour 2011. Electrical, mechanical and thermal properties of polyvinyl chloride composites filled with aluminum powder. *Materials & Design* 32: 62–68.

Bora, T., B.L. McKinney, H. Faust, 1994. *Polyester-Polyolefins Blends Containing a Functionalised Elastomer*, US Patent.

Boudenne, A., L. Ibos, M. Fois, E. Gehin, J.-C. Majeste, 2004. Thermophysical properties of polypropylene/aluminum composites. *Journal of Polymer Science Part B: Polymer Physics* 42(4): 722–732.

Chen, Y., R. Das, M. Battley, 2014. Modelling of closed-cell foams incorporating cell size and cell wall thickness variations. In *Proceedings of the 11th World Congress on Computational Mechanics*: 20–25.

Danes, F., B. Garnier, T. Dupuis, 2003. Predicting, measuring, and tailoring the transverse thermal conductivity of composites from polymer matrix and metal filler. *International Journal of Thermophysics* 24(3): 771–784.

Dominici, F., F. Sarasini, F. Luzi, L. Torre, D. Puglia, 2020. Thermomechanical and morphological properties of poly (ethylene terephthalate)/anhydrous calcium terephthalate nanocomposites. *Polymers* 12(2): 276.

Dutta, A.L., R.P. Singh, A.K. Misra, 1992. Electrical conduction and ultrasound wave propagation in particulate and short fiber composites of PVC. I: PVC-Cu/Al particulate composites. *Journal of Vinyl Technology* 14: 33–42.

Fu, Q., G. Wang, 1992. Polyethylene toughened by rigid inorganic particles. *Polymer Engineering & Science* 32(2): 94–97.

Gorassi, G., S.D. Ambrosio, G. Patimo, R. Pantani, 2014. Hybrid clay-carbon nanotube/PET composites: Preparation, processing, and analysis of physical properties. *Journal of Applied Polymer Science* 131: 40441.

Gu, H., AlFayez, F., Yang, L., Ahmed, T., Bashir, Z. Powder bed fusion of aluminium—Poly(ethylene terephthalate) hybrid powder: Process behaviour and characterization of printed parts. Addit. Manuf. 2022, 51, 102616

Huang, J.-C. 2002. Carbon black filled conducting polymers and polymer blends. *Advances in Polymer Technology: Journal of the Polymer Processing Institute* 21(4): 299–313.

Irina, D., E. Kuvardina, V. Krasheninnikov, S. Lomakin, I. Tchmutin, S. Kuznetsov, 2010. The effect of multiwalled carbon nanotube dimensions on the morphology, mechanical, and electrical properties of melt mixed polypropylene-based composites. *Journal of Applied Polymer Science* 117: 259–272.

Krishnaiah, A., U. Chakkingal, H.S. Kim, 2008. Mechanical properties of commercially pure aluminium subjected to repetitive bending and straightening process. *Transactions of the Indian Institute of Metals* 61(2): 165–167.

Li, Z., G. Luo, F. Wei, Y Huang, 2006. Microstructure of carbon nanotubes/PET conductive composites fibers and their properties. *Composites Science and Technology* 66: 1022–1029.

Mamunya, Y.P., V.V. Davydenko, P. Pissis, E.V. Lebedev, 2002. Electrical and thermal conductivity of polymers filled with metal powders. *European Polymer Journal* 38: 1887–1897.

MayPat, A., F. Avilés, P. Toro, M. Yazdani-Pedram, J.V. Cauich-Rodríguez, 2012. Mechanical properties of PET composites using multiwalled carbon nanotubes functionalized by inorganic and itaconic acids. *eXPRESS Polymer Letters* 6(2): 96–106.

Michler, G.H., H.-H.K.-B. von Schmeling, 2013. The physics and micro-mechanics of nano-voids and nanoparticles in polymer combinations. *Polymer* 54: 3131–3144.

Nakamura, S., K. Saito, G. Sawa, K. Kitagawa, 1997. Percolation threshold of carbon black-polyethylene composites. *Japanese Journal of Applied Physics* 36(8): 5163.

Nicodemo, L., L. Nicolais, 1983. Mechanical properties of metal/polymer composites. *Journal of Materials Science Letters* 2(5): 201–203.

Osman, A.F., M. Mariatti, 2006. Properties of aluminum filled polypropylene composites. *Polymers and Polymer Composites* 14(6): 623–633.

Pearson, R.A., A.F. Yee, 1993. Toughening mechanisms in thermoplastic-modified epoxies: 1. Modification using poly (phenylene oxide). *Polymer* 34(17): 3658–3670.

Pecorini, T.J., D. Calvert, 2000. The role of impact modifier particle size and adhesion on the toughness of PET. *ACS Symposium Series* 759, Chapter 9: 141–158.

Pinto, G., A. Jiménez-Martín, 2001. Conducting aluminum-filled nylon 6 composites. *Polymer Composites* 22(1): 65–70.

Ren, D., S. Zheng, F. Wu, W. Yang, Z. Liu, M. Yang, 2014. Formation and evolution of the carbon black network in polyethylene/carbon black composites: Rheology and conductivity properties. *Journal of Applied Polymer Science* 131(7): 39953.

Ronkay, F., B. Molnár, F. Szalay, D. Nagy, B. Bodzay, I.E. Sajó, K. Bocz, 2019. Development of flame-retarded nanocomposites from recycled PET bottles for the electronics industry. *Polymers* 11(2): 233.

Sahai, R.S.N., N. Pawar, 2017. Studies on mechanical properties of mica filled polyphenylene oxide composite with coupling agent. *Asian Journal of Applied Science and Technology* 1(7): 153–157.

Sandler, J., J. Kirk, I. Kinloch, M. Shafer, A. Windle, 2003. Ultra-low electrical percolation threshold in carbon-nanotube-epoxy composites. *Polymer* 44: 5893–5899.

Shabafrooz, V., S. Bandla, M. Allahkarami, J.C. Hanan, 2018. Graphene/polyethylene terephthalate nanocomposites with enhanced mechanical and thermal properties. *Journal of Polymer Research* 25(12): 1–12.

Tanrattanakul, V., W.G. Perkins, F.L. Massey, A. Moet, A. Hiltner, E. Baer, 1997. Fracture mechanisms of poly(ethylene terephthalate) and blends with styrene-butadiene-styrene elastomers. *Journal of Materials Science* 32: 4749–4758.

Tony, M., P. Pötschke, P. Halley, M. Murphy, D. Martin, S.E.J. Bell, G.P. Brennan, D. Bein, P. Lemoine, J.P. Quinn, 2005. Polyethylene multiwalled carbon nanotube composites. *Polymer* 46: 8222–8232.

Van den Brand, J., O. Blajiev, P.C.J. Beentjes, H. Terryn, J.H.W. De Wit, 2004. Interaction of ester functional groups with aluminum oxide surfaces studied using infrared reflection absorption spectroscopy. *Langmuir* 20(15): 6318–6326.

Wang, J., F. Song, Y. Ding, M. Shao, 2020. The incorporation of graphene to enhance mechanical properties of polypropylene self-reinforced polymer composites. *Materials & Design* 195: 109073.

Wei, G.-X., H.-J. Sue, J. Chu, C. Huang, K. Gong, 2000. Toughening and strengthening of polypropylene using the rigid-rigid polymer toughening concept Part II Toughening mechanisms investigation. *Journal of Materials Science* 35(3): 555–566.

Wei, P., S. Bai, 2015. Fabrication of a high-density polyethylene/graphene composite with high exfoliation and high mechanical performance via solid-state shear milling. *RSC Advances* 5(114): 93697–93705.

Wetzel, B., Haupeiit, F., Friedrich, K., Zhang, M.Q., Rong, M.Z., 2002. Impact and wear resistance of polymer nanocomposites at low filler content. *Polymer Engineering and Science* 42: 1919–1927.

Index

For Product Safety Concerns and Information please contact our EU representative GPSR@taylorandfrancis.com
Taylor & Francis Verlag GmbH, Kaufingerstraße 24, 80331 München, Germany

www.ingramcontent.com/pod-product-compliance
Lightning Source LLC
Chambersburg PA
CBHW080932220326
41598CB00034B/5760

* 9 7 8 1 0 3 2 3 8 5 4 9 5 *